Nickel and the Skin: Immunology and Toxicology

Editors

Howard I. Maibach, M.D.
Professor
Department of Dermatology
University of California at San Francisco
San Francisco, California

and

Torkil Menné, M.D., Ph.D.
Consultant Dermatologist
Department of Dermatology
Gentofte Hospital
Hellerup, Denmark

CRC Press, Inc.
Boca Raton, Florida

Library of Congress Cataloging-in-Publication Data

Nickel and the skin: immunology and toxicology/editors, Howard
I. Maibach and Torkil Menné.
p. cm.
Includes bibliographies and index.
ISBN 0-8493-6976-2
1. Nickel--Toxicology. 2. Dermatotoxicology. 3. Contact
dermatitis. I. Maibach, Howard I. II. Menné, Torkil.
[DNLM: 1. Dermatitis, Contact--etiology. 2. Nickel--adverse
effects. 3. Nickel--immunology. WR 175 N632]
RL803.N53 1989
616.97′3--dc19
DNLM/DLC
for Library of Congress 88-19378
CIP

Direct all inquiries to CRC Press, Inc., 2000 Corporate Blvd., N.W., Boca Raton, Florida, 33431.

International Standard Book Number 0-8493-6976-2

Library of Congress Card Number 88-19378
Printed in the United States

PREFACE

Eczematous sensitivity to nickel is now the most common cause of allergic contact dermatitis throughout the industrialized world. In 1880 the first patent on the nickel plating of metal was issued. Blascko, 6 years later, published his first report on occupational dermatitis from nickel, the so-called nickel-itch. In the following 50 years nickel dermatitis remained an occupational allergy seen in the nickel-plating industry. When the itching eruptions from nickel occurred in industry they were thought to be of an irritating nature. On the whole, nickel dermatitis was considered an uncommon disease, which is evident from the fact that it was hardly mentioned at the World Congress of Dermatology in 1930.

In 1935 nickel irritation was discussed at the meeting of the Nordic Dermatological Association. The proceedings of that congress were published and included the discussion which revealed that the prevalent idea of nickel dermatitis was that it was due to an eruption caused by mechanical irritation.

The imports of nickel to the European continent decreased during the Second World War, and at the same time the incidence of nickel dermatitis declined. Since the end of the Second World War nickel dermatitis has been on a steady increase. It now became universally recognized that nickel dermatitis was a consumer's dermatitis. The number of occupationally provoked cases decreased due to improved methods of occupational hygiene.

The major sources of sensitization were nickel-plated clasps, buttons, and female suspenders. It was hoped that if nickel-plated suspenders were replaced by plastic material, the dermatitis would become much rarer. This was not so. New fashions introduced new sources of sensitization, e.g., metal buttons, zip fasteners, and ear clasps.

Over the last 10 years nickel in costume jewelery has taken the lead as the major sensitizer. The introduction of less sensitizing alloys may possibly reverse the trend of nickel allergy. In spite of the differences both in fashions and industrial consumption, nickel sensitivity has steadily increased, so it is now by far the most common allergen in women tested in Eastern and Western Europe.

Nickel sensitivity predisposes for later industrial dermatitis and for that reason insertion of nickel earrings in school-age girls may decide the ultimate occupation/industrial prognoses of these girls. A person with nickel sensitivity may develop recurrent hand eczema in varying degrees or may even be incapacitated from chronic hand eczema.

During the last 40 years consumer sensitivity has become a most serious occupational disorder. Therefore, this book is well timed and an important source of information not only for dermatologists, but for all occupational physicians. The chapters cover all aspects of nickel sensitivity and should be a stimulus for further research aimed at reducing the frequency of sensitization.

Niels Hjorth
Gentofte Hospital
Denmark

EDITORS

Howard Maibach, M.D. is a Professor in the Department of Dermatology at the University of California Medical School, San Francisco.

He received his M.D. at Tulane University and clinical and research training in Dermatology at the University of Pennsylvania. He has been at the University of California since that time.

He is a member of the Society of Investigative Dermatology, Society of Toxicology, International Contact Dermatitis Research Group, and several other societies.

His research includes dermatopharmacy, dermatotoxicology, contact dermatitis, and psoriasis. He is the author of 750 articles, and the author and/or editor of 33 books.

Torkil Menné, M.D., Ph.D. is Assistant Professor in the Department of Dermatology, Gentofte Hospital, Copenhagen, Denmark.

He obtained his M.D. from the University of Copenhagen in 1971 and his Ph.D. in 1983 at the same university for investigations on genetic and epidemiologic aspects of nickel dermatitis.

Dr. Menné is a member of the Danish Dermatological Society, American Academy of Dermatology, President for the Danish Contact Dermatitis Research Group, and a member of the European Environmental Contact Dermatitis Research Group.

He has presented over ten invited lectures at international meetings and has published more than 100 research and review papers. His current major research interest is environmental dermatology.

CONTRIBUTORS

Ole Andersen, Ph.D.
Assistant Professor
Department of Environmental Medicine
Odense University
Odense, Denmark

Desmond Burrows, M.D., F.R.C.P.
Consultant Dermatologist
Department of Dermatology
Royal Victoria Hospital
Belfast, Northern Ireland

Jette Christophersen, Ph.D.
Senior Registrar
Department of Dermatology
Gentofte Hospital
Hellerup, Denmark

M. A. Cooke, M.B.
Visiting Professor
Health and Safety Unit
Department of Mechanical and Production Engineering
Aston University
Birmingham, England

R. A. Cooke, M.B.
Chief Medical Officer
G.K.N. plc
Birmingham, England

Bjorn Edman, B.A.
Department of Dermatology
Malmö General Hospital
Malmö, Sweden

Torkel Fischer, M.D.
Assistant Professor
Department of Occupational Dermatology
University Hospital
Uppsala, Sweden

G. N. Flint, B.Sc.
Consultant
c/o INCO Europe, Ltd.
Swansea, Wales, U.K.

Philippe Grandjean, M.D.
Professor
Department of Environmental Medicine
Odense University
Odense, Denmark

Anders Green, Ph.D.
Registrar
Internal Medical Department C
Odense University Hospital
Odense, Denmark

Arne Hensten-Pettersen, Dr. Odont.
Senior Scientist
Biological/Clinical Division
NIOM
Scandinavian Institute of Dental Materials
Haslum, Norway

Niels V. Holm, Ph.D.
Associate Professor
Institute of Medical Genetics
Odense University
Odense, Denmark

Lars Janzon, M.D.
Professor
Department of Epidemiology
Community Health Sciences University
Malmö General Hospital
Malmö, Sweden

K. Lammintausta
Department of Dermatology
University of California
San Francisco, California

Howard Maibach, M.D.
Professor
Department of Dermatology
University of California Hospital
San Francisco, California

Clifford McMillan, B.Sc., M.R.C.P.
Consultant Dermatologist
Department of Dermatology
City Hospital
Belfast, Northern Ireland

Torkil Menné, Ph.D.
Consultant Dermatologist
Department of Dermatology
Gentofte Hospital
Hellerup, Denmark

Lindsay G. Morgan, M.B.
Doctor
Medical Department
INCO Europe, Ltd.
Swansea, Wales, U.K.

Gitte Dalsgaard Nielsen, Ph.D.
Research Fellow
Department of Environmental Medicine
Odense University
Odense, Denmark

Rik J. Scheper, Ph.D.
Doctor
Department of Pathology
Free University Hospital
Amsterdam, Netherlands

F. William Sunderman, Jr., M.D.
Professor
Departments of Laboratory Medicine and Pharmacology
University of Connecticut Medical School
Farmington, Connecticut

Ingrid M. V. van Hoogstraten, M.Sc.
Department of Pathology
Free University Hospital
Amsterdam, Netherlands

Niels K. Veien, M.D.
The Dermatology Clinic
Aalborg, Denmark

Mary Von Blomberg, Ph.D.
Doctor
Department of Pathology
Free University Hospital
Amsterdam, Netherlands

Karel J. J. Vreeburg, M.D.
Department of Biomaterials
Free University
Amsterdam, Netherlands

Jan E. Wahlberg, M.D.
Professor
Department of Occupational Dermatology
Karolinska Hospital
Stockholm, Sweden

Kaare Weismann, M.D., Ph.D.
Consultant Dermatologist
Department of Dermatology
Bispebjerg Hospital
Copenhagen, Denmark

D. S. Wilkinson, M.D.
Consultant Dermatologist
Chilten Hospital
Buckinghamshire, England

J. D. Wilkinson, M.D., F.R.C.P.
Consultant Dermatologist
Department of Dermatology
Wycombe General Hospital
Buckinghamshire, England

Ruixi Yang, M.D.
Department of Dermatology
Institute for Industrial Hygiene and Occupational Disease
Shanghai, People's Republic of China

TABLE OF CONTENTS

Chapter 1
Chemistry, Analysis, and Monitoring 1
F. W. Sunderman, Jr.

Chapter 2
Human Nickel Exposure and Chemobiokinetics 9
P. A. Grandjean, G. D. Nielsen, and O. Andersen

Chapter 3
Carcinogenicity 35
M. A. Cooke and R. A. Cooke

Chapter 4
Nickel Alloys and Coatings: Release of Nickel 45
L. G. Morgan and G. N. Flint

Chapter 5
Recent Advances in Immunology of Nickel 55
Rik Scheper, M. von Blomberg, K. J. J. Vreeburg, and I. M. V. van Hoogstraten

Chapter 6
Nickel:Animal Sensitization Assays 65
J. Wahlberg

Chapter 7
Nickel:Human Sensitization Assays 75
R. Yang and H. I. Maibach

Chapter 8
In Vitro Testing in Nickel Contact Hypersensitivity 79
C. McMillan and D. Burrows

Chapter 9
Clinical Concepts in Nickel Testing 91
K. Lammintausta and H. I. Maibach

Chapter 10
Genetic Aspects of Nickel Sensitization 101
T. Menné and N. V. Holm

Chapter 11
Epidemiology of Nickel Dermatitis 109
T. Menné, J. Christophersen, and A. Green

Chapter 12
Occupational Nickel Dermatitis 117
T. Fischer

Chapter 13
Nickel Allergy and Hand Eczema 133
D. S. Wilkinson and J. D. Wilkinson

Chapter 14
Nickel Dermatitis: Its Relationship to Food and Experimental Oral Challenge 165
N. K. Veien

Chapter 15
Nickel Allergy and Drug Interaction 179
K. Weismann and T. Menné

Chapter 16
Nickel Allergy and Orthopedic Prostheses 187
J. D. Wilkinson

Chapter 17
Nickel Allergy and Dental Treatment Procedures 195
A. Hensten-Pettersen

Chapter 18
Social and Demographic Aspects of Nickel Contact Allergy 207
Bjorn Edman and L. Janzon

Chapter 19
Nickel Dermatitis: A Preventable Health Problem 215
T. Menné and H. I. Maibach

Index 217

Chapter 1

CHEMISTRY, ANALYSIS, AND MONITORING

F. William Sunderman, Jr., M.D.

TABLE OF CONTENTS

I. Introduction 2

II. Physicochemical Properties 2

III. Oxidation States 2

IV. Radioisotopes 2

V. Chemical Compounds 3

VI. Analytical Techniques 3

VII. Electrothermal Atomic Absorption Spectrophotometry 4

VIII. Nickel Carbonyl Analysis 4

IX. Specimens for Biological Monitoring 4

X. Reference Values 5

XI. Quality Assurance 5

XII. Summary 6

References 7

I. INTRODUCTION

A brief synopsis of the chemistry and analytical technology of nickel, in relation to biological monitoring of nickel exposures, is provided in this chapter,. The discussions are condensed from recent articles by the author and his collaborators.[1-9] For additional information, consult comprehensive monographs on nickel,[10-15] as well as reviews on analytical toxicology of nickel[16-18] and biological monitoring of nickel exposures.[19-22]

II. PHYSICOCHEMICAL PROPERTIES

In 1751, Axel Frederik Cronstedt, a Swedish mineralogist, identified a previously unrecognized metallic element as a constituent of a common mineral, ''kupfernickel'' (now known as ''Gersdorffite'', NiAsS). In Germanic folklore, ''nickel'' referred to a dwarf, devil, or scamp (''Old Nick''); thus ''kupfernickel'' denoted ''Old Nick's copper''. Cronstedt named the new element ''nickel'' and described some of its physical and chemical properties. Nickel is a silver-white, hard, malleable, ductile metal that maintains high luster and is a fair conductor of electricity and heat. Nickel is ferromagnetic at ambient temperatures and becomes paramagnetic at temperatures above its Curie point (353°C). Nickel is relatively resistant to corrosion in many acids, salts, and alkalies; in fresh and salt water; and in wet and dry gases, accounting for the widespread use of nickel in ferronickel alloys (e.g., stainless steel), cupronickel alloys (e.g., Monel metal), and precious metal alloys (e.g., German silver).

The atomic number of nickel is 28 and its atomic weight is 58.71, comprising a mixture of five stable isotopes with atomic weights of 58, 60, 61, 62, and 64. Metallic nickel exists in two principal crystalline phases, ''α-nickel'' with an hexagonal lattice and ''β-nickel'' with a cubic lattice. The α- to β-phase transition occurs upon heating above 250°C. The melting point of metallic nickel is 1453°C, its boiling point is 2732°C, and the specific gravity of nickel at 25°C is 8.902.

III. OXIDATION STATES

Oxidation states of nickel include −1, 0, +1, +2, +3, and +4, but the prevalent valences are 0 (in nickel metal and its alloys) and +2 (in most nickel salts and stable inorganic compounds). Nickel atoms contain unpaired electrons in two outer 3d orbitals and thus undergo changes in oxidation state that involve one electron. Complexation with peptides can reduce the redox potential of the Ni^{3+}/Ni^{2+} couple from 4.2 V to values from 0.7 to 1.0 V, enabling stable Ni^{3+} complexes to form under certain biological conditions.[23,24] The oxidation-reduction properties of the Ni^{3+}/Ni^{2+} couple may account for the biological roles of nickel in enzymes and coenzymes, such as ureases, hydrogenases, dehydrogenases, and factor F_{430}, which have been identified in plants and microorganisms, and may initiate the free-radical reactions (e.g., lipid peroxidation, DNA-nucleoprotein cross-linking) that are speculated to be involved in the pathogenesis of nickel toxicity and carcinogenesis.[25]

IV. RADIOISOTOPES

The seven radioisotopes of nickel include ^{56}Ni (γ, $T_{1/2}$ = 6.1 d), ^{57}Ni ($\beta + \gamma$, $T_{1/2}$ = 36 h), ^{59}Ni (γ, $T_{1/2} = 8 \times 10^4$ years), ^{63}Ni (β, $T_{1/2}$ = 92 years), ^{65}Ni ($\beta + \gamma$, $T_{1/2}$ = 2.5 h), ^{66}Ni (β, $T_{1/2}$ = 55 h), and ^{67}Ni ($\beta + \gamma$, $T_{1/2}$ = 50 s).[1,10] Two of these isotopes are practical as radiotracers for chemical and biological studies. ^{57}Ni is advantageous for clinical investigations and diagnostic applications in nuclear medicine, since it has a relatively short half-life (36 h) and high-energy γ-radiation (1378 keV), which can readily be measured in the presence of the low-energy radiation of its ^{57}Co daughter.[1,26] ^{63}Ni is well suited for *in vitro* studies in cell and organ cultures and *in vivo* studies in experimental animals because of its (1) low energy β-emission (67 keV), which is intermediate between ^{3}H and ^{14}C, (2) long half-life (92 years), and (3) economical availability at high specific activity (up to 3.87 kCi/g-atom). The soft β-emission of ^{63}Ni is readily counted by liquid scintillation spectrometry and provides autoradiograms with exceptionally high resolution.[1]

V. CHEMICAL COMPOUNDS

Nickel monoxide (NiO) exists in two major forms with different chemical properties. Black NiO is chemically reactive and forms nickel salts upon contact with mineral acids, whereas green NiO is relatively inert and refractory to solubilization in dilute acids.[7,14,15] Numerous nickel sulfides are known, including crystalline sulfides (βNiS, αNi_3S_2), which are highly carcinogenic in experimental animals, and amorphous NiS, which has low carcinogenic activity.[27] Nickel chloride ($NiCl_2$), sulfate ($NiSO_4 \cdot 6H_2O$), ammonium sulfate [$Ni(NH_4)_2(SO_4)_2$], nitrate [$Ni(NO_3)_2 \cdot 6H_2O$], carbonate ($NiCO_3$), hydroxide [$Ni(OH)_2$], and acetate [$Ni(COOCH_3)_2$] are the salts of greatest commercial importance. Nickel carbide (NiC_3) is a heat-resistant compound that is used to coat the turbine vanes of high-compression jet engines. Nickel salts of fatty acids (e.g., nickel laurate) are used in paper coatings for pressure-sensitive credit card forms. Nickel carbonyl ([$Ni(CO)_4$] is a colorless liquid that boils at 43°C and decomposes at 60°C, producing carbon monoxide and finely powdered nickel. Nickel carbonyl decays spontaneously in air with a half-time of approximately 30 min.[28] Owing to its volatility, lack of strong odor, and propensity for inadvertent formation, nickel carbonyl is generally considered the most toxic compound of nickel that is encountered in the workplace.[7]

VI. ANALYTICAL TECHNIQUES

Current techniques for trace analysis of nickel include[2,3,16,18]

1. High-performance liquid chromatography of nickel complexes, with spectrophotometric or fluorometric detection
2. Gas chromatography of volatile nickel chelates, with electron-capture detection
3. X-ray fluorescence or particle-induced X-ray emission
4. Neutron or charged-particle activation analysis
5. Inductively coupled plasma emission with stable isotope-dilution mass spectrometry
6. Electrothermal atomic absorption spectrophotometry (EAAS)
7 Differential pulse absorption voltametry (DPAV)

EAAS and DPAV are practical methods that provide the requisite sensitivity for measurements of nickel concentrations in biological samples. The detection limits for nickel determinations by EAAS analysis with Zeeman background correction are approximately 0.45 μg/l for urine, 0.1 μg/l for whole blood, 50 ng/l for serum or plasma, and 10 ng/g (dry weight) for tissues, food, or feces.[29-32] Greater analytical sensitivity for nickel can be achieved by DPAV analysis, using mercury electrodes sensitized with dimethylglyoxime, furildioxime, or dithiocarbamate.[17,33,34] For example, DPAV analyses using a dimethylglyoxime-sensitized mercury electrode provide detection limits of approximately 1 ng/l for nickel determinations in whole blood, urine, saliva, and tissue homogenates.[33] However, current DPAV methods for nickel analysis are more cumbersome and time-consuming than EAAS procedures.

VII. ELECTROTHERMAL ATOMIC ABSORPTION SPECTROPHOTOMETRY

The EAAS procedure of Brown et al.,[35] which was developed through collaborative interlaboratory trials by biochemists in 13 nations,[36-38] has been accepted as a reference method by the International Union of Pure and Applied Chemistry (IUPAC) and the International Agency for Research on Cancer (IARC).[5,35] This procedure involves

1. Sample digestion with mixed nitric, sulfuric, and perchloric acids
2. Chelation of nickel with ammonium tetramethylenedithiocarbamate
3. Extraction of *bis*(1-pyrrolidinecarbodithioato)-nickel into 4-methyl-2-pentanone
4. Measurement of nickel in the solvent extract by EAAS

Three routine procedures for EAAS analysis of various biological materials, described by Sunderman et al.,[29-32] yield analytical results that agree closely with measurements by the IUPAC/IARC reference procedure.[35]

Acid digestion technique — Organic constituents of the sample are oxidized with mixed nitric, sulfuric, and perchloric acids prior to direct EAAS analysis (without chelation and solvent extraction).[30,32] This technique can be used to determine nickel in tissues, diets, and feces, as well as biological samples containing nickel that would be incompletely released from macromolecules after acidification with nitric acid (e.g., nickel in the active site of urease), or nickel in a volatile form that would be lost by vaporization during the preliminary ashing steps of EAAS analysis (e.g., nickel in a Tris-Ni[II] complex).[39]

Protein-precipitation technique — Proteins are precipitated with nitric acid and heat prior to EAAS analysis.[29,32] This rapid technique can be used to measure nickel concentrations in serum, plasma, whole blood, saliva, bile, and cerebrospinal fluid.

Direct technique — The sample is merely acidified with nitric acid.[31,32] This technique can be used to determine nickel in urine, water, and protein-free aqueous samples.

VIII. NICKEL CARBONYL ANALYSIS

Nickel carbonyl can be measured quantitatively in air or exhaled breath by gas chromatography, chemiluminescence, or EAAS.[40-42] The chemiluminescent nickel carbonyl detector, developed by Stedman et al.,[41] is a portable unit for industrial use. By action of ozone and carbon monoxide, nickel carbonyl is converted to an excited state of nickel oxide, $NiO^{\cdot}$, which emits a photon during decay to the ground state. Detection of photon emission is achieved with a specially designed chemiluminometer. A sensitive technique for determination of nickel carbonyl in air has been developed by Filkova,[42] based upon sampling filtered air through a heated graphite tube to deposit nickel in the tube and measurement of the deposited nickel by EAAS analysis.

IX. SPECIMENS FOR BIOLOGICAL MONITORING

Advantages and disadvantages of various biologicals materials that are sampled to monitor human exposures to nickel compounds are summarized in Table 1.[6,9,19,21] Urine and serum are the most convenient samples for biological monitoring. Urine specimens have the drawbacks of matrix variability and fluctuating specific gravity, but urine is usually preferable to serum, since collection of urine is noninvasive, while collection of blood for nickel analysis requires venepuncture with plastic intravenous cannulae to avoid nickel contamination from steel needles. Nickel concentrations are generally higher in urine than in serum, enhancing the accuracy of analysis. Determinations of nickel concentrations in dilute urine specimens (creatinine concentrations <0.3 g/l or specific gravity <1.010) should be repeated

Table 1
ADVANTAGES AND DISADVANTAGES OF VARIOUS SPECIMENS FOR NICKEL ANALYSES TO MONITOR HUMAN EXPOSURES TO NICKEL AND ITS COMPOUNDS

Specimens	Advantages	Disadvantages
Whole blood	Collection is convenient; centrifugation unnecessary; Ni contamination of heparin must be avoided	Venepuncture needed; blood may clot during storage
Serum	Collection is convenient; analysis is reliable	Venepuncture and centrifuge needed
Urine	No discomfort; analysis is reliable; Ni conc higher than in serum or whole blood	Dust contamination; matrix variability; dilute urines are unsatisfactory
Parotid saliva	Minimal discomfort; analysis is reliable	Collection is slow; special equipment; validity uncertain
Scalp hair	No discomfort; Ni conc does not fluctuate rapidly	Sampling is variable; contamination from dust, dyes, sweat; validity uncertain
Nasal mucosal biopsy	Correlates with chronic exposure to insoluble dust	Discomfort and risk of bleeding or infection; difficult to analyze

upon more concentrated urine specimens. Results of urine nickel analyses are generally reported as unadjusted nickel concentrations, with optional provision of supplemental values normalized on the basis of creatinine or specific gravity, according to the preference of the responsible physician.[6] Serum should obviously be used to monitor nickel exposures from hemodialysis, since little or no urine is available. Fecal nickel provides the most reliable index of oral exposures to nickel,[43] and expired breath may be useful for detection of nickel carbonyl following inhalation of this toxic compound[40] Analyses of nickel in other media, such as tissue samples (obtained by biopsy or autopsy), whole blood, saliva, sweat, milk, or hair may be useful for investigative or medicolegal purposes.[6,9,19,21,44]

X. REFERENCE VALUES

As a result of lower detection limits, improved analytical techniques, and better procedures to minimize nickel contamination from needles, containers, utensils, and reagents, the reference values for nickel concentrations in specimens from nonexposed persons have diminished substantially during the past 15 years.[45] Reference values for nickel concentrations in body fluids, excreta, and tissues from healthy adult persons, without occupational or iatrogenic exposures to nickel compounds, are listed in Table 2, based upon measurements in the author's laboratory.

XI. QUALITY ASSURANCE

Reference materials for nickel analyses have been described in several recent studies.[46-49] Specific recommendations for specimen collection, transport, storage, contamination control, recovery tests, and quality assurance of nickel analyses have been outlined by Sunderman et al.[6] Accurate measurements of nickel concentrations in body fluids require:

Table 2
NICKEL CONCENTRATIONS IN BODY FLUIDS, EXCRETA, AND TISSUES OF NONEXPOSED PERSONS[3,8,44]

Specimen	Subjects	Nickel concentrations Mean ± SD	Nickel concentrations Range	Units
Whole blood	30	0.34 ± 0.28	<0.05—1.05	μg/l
Serum	30	0.28 ± 0.24	<0.05—1.08	μg/l
Urine (spot)	34	2.0 ± 1.5	0.5—6.1	μg/l
		2.0 ± 1.6	0.4—6.0	μg/g creatinine
Urine (24 h)	50	2.2 ± 1.2	0.7—5.2	μg/l
		2.6 ± 1.4	0.5—6.4	μg/d
Feces (72 h)	10	14.2 ± 2.7	10.8—18.7	μg/g (dry wt)
		258 ± 126	80—540	μg/d
Scalp hair	20	220 ± 80	130—510	μg/kg
Arm sweat	33	52 ± 36	7—180	μg/l
Lung	9	173 ± 94	71—371	μg/kg (dry wt)
Kidney	10	62 ± 43	19—171	μg/kg (dry wt)
Liver	10	50 ± 31	11—102	μg/kg (dry wt)
Heart	9	54 ± 40	10—110	μg/kg (dry wt)
Spleen	10	37 ± 31	9—95	μg/kg (dry wt)

1. A clean, well-equipped, laboratory room that is solely devoted to trace element analysis
2. An analyst who is skilled in trace element determinations and who has sufficient time, patience, and motivation to perform the measurements meticulously and to tend to preventive maintenance
3. An electrothermal atomic absorption spectrophotometer, differential pulse polarograph, or comparable instrument for trace analysis, which is kept in electronic and mechanical adjustment
4. Auxilary staff to acid-wash the containers and utensils and to assist in specimen collection, identification, transport, storage, record keeping, and reporting
5. Availability of reference materials to monitor analytical accuracy and precision
6. Participation in interlaboratory programs for harmonization and quality assurance of nickel analyses
7. Supervision by a clinical pathologist, clinical chemist, or toxicologist who is concerned and knowledgeable about the accuracy, reliability, and clinical interpretation of the analytical results

Interpretations of nickel concentrations in body fluids require information about the routes, sources, and durations of possible nickel exposures, the chemical and physical properties of the nickel compounds that may be involved, and relevant clinical and physiological data (e.g., renal function tests).

XII. SUMMARY

The chemistry and analysis of nickel, in relation to biological monitoring of human exposures to nickel alloys and compounds, has been summarized in this chapter. EAAS and DPAV are currently the most practical and reliable instrumental techniques to measure nickel in biological materials. Effective contamination control during specimen collection, processing, and analysis, as well as a comprehensive program for quality assurance, are necessary to obtain clinically useful measurements of nickel concentrations in body fluids, excreta, and tissue samples.

REFERENCES

1. **Kasprzak, K. S. and Sunderman, F. W., Jr.,** Radioactive ^{63}Ni in biological research, *Pure Appl. Chem.,* 51, 1375, 1979.
2. **Sunderman, F. W., Jr.,** Analytical biochemistry of nickel, *Pure Appl. Chem.,* 52, 527, 1980.
3. **Sunderman, F. W., Jr.,** Nickel, in *Techniques and Instrumentation in Analytical Chemistry,* Vercruysse, A., Ed., Plenum Press, New York, 1984, 279.
4. **Sunderman, F. W., Jr.,** Nickel, in *Handbook on the Toxicology of Metals,* Sigel, H. and Seiler, H. G., Eds., Marcel Dekker, New York, 1986, 453.
5. **Sunderman, F. W., Jr.,** Determination of nickel in body fluids, tissues, excreta, and water, in *Environmental Carcinogens — Selected Methods of Analysis,* Vol. 8, O'Neill, I. K., Ed., International Agency for Research on Cancer, Lyon, France, 1986, 319.
6. **Sunderman, F. W., Jr., Aitio, A., Morgan, L. M., and Norseth, T.,** Biological monitoring of nickel, *Toxicol. Ind. Health,* 2, 17, 1986.
7. **Sunderman, F. W., Jr. and Oskarsson, A.,** Nickel, in *Metals and Their Compounds in the Environment,* Merian, E., Ed., VCH Publishers, Weinheim, West Germany, in press.
8. **Sunderman, F. W., Jr.,** Nickel, in *Handbook of Clinical Chemistry,* Werner, M., Ed., CRC Press, Boca Raton, FL, in press.
9. **Sunderman, F. W., Jr.,** Nickel, in *Biological Monitoring of Toxic Metals,* Clarkson, T., Friberg, L., Nordberg, G., and Sager, P., Eds., Plenum Press, New York, 1988, 265.
10. **Nriagu, J. O., Ed.,** *Nickel in the Environment,* John Wiley & Sons, New York, 1980.
11. **Brown, S. S. and Sunderman, F. W., Jr., Eds.,** *Nickel Toxicology,* Academic Press, London, 1980.
12. **Sunderman, F. W., Jr., Ed.,** *Nickel in the Human Environment,* International Agency for Research on Cancer, Lyon, France, 1984.
13. **Brown, S. S. and Sunderman, F. W., Jr., Eds.,** *Progress in Nickel Toxicology, Blackwell Scientific,* Oxford, 1985.
14. **Anon.,** Health Assessment Document for Nickel, U.S. Environmental Protection Agency, Research Triangle Park, NC, 1985.
15. **Grandjean, P.,** Health Effects Document on Nickel, Ontario Ministry of Labour, Toronto, 1986.
16. **Stoeppler, M.,** Analysis of nickel in biological materials and natural waters, in *Nickel in the Environment,* Nriagu, J. O., Ed., John Wiley & Sons, New York, 1980, 661.
17. **Stoeppler, M.,** General analytical aspects of the determination of lead, cadmium, and nickel in biological fluids, in *Analytical Techniques for Heavy Metals in Biological Fluids,* Facchetti, S., Ed., Elsevier, Amsterdam, 1981, 133.
18. **Schaller, K. H., Stoeppler, M., and Raithel, H. J.,** The analytical determination of nickel in biological matrices: summary of present knowledge and experience, *Staub. Reinhalt. Luft,* 42, 137, 1982.
19. **Aitio, A.,** Biological monitoring of occupational exposure to nickel, in *Nickel in the Human Environment,* Sunderman, F. W., Jr., Ed., International Agency for Research on Cancer, Lyon, France, 1984, 497.
20. **Mushak, P.,** Nickel metabolism in health and disease, in *Clinical Laboratory Annual,* Homburger, H. H. and Batsakis, J. G., Eds., Appleton-Century-Crofts, New York, 1984, 249.
21. **Norseth, T.,** Chromium and nickel, in *Biological Monitoring and Surveillance of Workers Exposed to Chemicals,* Aitio, A., Riihimaki, V., and Vainio, H., Eds., Hemisphere, Washington, D.C., 1984, 49.
22. **Norseth, T.,** Nickel, in *Handbook on the Toxicology of Metals,* Vol. 2, Friberg, L., Nordberg, G., and Vouk, V. B., Eds., Elsevier, Amsterdam, 1986, 462.
23. **Nieboer, E., Stetsko, P. I., and Hin, P. Y.,** Characterization of the Ni[III]/Ni[II] redox couple for the Ni[II]-complex of human serum albumin, *Ann. Clin. Lab. Sci.,* 14, 409, 1984.
24. **Cross, J. E., Hughes, D. M., and Williams, D. R.,** Critical review of the evidence for nickel[III] in animals and man, in *Progress in Nickel Toxicology, Brown, S. S. and Sunderman, F. W., Jr., Eds.,* Blackwell Scientific, Oxford, 1985, 109.
25. **Sunderman, F. W., Jr.,** Lipid peroxidation as a mechanism of acute nickel toxicity, *Toxicol. Environ. Chem. Rev.,* 15, 59, 1987.
26. **Nielsen, G. D., Andersen, O., Jensen, M., and Grandjean, P.,** Gastrointestinal nickel absorption: a new experimental model using the gamma-emitting isotope ^{57}Ni, *Abstr. UOEH Int. Symp. Bio- and Toxico-Kinetics of Metals,* University of Occupational and Environmental Health, Kitakyushu, Japan, 1986.
27. **Sunderman, F. W., Jr.,** Recent progress in nickel carcinogenesis, *Toxicol. Environ. Chem. Rev.,* 8, 235, 1984.
28. **Stedman, D. H., Pearson, R., and Yalvac, E. D.,** Nickel carbonyl: decomposition in air and related studies, *Science,* 208, 1029, 1980.
29. **Sunderman, F. W., Jr., Crisostomo, C., Reid, M. C., Hopfer, S. M., and Nomoto, S.,** Rapid analysis of nickel in serum and whole blood by electrothermal atomic absorption spectrophotometry, *Ann. Clin. Lab. Sci.,* 14, 232, 1984.

30. **Sunderman, F. W., Jr., Marzouk, A., Crisostomo, C., and Weatherby, D. R.,** Electrothermal atomic absorption spectrophotometry of nickel in tissue homogenates, *Ann. Clin. Lab. Sci.,* 15, 299, 1985.
31. **Sunderman, F. W., Jr., Hopfer, S. M., Crisostomo, C., and Stoeppler, M.,** Rapid analysis of nickel in urine by electrothermal atomic absorption spectrophotometry, *Ann. Clin. Lab. Sci.,* 16, 219, 1986.
32. **Sunderman, F. W., Jr.,** Nickel analysis by electrothermal atomic absorption spectrophotometry, *Methods Enzymol.,* 158, 382, 1988.
33. **Ostapczuk, P., Valenta, P., Stoeppler, M., and Nurnberg, H. W.,** Voltammetric determination of nickel and cobalt in body fluids and other biological materials, in *Chemical Toxicology and Clinical Chemistry of Metals,* Brown, S. S. and Savory, J., Eds., Academic Press, London, 1983, 61.
34. **Uto, M., Itoh, Y., and Sugawara, M.,** Differential pulse polarographic determination of nickel as water-soluble dithiocarbamate, *Fresenius Z. Anal. Chem.,* 321, 68, 1985.
35. **Brown, S. S., Nomoto, S., Stoeppler, M., and Sunderman, F. W., Jr.,** IUPAC reference method for analysis of nickel in serum and urine by electrothermal atomic absorption spectrophotometry, *Clin. Biochem.,* 14, 295, 1981.
36. **Brown, S. S., DiMichel, A. D., and Sunderman, F. W., Jr.,** Interlaboratory comparison of analyses of serum and urine for nickel by the IUPAC provisional reference method, in *Nickel Toxicology,* Brown, S. S. and Sunderman, F. W., Jr., Eds., Academic Press, New York, 1980, 167.
37. **Brown, S. S.,** Harmonization of trace metal analysis in clinical biochemistry: a case study, in *Proc. 11th Int. Congr. Clinical Chemistry,* Kaiser, E., Gabl, F., Müller, M. M., and Bayer, M., Eds., de Gruyter, Berlin, 1982, 1107.
38. **Sunderman, F. W., Jr., Brown, S. S., Stoeppler, M., and Tonks, D. B.,** Interlaboratory evaluations of nickel and cadmium analyses in body fluids, in *Collaborative Interlaboratory Studies in Chemical Analysis,* West, T. S. and Egan, H., Eds., Pergamon Press, London, 1982, 25.
39. **Grove, T. H. and Sunderman, F. W., Jr.,** Interference by Tris(hydroxymethylamino)methane in nickel analysis by electrothermal atomic absorption spectrometry, *Ann. Clin. Lab. Sci.,* 8, 495, 1978.
40. **Sunderman, F. W., Jr., Rozel, N. O., and Clark, R. J.,** Gas chromatography of nickel carbonyl in blood and breath, *Arch. Environ. Health,* 16, 836, 1968.
41. **Stedman, D. H., Tammaro, D. A., Branch, D. K., and Pearson, R.,** Chemiluminescence detector for the measurement of nickel carbonyl in air, *Anal. Chem.,* 51, 2340, 1979.
42. **Filkova, L.,** *Analysis of Nickel Carbonyl in Air,* Institute of Hygiene and Epidemiology, Prague, 1985.
43. **Hassler, E., Lind, B., Nilsson, B., and Piscator, M.,** Urinary and fecal elimination of nickel in relation to air-borne nickel in a battery factory, *Ann. Clin. Lab. Sci.,* 13, 217, 1983.
44. **Rezuke, W. N., Knight, J. A., and Sunderman, F. W., Jr.,** Reference values for nickel concentrations in human tissues and bile, *Am. J. Ind. Med.,* 11, 419, 1987.
45. **Versieck, J.,** Trace elements in body fluids and tissues, *CRC CRit. Rev. Clin. Lab. Sci.,* 22, 97, 1985.
46. **Veillon, C., Lewis, S. A., Patterson, K. Y., Wolf, W. W., Harnly, J. M., Versieck, J., Vanballenberghe, L., Cornelis, R., and O'Haver, T. C.,** Characterization of a bovine serum reference material for major, minor, and trace elements, *Anal. Chem.,* 57, 2106, 1985.
47. **Kiilunen, M., Jarvisalo, J., Makitie, O., and Aitio, A.,** Analysis, storage stability, and reference values for urinary chromium and nickel, *Int. Arch. Occup. Environ. Health,* 59, 43, 1987.
48. **Christensen, J. M., Ihnat, M., Stoeppler, M., Thomassen, Y., Veillon, C., and Wolynetz, M.,** Human body fluids: IUPAC proposed reference materials for trace element analysis, *Fresenius Z. Anal. Chem.,* 326, 639, 1987.
49. **Valkonen, S., Jarvisalo, J., and Aitio, A.,** Lyophilized and natural urine specimens in the quality control of analyses of toxic metals, *Ann. Clin. Lab. Sci.,* 17, 145, 1987.

Chapter 2

HUMAN NICKEL EXPOSURE AND CHEMOBIOKINETICS

P. Grandjean, G. D. Nielsen, and O. Andersen

TABLE OF CONTENTS

I. Introduction 10

II. Nickel Exposure Sources 10
- A. Environmental Exposures of the Skin 10
- B. Occupational Exposures of the Skin 11
- C. Oral Exposure 12
- D. Pulmonary Exposure 18
- E. Iatrogenic Exposures 19

III. Chemobiokinetics 20
- A. Absorption 20
  - 1. Dermal Absorption 20
  - 2. Intestinal Absorption 21
  - 3. Pulmonary Deposition, Clearance, and Absorption 22
- B. Systemic Distribution 23
- C. Organ Distribution 23
- D. Cellular Distribution 24
- E. Excretion 24

IV. Diagnostic Tests for Exposure 25

V. Prevention 26
- A. Preventive Strategies 26
- B. Cutaneous Exposures 26
- C. Dietary Restrictions 26
- D. Pulmonary Exposures 27

VI. Conclusions 27

Acknowledgment 28

References 28

I. INTRODUCTION

Nickel is a transition-group metal which is used in a large number of alloys and chemical compounds. It is the fourth most-used metal, and only iron, chromium, and lead are produced in larger amounts. In the environment, nickel constitutes about 0.008% of the earth's crust and ranks as the 24th element in order of abundance. Nickel has therefore been detected in different media in all parts of the biosphere. Thus, humans are constantly exposed to this ubiquitous element, although in variable amounts. The average natural nickel exposure from food in the past has probably been somewhat, but not much, below current levels.

Human nickel exposures are of concern for several reasons. Nickel is an essential nutrient for several species of laboratory animals[1] and may also be essential for humans. The possible nickel requirement is probably very low. On the other hand, inorganic nickel may cause acute toxicity at high dose levels.[2] However, the major emphasis today is placed on allergenic and carcinogenic effects. As reviewed elsewhere in this volume, nickel is a frequent cause of contact dermatitis. Although skin contact is of major significance in this regard, provocation experiments suggest a possible significance of oral nickel intake as a cause of eczema exacerbation or development in nickel-sensitive individuals.[3] Further, case reports suggest the possible induction of respiratory allergy in the form of asthma following inhalation of soluble nickel compounds,[4] but such cases appear to be rare. In addition, increased respiratory cancer risk after exposure to sparingly soluble nickel compounds has been firmly demonstrated, and increased gastrointestinal cancer risk after exposure to soluble nickel compounds has been suggested.[5]

For these reasons, environmental and occupational nickel exposure needs to be controlled. Our present knowledge about human nickel exposures, the fate of nickel in the human body, and the potentials for prevention are reviewed in this chapter.

II. NICKEL EXPOSURE SOURCES

A. Environmental Exposures of the Skin

Since 1940, the incidence of nickel allergy has increased considerably in parallel with the augmented use of this metal in everyday articles.[6] Today almost half of the nickel production in the world is used for stainless steel, while about 10% is used in alloy steels. Cupronickel and other nonferrous, high-nickel alloys account for 20% of the production, electroplating and certain foundry uses for 10%, and a smaller amount is used for batteries, catalysts, chemicals, etc.[7]

Sensitization is mainly a result of daily skin contact with corrosive metal objects containing nickel. Such everyday objects include jewelry made of German silver (containing 10 to 20% nickel), especially earrings for pierced ears, nickel-plated spectacle frames, buttons for blue jeans, clasps, buckles, zippers, etc.; even jewelry made of white gold with 2 to 15% nickel may cause eczema.[8-10] Common sources of nickel dermatitis are listed in Table 1.

The amount of nickel released from such objects depends on the corrosion resistance of the object and the presence of sweat and other corrosive fluids in the environment. Due to the chloride content and the relatively low pH, sweat may dissolve nickel. Experiments with synthetic sweat and physiological salt solutions have shown that nickel may be leached from various materials. A particular group of people called "rusters", with an abnormally high chloride concentration in their sweat, are likely to release even more nickel.[11] After a 1-week submersion in a sweat solution at room temperature, milligram quantities were leached from coins, and even smaller amounts were released from stainless-steel materials containing about 10% nickel, although stainless steel is known to be very resistant to corrosion due to the protecting effect of chromium oxides on the surface.[12] More recently, the dissolution of nickel in synthetic sweat was examined for 15 different metals and alloys.[13] A high corrosion

Table 1
SITES AND NICKEL SOURCES OF CONTACT DERMATITIS

Location	Nickel source causing dermatitis
Scalp	Hairpins, curlers, bobby pins
Eyelids	Eyelash curler
Earlobes	Earrings
Back of ears	Eyeglass frames
Sides of face	Bobby pins, curlers
Lips	Metal pins held in mouth, metal lipstick holder
Neck	Clasp of necklace, zipper
Upper chest	Medallions, metal identification tags, buttons
Axilla	Zipper
Breast	Wire support of bra cup
Belly	Buttons on jeans
Palms	Handles of doors, handbags
Fingers	Thimbles, needles, scissors, coins, pens
Wrists	Watchbands, bracelets
Arms	Bracelets
Ankle	Bracelets
Dorsum of foot	Metallic eyelets of shoes
Plantar aspect	Metal arch support
Postoperative sites	Screws, bolts, plates in orthopedic implants

Modified from Fisher.[8]

was documented for nickel-plated items (both electrolytically and chemically plated), nickel-iron alloy (65% nickel), German silver (10 to 20% nickel), Monel 400 alloy (66% nickel), Nicrobraze LM alloy (83% nickel), and coin alloy (25% nickel). Nine of the materials examined caused medium to severe degrees of allergic skin reactions, and the amount of nickel dissolved by corrosion was related to the degree of allergic reaction.

An allergic patient may successfully use Fisher's dimethylglyoxime spot test to identify free nickel ions. One drop of a 1% solution of dimethylglyoxime in ethanol and one drop of 10% ammonia are placed on the object; the fluids are mixed and rubbed onto the object with a small cotton swab. The appearance of a red complex will indicate dissolution of nickel from the object.[8]

The nickel concentration used when assessing the presence or absence of nickel sensitivity is usually 5% as $NiSO_4$-H_2O. However, the average threshold for provocation of a positive response in a group of nickel-allergic patients was about one tenth of that concentration (5 g/l) in distilled water and even lower when administered in petrolatum.[14] Further, a 6-h exposure of forearm skin to nickel concentrations of 12 mg/l may result in an allergic reaction in allergic individuals. A tentative safety limit for nickel solutions which come into contact with skin has therefore been proposed at 1 μmol/l, i.e., 60 μg/l.[15] Also, alloys with a nickel release to synthetic sweat above 1 μg/cm²/week gave a strong patch test reaction in nickel-sensitive patients, whereas those with a release below 0.5 μg/cm²/week generally showed a weak reaction.[13] Unfortunately, detailed quantitative data which indicate dose-response relationships and possible threshold levels are not available.

B. Occupational Exposures of the Skin

Nickel allergy has often been associated with occupational exposure. As reviewed by

Marcussen,[6] many cases of skin disease in nickel platers were reported before 1930 from large industrial centers such as Berlin, Kiel, Leningrad, and Toronto. The disease mainly occurred in workers occupied directly with the nickel baths. In some instances, the picture was complicated by irritative eczema and pyoderma, but the concept of allergic contact dermatitis was not yet known at that time. During the period from 1930 to 1940, when hygienic improvements were introduced in the industry, most of these cases disappeared. Later, nickel-associated skin disease was reported from several occupations. One study found that 4% of 621 eczema cases were related to nickel plating, 9.5% to other occupations, and 86.5% were apparently not associated with present occupation. Occupations with increased risk included sewer workers, tailors, hairdressers, waitresses, workers from the metal industry, cashiers, shop assistants, nurses, and physicians, but nickel sensitization in most of these cases had probably not occurred as a direct result of the occupation.[6]

With regard to specific occupations, Wahlberg[16] investigated a group of 35 hairdressers with hand eczema and found that 14 (40%) were sensitive to nickel when patch-tested. Marks and Cronin[17] compared a group of hairdressers with hand eczema to a matched group of housewives with hand eczema and found a significantly higher prevalence of nickel allergy among the hairdressers. The reason for this difference may be a combination of jewelry worn and handling at work of nickel-containing objects under conditions where the barrier function of the skin may be weakened.[18]

Among 853 hard-metal workers, nickel sensitivity was found in 2 men and 38 women. A jewelry dermatitis in 88% of the nickel-sensitive individuals had developed prior to employment in the hard-metal industry or before the appearance of hand eczema, but 40% subsequently developed severe hand eczema, generally 6 to 12 months after first employment, and 95% of the nickel-allergic women had pierced ear lobes.[19]

At an electroforming plant with heavy exposure to nickel, 13 of 27 workers had nickel allergy as determined by patch testing. Improvements in industrial hygiene led to an immediate decrease in the incidence of dermatitis.[20]

A high incidence of nickel allergy with hand eczema among janitors found in a 10-year study by Fregert[21] may be due to nickel concentrations increasing in the water during successive stages of cleaning; a high level of 90 μg/l was found in water from used cloths.[22]

Although many patients with an occupational nickel-related contact dermatitis may have been sensitized elsewhere, working conditions frequently provoke the outbreak of eczema. These cases may not necessarily be recognized as occupationally related and are frequently not reported as such. During the period 1966 to 1980, only 83 claims were allowed by the Ontario Workmen's Compensation Board for contact dermatitis, an average of 5 to 6 cases per year.[23] These claims occurred in a few specific nickel-exposed trades only, and other cases of nickel-induced contact eczema may be found in the larger number of other eczema claims allowed. This number appears to be very low for a province with the largest nickel production in the world, compared to an annual average of about 20 applications for permanent disability pension because of occupational contact dermatitis due to nickel in Denmark,[24] a country with fewer inhabitants, However, these statistics may be influenced by a tendency of nickel-sensitive patients to avoid jobs with nickel exposure or rapidly leave the job if reactions occur.

C. Oral Exposure

Nickel content of various food items may span a very wide range, as shown in Tables 2 and 3. The highest nickel contents (above 1 mg/kg fresh weight) were found in cocoa and chocolate, soy beans, soy products and other dried legumes, different nuts, oatmeal, and buckwheat.[25,26]

The variations in nickel concentrations of different types of fruits, roots, and vegetables may be due to genetic factors of the species rather than to environmental factors, such as

Table 2
NICKEL CONTENT IN FOODS AND NICKEL INTAKE IN THE DANISH AVERAGE DIET

	Nickel content (μg/g)			Consumption		Nickel intake	
	n	Range	Mean	g/person/day	% of total	μg/person/day	% of total
Milk products	94			514	24.5	12.7	8.7
Full milk	63	BDL—0.13	0.02	332	15.8	6.64	4.5
Skim milk			0.02	107	5.1	2.14	1.5
Ymir			0.01	10	0.5	0.10	0.1
Yogurt	3	0.004—0.03	0.01	10	0.5	0.10	0.1
Cream	3	0.01—0.04	0.03	26	1.2	0.78	0.5
Cheese	25	0.02—0.34	0.10	29	1.4	2.90	2.0
Meat, fish, egg	869			211	10.1	7.2	4.9
Beef	32	0.01—0.03	0.02	42	2.0	0.84	0.6
Pork	20	<0.02—0.02	0.02	85	4.0	1.70	1.2
Chicken	9	0.02—0.24	0.11	11	0.5	1.21	0.8
Lamb	12	<0.02—0.02	0.02	2	0.1	0.04	0.03
Liver, kidney	108	0—0.94	0.11	5	0.2	0.55	0.4
Fish	658	0.005—0.303	0.04	40	1.9	1.60	1.1
Egg	30	0.01—0.35	0.05	26	1.2	1.30	0.9
Roots and vegetables	228			298	14.2	43.2	29.5
Potatoes	45	BDL—0.44	0.14	174	8.3	24.4	16.6
Carrots	17	<0.01—0.16	0.04	10	0.5	0.40	0.3
Celery root	8	0.04—0.1	0.06	3	0.1	0.18	0.1
Beet root	7	0.01—0.3	0.12	3	0.1	0.36	0.2
Cabbage	31	0.01—0.63	0.17	20	1.0	3.40	2.3
Cauliflower	5	0.03—1.0	0.3	10	0.5	3.00	2.0
Kale	2	0.15—0.24	0.20	2	0.1	0.40	0.3
Salad	21	BDL—1.4	0.36	4	0.2	1.44	1.0
Spinach	15	0.02—2.99	0.52	2	0.1	1.04	0.7
Leeks	7	0.3—0.2	0.11	2	0.1	0.22	0.2
Asparagus	1	—	0.42	1	0.05	0.42	0.3
Onions	16	0.3—0.42	0.08	18	0.9	1.44	1.0
Cucumber	8	0.01—0.11	0.04	13	0.6	0.52	0.4
Tomatoes	21	0.01—0.25	0.07	26	1.2	1.82	1.2
Peas	24	0.13—0.8	0.42	10	0.5	4.20	2.9

Table 2 (continued)
NICKEL CONTENT IN FOODS AND NICKEL INTAKE IN THE DANISH AVERAGE DIET

	Nickel content (μg/g)			Consumption		Nickel intake	
	n	Range	Mean	g/person/day	% of total	μg/person/day	% of total
Fruits	153			140	6.7	7.5	5.1
Apples	11	BDL—0.03	0.01	51	2.4	0.51	0.3
Pears	10	0.07—0.42	0.14	5	0.2	0.70	0.5
Plums	10	0.03—0.20	0.12	4	0.2	0.48	0.3
Currants	13	0.01—0.2	0.06	7	0.3	0.42	0.3
Strawberries	9	0.03—0.08	0.05	5	0.2	0.25	0.2
Rhubarb	10	0.01—0.22	0.13	2	0.1	0.26	0.2
Grapes	4	0.01—0.04	0.02	5	0.2	0.10	0.1
Raisins	3	0.02—0.04	0.03	4	0.2	0.12	0.1
Citrus fruits	3	0.01—0.04	0.03	27	1.3	0.81	0.6
Bananas	4	0.01—0.03	0.02	15	0.7	0.30	0.2
Canned fruits	65	0.02—1.36	0.31	11	0.5	3.41	2.3
Juice	11	0.01—0.17	0.04	4	0.2	0.16	0.1
Meal, grain, and bread	81			188	9.0	36.0	24.6
Wheat flour	32	0.03—0.3	0.13	113	5.4	14.7	10.0
Rye flour	15	0.03—0.3	0.1	62	3.0	6.20	4.2
Oatmeal	18	0.80—4.7	1.76	8	0.4	14.1	9.6
Rice	16	0.08—0.45	0.21	5	0.2	1.05	0.7
Potato flour			0.63	10	0.5	6.30	4.3
Fats	17			92	4.4	19.7	13.4
Butter	4	0.03—0.2	0.1	23	1.1	2.3	1.6
Fat			0.02	19	0.9	0.38	0.3
Margarine	13	0.2—2.5	0.34	50	2.4	17.0	11.6
Sugar	22	0.01—0.09	0.05	132	6.3	6.6	4.5
Beverages	94			498	23.7	6.1	4.8
Beer	14	0.003—0.02	0.01	295	14.1	2.95	2.0
Light beer			0.01	40	1.9	0.40	0.3
Wine	21	0—0.14	0.06	15	0.7	0.90	0.6
Fortified wine and spirits	4	0.08—0.17	0.10	15	0.7	1.50	1.0

Mineral water and soda water	37	0.002—0.01	0.01	133	6.3	1.33	0.9
Tea	9	0.02—0.09	0.05	1	0.5	0.05	0.03
Coffee	9	0.006—0.04	0.02	15	0.7	0.30	0.2
Total	1558			2099	100.0	146.6	100.0

Note: BDL = below detection limit.

Modified from Flyvholm et al.[26]

Table 3
NICKEL CONTENT IN FOOD NOT INCLUDED IN THE DANISH AVERAGE

	Range (μg/g)	n	Mean (μg/g)
Cocoa	8.2—1.2	7	9.8
Soy beans	4.7 5.9	3	5.2
Soy products	1.08—7.8	7	5.1
Walnuts	—	1	3.6
Peanuts	1.6—3.9	2	2.8
Oats	0.33—4.8	37	2.3
Buckwheat	1.3—2.8	3	2.0
Bitter chocolate	1.3—2.7	7	1.9
Hazelnuts	0.66—2.3	12	1.8
Dried legumes	0.57—3.3	17	1.7
Almonds	1.2—1.3	5	1.3
Pistachio nuts	—	1	0.8
Milk chocolate	0.4—1.2	11	0.7
Prunes	0.5—0.8	3	0.6
Whole meal, white bread	0.20—0.51	41	0.49
Canned beans	0.33—0.61	8	0.45
Beans (frozen and fresh)	0.15—1.0	23	0.40
Whole wheat	0.1—0.8	85	0.33
Endive	0.07—0.66	10	0.31
Broccoli	0.03—0.8	9	0.28
Canned peas	0.1—0.53	11	0.28
White bread	0.02—0.31	65	0.27
Maize	0.1—0.3	3	0.2
Preserved canned meat	n.d.—1.04	221	0.18
Sprouts	0.02—0.80	19	0.16
Mushrooms	0—0.61	30	0.16
Rye bread	0.1—0.28	10	0.13
Shellfish and crayfish	0.08—0.14	3	0.10
Red cabbage	0.05—0.1	3	0.1
Herring	0.01—0.02	6	0.20
Total	—	663	—

Note: n.d. = not detected.

Modified from Flyvholm et al.[26]

variations in nickel content in the farm soil and sedimentation from the air. Although farm soils contain nickel concentrations in the range 0.003 to 0.1%,[27] this wide variation is not reflected in the nickel levels in crops grown on the specific soils. Because of a relatively tight binding of nickel to soil colloids, the nickel levels in soil are only slowly changeable. The mean residence time for nickel in soil has been estimated at about 2400 years.[28,29]

Seafood may be quite high in nickel, and oysters, other shellfish, and salmon may approach or exceed a nickel level of 1 mg/kg.[27,30-32] However, fish and meat products normally contain much less nickel, with the possible exception of kidneys[26,32,32] (Tables 2 and 3).

In addition, various fats may contain much nickel, probably related to the use of nickel catalysts for commercial hydrogenation. Margarine normally contains less than 0.2 g Ni/kg·3[3] but levels up to 6 mg/kg have been found.[27]

Daily oral intake of nickel has been examined by various methods. Total diets have been studied by the duplicate portion method. A study of hospital diets in the U.S. showed that the general diet contained 160 μg/d, and special diets varied by less than 40% from this level.[34] A daily intake of 290 μg was found in a balance study in the U.S.S.R.,[35] and another study stated that the daily intake was 0.5 mg without mentioning the method used.[36] As only a small fraction of the nickel in the diet will be absorbed in the gut, fecal nickel excretion will normally reflect dietary intakes (except in the case of occupational exposure). An American study showed an average excretion of 260 μg/d.[37]

Based on studies of major food items, Schroeder et al.[30] found 470 μg in a 1-d hospital diet and estimated the normal daily intake at 300 to 600 μg, although with different diets the nickel intake could vary between 100 and 900 μg/d. The latter diet included oatmeal, soybeans, cocoa, and nuts. These food items have a high nickel content.[38] Vegetables, cereals, and bread contributed most of the daily nickel intake of 150 μg in the average Danish diet.[38] However, even with a daily average nickel intake of 200 μg or less, three or four times that amount may be ingested by a vegetarian, particularly when chocolate and nuts are included in the diet.

Drinking water can be taken from groundwater or from surface water. Nickel levels in surface waters affected by geochemical processes vary considerably and exceed, in general, those found in groundwaters.[39] The nickel content of groundwater is usually low. The levels appear to be similar in raw, treated, and distributed municipal water.[40] The median level in Canada is below 2 μg/l, but high levels occur in Ontario.[40] In particular, municipal tap water near large open-pit mines averaged about 200 μg/l, while a control area showed an average of about 1 μg/l. This difference was also reflected in nickel levels in serum and urine of the inhabitants.[41] Other studies have suggested low background levels in drinking water, e.g., in Finland an average of about 1 μg/l.[42] In the German Democratic Republic, drinking water from groundwater averaged 10 μg/l, slightly below the amount present in surface water.[43] Groundwater pollution with soluble nickel compounds from a nickel-plating facility resulted in nickel levels up to 2,500 μg/l and a median for 12 wells of 180 μg/l.[44] Except for such unusual cases, the direct contribution of drinking water to the daily nickel intake is small. Very high nickel contents of water could conceivably be of significance to sensitive patients if skin exposure is more than a brief shower, but this possibility has not been examined.

On the other hand, drinking water may contain much nickel due to leaching from nickel-chromium-plated armatures. One study showed up to 490 μg/l,[45] and another up to 957 μg/l in "first draw" drinking water.[46] Ten used water faucets which were filled with 15 ml deionized water in an inverted position and left overnight for 16 h leached between 18 and 900 μg of nickel.[46]

Nickel may also be leached from kitchen utensils during food preparation. Kitchen utensils made of nickel may add more than 100 mg of nickel to the daily intake due to leaching processes during food preparation.[47] Less dissolution takes place from alloys containing nickel and chromium. One hour of boiling in an alloy container with a 15% nickel content may leach 60 μg of nickel per square decimeter of exposed surface.[47] Several micrograms of nickel were even released from stainless-steel saucepans during up to 1.5 h of boiling, the release being larger at a lower pH.[48]

In one study, 82 kitchen utensils were examined by boiling water containing 4% acetic acid in them (for three times 30 min), a standard test usually employed for examining metal releases from ceramic glazes.[49] The highest concentrations, as expressed in relation to the effective volume, were 1.16 g/l or 91 mg/dm^2 (a nickel-plated tea ball), 137 mg/l (an electric kettle), 52 mg/l or 12.5 mg/dm^2 (an eggbeater), and 26 mg/l (a blender). The electric kettle released 36 mg/l when water (without acetic acid) was used for the boiling test. Four stainless-steel saucepans and pots released an average of 49 μg/l to the acetic acid solution. Although some test conditions were somewhat extreme and not entirely realistic, this study demonstrated the occurrence of very high leaching levels.

The average contribution from leaching processes to the daily nickel intake is unknown. Cookware, kitchen utensils, and armatures may occasionally add 1 mg to daily intake, i.e., an amount which may be equivalent to the intake resulting from nickel in foods and beverages.

Leaching of nickel can also take place from nickel-containing dental alloys. One experiment suggested that the average daily nickel release from such materials to the saliva is about 4.2 $\mu g/cm^2$ of surface.[50] This source seems to be of limited importance.

Additional gastrointestinal exposure may occur after occupational exposure through the respiratory tract (as described below) because of pumonary clearance. Due to insufficient personal hygiene, poor work practices, or the lack of necessary facilities, additional amounts of nickel dusts may enter the body through the gastrointestinal tract. One study has documented these exposure routes in a battery factory where high fecal nickel levels were related to dusty work conditions.[51]

The significance of these oral intakes of nickel for the eczema status in nickel-sensitized patients it not yet clear. Several investigators have observed exacerbation of hand eczema after oral challenge of nickel-sensitive women with 0.6 to 5.6 mg nickel as nickel sulfate administered in a lactose capsule[3,38,52-55] (see Chapter 14). Increased lymphocyte transformation rate was seen after oral challenge of nickel-sensitive patients with 4 mg nickel as nickel sulfate.[56] On the other hand, intense handling of nickel-contaminated metal objects by these patients did not induce any visible eczematous activity,[52] thus indicating that worsening of hand eczema could be provoked by systemic rather than cutaneous exposure. However, the oral nickel challenge corresponds to at least the upper limit of normal daily nickel intake, and the nickel speciation may be different from that occurring in food and beverages.

D. Pulmonary Exposure

Nickel concentrations in the atmosphere at remote locations are about 1 ng/m^3, a level which would contribute a negligible amount to human absorption of nickel. Natural sources contribute an estimated 16% of current nickel levels in the atmosphere, while burning of residual and fuel oils, nickel mining and refining, and municipal waste incineration result in the majority of anthropogenic nickel emissions to the atmosphere.[57] Accordingly, suburban and rural areas usually exhibit air nickel levels from a few to several nanograms per cubic meter but average levels in cities in excess of 100 ng/m^3 have been documented; even higher community levels have been measured in Sudbury, Ontario.[27] With indoor levels somewhat lower than those outside, an exposure level of nickel would generally average about 20 ng/m^3 in urban areas and 10 ng/m^3 in rural areas.[58] These concentrations can result in a total exposure of 0.25 to 0.5 μg per person per day when inhaling 25 m^3 air per day.

Additional pulmonary intake may occur through smoking. One cigarette may contain about 2 μg of nickel,[59,60] of which perhaps 10% is released into the mainstream smoke, possibly in part as nickel carbonyl. Unfortunately, no recent information in this area is available. The pulmonary exposure from cigarettes could contribute 4 μg nickel if one smokes 20 cigarettes, obviously much more than that originating from an urban atmosphere. The exact contribution cannot be evaluated until information of nickel levels in tobacco and speciation in the smoke has been updated.

Pulmonary exposure to airborne nickel dust exists in a variety of nickel-producing and -using industries. Nickel levels in the mines are usually about 0.01 mg/m^3, while somewhat higher levels occur in the mills where the wet ore is processed, generally well below 1 mg/m^3; at the smelters, average nickel levels of 0.03 to 0.05 mg/m^3 occur near roasters, furnaces, and converters, but much higher levels have been measured in the matte casting and crushing area.[23] In a nickel refinery, the highest averages approach 0.2 mg/m^3 in the feed preparation area and at the converters, and additional exposures occur in iron and copper recovery plants.[23] The pulmonary exposure of inorganic nickel in nickel-producing industries occurs

Table 4
OCCUPATIONS WITH EXPOSURE TO NICKEL

Battery makers	Nickel miners
Ceramic makers	Nickel refiners
Coal gasification workers	Nickel smelters
Dyers	Oil dehydrogenators
Electroformers	Paint makers
Electroplaters	Sandblasters
Enamelers	Spark plug makers
Glassworkers	Spray painters
Ink makers	Stainless steel makers
Jewelers	Textile dyers
Magnet makers	Varnish makers
Metalworkers	Welders

mainly in the form of insoluble or sparingly soluble nickel compounds. An average nickel level of 0.2 mg/m^3 air may result in a daily pulmonary retention of about 1.6 mg.

Further, respiratory exposure to nickel occurs in several user industries. A number of occupations with nickel exposures have been identified by the National Institute for Occupational Safety and Health (NIOSH)[61] or identified by biological monitoring (Table 4). Limited evidence suggests that exposures at electroplating operations may be considerable, and an average of about 1 mg/m^3 has been reported.[23] Grinding, casting, handling of nickel powder, and welding may also cause high nickel exposures. Under circumstances of insufficient ventilation, stainless-steel welding,[62] or deseaming of steel with oxygen-gas torches[63] have resulted in nickel levels of several milligrams per cubic meter. Sandblasters may be exposed to dusts from old paints containing nickel and, additionally, to nickel containing abrasive materials. Coal slag has been recently introduced as a substitute for sand in abrasive blasting, and in 12 samples the nickel concentration averaged 40 μg/g.[64] Metal workers may be exposed to nickel dusts from various alloys,[65,66] and some exposure may occur through leaching into cutting oils.[67] Auto mechanics are exposed to nickel from welding fumes and perhaps through the skin from nickel in motor oils.[68] Nickel exposures occur in the production of nickel-cadmium batteries[51,65] and in glass production when nickel molds are used.[69] A questionnaire study indicated that nickel powder was used in at least 10% of various types of laboratories,[70] but the extent of possible exposures was not assessed. In dental laboratories, where nickel-containing alloys may be worked, exposures appear to be minimal.[71] Additional, occasionally unexpected exposures may occur in the case of bystanders with indirect nickel exposures.[66]

Several studies have documented increased nickel concentrations in the body fluids of welders.[66,72,73] Electroplaters and polishers are exposed to soluble nickel salts that may cause high urinary levels.[65,74-76] Increased nickel concentrations in plasma and urine have also been found in paint and pigment workers,[74,76] painters, and sandblasters.[66] Coal gasification workers showed only minimally increased nickel levels in urine.[65]

D. Iatrogenic Exposures

Nickel-containing alloys may be implanted in a patient for various purposes, such as joint prostheses, plates and screws for fractured bones, surgical clips, and steel sutures. The most commonly implanted alloys are stainless steel (usually 10 to 14% nickel) and Vitallium (cobalt-chromium alloys containing up to 2.5% nickel). The wire and probe leading to the heart muscle from a nickel-cadmium battery pacemaker may contain 35% nickel.[77] Many commonly used prosthetic heart valves implanted in patients with end-stage rheumatic heart disease contain nickel in the alloy which forms the metal framework.[78]

Leaching of nickel may occur due to slow corrosion of the implanted parts because of the chloride level and the low oxygen content. Further, pH may not always be neutral, e.g., in a wound during the healing process. Immediately after surgery, the pH drops to 5.5 and only in the course of 10 to 15 d does it again reach neutral.[79] Corrosion from implanted parts has been simulated in the laboratory. A 1-week immersion of alloys in physiological saline, whole blood, or plasma suggested initial leaching rates corresponding to several micrograms per square centimeter.[77] This release corresponds to the leaching from nickel alloys to saliva.[50] In a longer-term study, nickel release from steel powder into a Ringer's solution declined after a few days and settled at a very low rate of about 0.3 $ng/cm^2/d$.[80] Initial release was only about ten times that rate, i.e., about one thousandth of other reported release rates. This discrepancy may not be surprising, as corrosion is very dependent on the alloy surface, the test solution, and, as a result of these, the possibility of obtaining an electric potential. Using the lower figure for nickel leaching, 100 cm^2 of surface area of a hip prosthesis may release 10 μg of nickel per year. The higher release rates would suggest an even higher amount per day.

Significant concentrations have been detected in intravenous albumin solution with three commercial brands averaging 200 μg/l.[81] Additional, though small, amounts of nickel may be dissolved from intravenous needles. Leaching of nickel may contaminate intravenous fluids. In one severe incident, a nickel-plated stainless-steel water heater tank contaminated the dialysate and caused plasma nickel levels of 3 mg/l and acute nickel toxicity in patients treated by hemodialysis.[82] Even during normal operation, the average intravenous nickel uptake may be estimated at 100 μg per dialysis.[81]

Further, nickel released from a hemodialysis system was in a case study shown to cause cutaneous hypersensitivity when administered intravenously. An *in vitro* experiment was performed to demonstrate the ability of nickel to be dialyzed into blood through a standard hemodialysis system. A plasma nickel level 89% higher than that of the original dialysate was achieved with a single cycle of dialysis.[83]

The biological significance of nickel release from various implants has been questioned[55] (Chapter 16). However, failure of implanted heart valves has been blamed on nickel allergy.[78] Likewise, Evans et al.[84] suggested that metal sensitivity is a cause of bone necrosis and loosening of the prosthesis in total joint replacement, as 9 of 14 patients with loose prosthesis were found to be metal sensitive by skin patch testing. In 24 patients with intact prostheses, no sensitivity was demonstrated. Further, implanted stainless-steel screws in animals sensitized to nickel resulted in severe reactions with necrosis, increased number of inflammatory cells, and foreign-body giant cells, when compared to control animals.[85] Also, allergic dermatitis may develop after such internal exposure to nickel,[86,87] but sensitization may occur only after several years.

III. CHEMOBIOKINETICS

A. Absorption

Absorption of nickel in humans may occur after inhalation, ingestion, and dermal exposure. All three routes have proven important in relation to nickel toxicity in humans. They differ profoundly, however, with regard to the types of nickel compounds of major importance, the fraction of the exposure which is absorbed, and the toxic effects encountered. The different absorption routes may interplay. Thus, oral intake may cause dermatological response in individuals already sensitized by skin exposure, and gastrointestinal exposure may occur due to mucociliary clearance after pulmonary exposure.

1. Dermal Absorption

Nickel is absorbed through appendages of the skin (mainly sweat ducts and hair follicles) at a faster rate than through the epidermis.[88] However, the surface area of the appendages is very small. Thus, penetration through the skin is primarily determined by the diffusion rate through the horny layer. Lipophilic organic nickel compounds may pass the dermal barrier rapidly and cause systemic toxicity, but are not considered here.

Studies of human skin obtained from autopsies have documented that percutaneous transport of nickel occurs and that nickel was accumulated in Malpighi's layer, sweat glands, and blood vessels.[89,90] Binding to keratin has been demonstrated.[88] Nickel as $^{63}Ni^{2+}$ had accumulated in keratinaceous areas of stratum corneum and in hair sacs 1 h after application on the shaved skin of guinea pigs; after 4 h, nickel was present in the stratum granulosum and stratum spinosum, where the Langerhans cells are located.[91]

Skin injury or increased water content of the stratum corneum may significantly increase absorption through the skin. Some solvents and detergents may also increase percutaneous absorption. Sensitization of guinea pigs toward nickel due to repeated skin painting was augmented when the skin was painted with an aqueous solution of sodium dodecyl sulfate before the application of nickel.[92] However, skin painting with ointments containing a macromolecular cation exchange resin[93] or a chelating agent, such as ethylenediaminetetraacetic acid (EDTA), diethyldithiocarbamate (DDC), or dimethylglyoxime,[92,94] prevented the development of positive patch tests in nickel-sensitive individuals.

The Ni(II) ion is considered exclusively responsible for the immunological effects of nickel.[95] In order to induce an allergic response, nickel as a hapten must react with a protein in the skin to form a complete allergen which is then taken up by a macrophage for antigen presentation to a T-lymphocyte. The Langerhans cells of the skin are believed to be responsible for transporting the allergen to the T-lymphocytes in the peripheral lymph node, where antigen presentation takes place.[96-100]

When dermal exposure to nickel has led to sensitization, subsequent dermal application (or possibly oral exposure) to very low doses of nickel may provoke contact dermatitis in sensitized individuals. Prevention of occupational nickel dermatitis should therefore focus on the prevention of sensitization. This requires efficient protection of the skin and low exposure limits for soluble nickel compounds.

2. *Intestinal Absorption*

Published studies on intestinal nickel absorption in human volunteers or experimental animals relate to water-soluble salts of nickel. As the biliary excretion of nickel is very low,[101] intestinal absorption can be determined from urinary nickel elimination in combination with data on either oral intake or fecal elimination, provided that pulmonary exposure to nickel, e.g., due to occupational exposure or smoking, can be excluded. After a single oral dose of a soluble nickel compound, plasma concentration increases rapidly and decreases again rather rapidly, while urinary concentration increases at a lower rate and remains elevated for a somewhat longer period. Thus, to allow calculation of the minimal intestinal absorption after a single oral dose of nickel, cumulative urinary nickel excretion should be measured over several days.

In a number of such experiments, intestinal inorganic nickel absorption was found to be 1 to 5%.[37,102,103] Ho and Furst[104] calculated that 3 to 6% of ^{63}Ni was absorbed from the intestine following administration of 4, 16, or 64 mg/kg of nickel as $NiCl_2$ by gavage to rats.

However, intestinal nickel absorption can be influenced by a number of factors. While a 5% absorption was seen with nickel added to scrambled eggs, 8 to 50% of nickel was absorbed if ingested as aqueous solution after an overnight fast.[105] Also, fasting human volunteers, given a single oral dose of nickel sulfate, excreted 4 to 16% of the dose in urine within 24 h,[3] while cumulative urinary excretion in nonfasting volunteers given a single oral dose (5.6 mg nickel as $NiSO_4$-$6H_2O$) indicated intestinal absorption of 3%.[106]

The increase in plasma nickel observed after a single oral dose (5 mg nickel as $NiSO_4$-

$6H_2O$) given to fasting human volunteers may be suppressed by giving the nickel dose with a meal or several different beverages.[107] Although calculation of intestinal absorption was not possible in this study, it demonstrates that dietary constituents (most likely macromolecular chelators such as "fibers", nonabsorbable, low molecular weight chelators such as phytate, and other metal ions competing for absorption) can reduce gastrointestinal nickel absorption.

Nickel naturally incorporated within food items also increases urinary nickel excretion. A diet containing 850 μg nickel resulted in an increased nickel excretion corresponding to slightly less than 1% of the amount in diet supplements, indicating a lesser fractional absorption for nickel in food items as compared to inorganic nickel.[108] Further, urinary nickel excretion in 14 nickel-dermatitis patients was found to be 1.5 (range 0.4 to 3.6) μg/d, and 2 weeks after restricting the oral nickel intake by avoiding a number of nickel-rich food items urinary excretion decreased to 1.0 (range 0.2 to 2.1) μg/d.[109]

Solomons et al.[107] also found that EDTA very efficiently prevented intestinal absorption of nickel. Other investigations have demonstrated that disulfiram and its metabolite DDC increase intestinal absorption of nickel, probably by forming a lipophilic NI-DDC complex.[110,111] Following oral ^{57}Ni exposure of mice, residual body burden of nickel after 22 and 48 h was increased severalfold in groups receiving DDC either p.o. or i.p. as compared to controls.[108]

Collectively, the available data indicate that only a few percent of nickel ingested with food is absorbed in the intestinal tract and that the absorption is even less when nickel is incorporated in food items. Intestinal absorption may, however, be considerably elevated by fasting.

3. Pulmonary Deposition, Clearance, and Absorption

In the occupational setting, pulmonary exposure is usually the major route of nickel absorption. Pulmonary deposition, clearance, and absorption of nickel particulates follow general principles of lung dynamics. Unfortunately, current knowledge relating to nickel mainly stems from studies on experimental animals, e.g., the distribution and elimination kinetics of intratracheally instilled nickel compounds in rats and mice.

The lung clearance after exposure of mice to a $NiCl_2$ aerosol indicated first-order clearance kinetics. About 28% of the dose was still present in the lungs 4 d postexposure.[112] This exposure method resembles conditions encountered in occupational exposures. As the results are somewhat at variance with those obtained with intratracheal instillation, they may therefore question the validity of the high clearance rates observed in the instillation experiments. Wehner and Craig[113] exposed groups of Syrian golden hamsters to 73.5 mg/m^3 of a NiO aerosol with a particle size of 1 to 2.5 μm mass median aerodynamic diameter (MMAD). Pulmonary deposition was 16.9% after 2 d of exposure, and after 45 and 100 d about 45 and 30%, respectively, of the NiO originally deposited was still present in the lungs. Kodama et al.[114] exposed groups of rats to aerosols of green NiO (MMAD = 1.2 μm) at low and medium dose levels (0.6 and 8.0 mg/m^3) for 140 to 147 h or 216 h. Nickel determination in lung tissue indicated a 14.5 and 19.8% deposition of NiO at the medium and low doses, respectively. About 10% of the dose of NiO deposited at the end of exposure had been cleared after 1 year at the medium dose, while $^2/_3$ of the low dose was cleared. The annual clearing rate was 120 and 90 μg NiO at the medium and low doses, respectively. Tanaka et al.[115] calculated that the biological half-life of pulmonarily deposited NiO depended on particle size, being 11.5 months for NiO with MMAD = 1.2 μm and 21 months for particles with MMAD = 4 μm.

Available evidence supports the following general conclusions on pulmonary absorption of nickel: (1) deposition in the airways depends on the physical characteristics of the particles inhaled, (2) mucociliary clearance leads to subsequent gastrointestinal transfer where some absorption may take place, (3) sparingly soluble compounds may be retained in the lungs for long periods of time, and (4) even after pulmonary deposition of large doses of insoluble nickel compounds, the nickel concentration in blood and other body organs is very low.

B. Systemic Distribution

Immediately after absorption, nickel is transported by the blood and retained by various tissues or excreted, mainly into the urine. During this initial systemic distribution phase, the largest amount of blood nickel is present in serum.[116] Between 2 and 24 hr after injection of 2 mg $NiCl_2$ in 0.1 ml saline into the thigh of female golden hamsters, the serum-nickel concentration was much higher than that in red blood cells, which again was higher than the nickel concentration in white blood cells. Between 6 and 24 h postinjection, the serum-nickel concentration was reduced about 40-fold.

Christensen and Lagesson[106] administered 5.6 mg of nickel as $NiSO_4{\cdot}6H_2O$ in lactose to eight volunteers and followed the whole-blood and serum concentrations, as well as the urinary excretion, for 3 d. Again, the largest part of blood-nickel was present in serum. Serum- and blood-nickel concentrations showed a very high positive correlation ($r = 0.99$). The nickel half-life in serum was 11 h (one-compartment model). The serum nickel concentration and urinary nickel excretion also showed a high positive correlation ($r = 0.98$). This study indicates that, during oral exposure to soluble nickel, urine seems to be the most useful sample to analyze.

Nickel in serum is chelated by histidine (His), serum albumin (SA), or both in a ternary complex.[117-121] The average logs (association constants) for the Ni-human serum albumin (HSA) complex and the ternary Hi-HSA-His complex are about 9.57 and 16.23, respectively.[122] NMR spectroscopy studies[123,124] have indicated that the nickel-binding site of HSA is formed by the three N-terminal amino acids Asp^1-Ala^2-His^3, which seem to be solely responsible for chelation of nickel in a pentacoordinate structure.

A macroglobulin "nickeloplasmin" has been demonstrated to bind nickel in rabbit serum,[125] and a glycoprotein binds a minor fraction of nickel in human serum.[2,126] These nickel-binding proteins may be associated with the essentiality of nickel but are not likely to be of importance in nickel homeostasis at higher serum nickel concentrations, such as are encountered during occupational exposures, where His and SA binding most likely dominate.[126]

Nickel levels in intercellular fluid have been studied in groups of nickel-allergic patients and controls. The results showed that although dietary intake and urinary nickel excretion levels were similar for the two groups, the nickel-allergic group had a significantly ($p < 0.05$) lower nickel concentration (median 9.9 μg/l) than the controls (median 2.7 μg/l) in the intercellular fluid, possibly due to cell binding or uptake.[127]

C. Organ Distribution

Nickel seems to be distributed to all organs, in pregnant females even to fetuses. However, the kidney is the only organ immediately accumulating appreciable amounts of nickel. Smaller amounts are deposited in the lungs (inhalation exposures may obviously lead to much higher levels). The distribution of nickel to major lymphoid organs, such as the thymus, non-hilar lymph nodes, and spleen, is only sparsely investigated. Although no organ seems to constitute a specific storage depot for nickel, experimental animal studies have indicated more rapid excretion of nickel from kidneys than from other organs after 1 d and continued retention of nickel in the lungs and skin. Also, the evidence from these studies indicates that the distribution of absorbed nickel is relatively independent of the route of entry.

However, the chemobiokinetics of nickel can be significantly changed by administration of chelating agents, such as thiuram sulfides and DDC.[128,129] Administration of clinically relevant doses of DDC (p.o. or i.p. in mice) resulted in a considerably changed organ distribution compared to controls. At 48 h the brain deposition was at least 100-fold above the control levels, and deposition in kidneys, liver, and lungs was also increased.[130] This finding may be of significance, as disulfiram, which is used in alcohol aversion therapy and sometimes also in treatment of nickel-related eczemas, is metabolized to two molecules of DDC.

D. Cellular Distribution

In the immunologic response, as briefly mentioned above, the Langerhans cells of the skin engulf and transport nickel to peripheral lymph nodes where it is presented to T-lymphocytes. Similarly, antigen-presenting macrophages present in the gastrointestinal and pulmonary tissues could theoretically transport nickel to T-lymphocytes in lymphoid tissues near the intestinal tract (Peyer's patches) and the lungs.

The mechanism for presentation of nickel to T-lymphocytes is not known in detail, but the T-lymphocytes in allergic individuals probably have a nickel-specific receptor. Thus, T-lymphocytes for nickel-sensitized individuals incubated with Ni^3S_2 were found to bind considerable amounts of nickel to the cell membrane, while this was only the case with very few cells from nonsensitized persons.[131] This observation may offer an explanation of the comparatively low nickel concentrations found in cell-free intercellular fluid of nickel-sensitized individuals when compared to controls.[127]

Several cell types are capable of phagocytosis or endocytosis under certain conditions. Thus, fibroblasts may phagocytize particles, including crystalline nickel compounds.[132,133] Also, bronchial epithelial cells are capable of phagocytic uptake of particles.[134] These processes may lead to intracellular solubilization of nickel. Kuehn and Sunderman[135] found that several particulate forms of sparingly soluble nickel compounds were solubilized more rapidly in serum and renal cytosol than in water. Cellular uptake followed by cytoplasmic dissolution may suggest a possible explanation for the carcinogenicity of sparingly soluble crystalline nickel compounds, which are rapidly phagocytized, as compared to amorphous nickel compounds of similar low solubility.[132,133,136,137]

Internalized nickel particles are preferentially localized close to the nuclear envelope. However, unsolubilized particles do not enter the nucleus.[137] While intracellular nickel is mainly found in extranuclear organelles after exposure to soluble nickel,[138] exposure to crystalline Ni_3S_2 or NiS results in high concentrations of nickel in the nucleus.[139,140] Phagocytized particulate nickel is solubilized relatively rapidly[141] and then binds strongly to nuclear material.[137] Sen and Costa[142] observed selective fragmentation of the heterochromatic long arms of the X-chromosome after Ni^{2+} exposure due to the generation of high cytoplasmic nickel concentrations during dissolution of NiS particles. Nickel also binds to isolated chromatin and to chromatin in isolated nuclei *in vitro*.[143] Nickel is capable of interacting with the tubulin-microtubule system, thus affecting the function of the spindle.[144-146]

E. Excretion

Excretion of systemically absorbed nickel may take place through sweat[147,148] and, as discussed below, feces. However, these routes are normally of minor importance as compared to urinary nickel excretion. Again, most of the evidence originates from experimental animal studies, as the human studies already referred to only monitored urinary excretion and not all potential excretion routes.

After administration of $^{63}NiCl_2$ to rats, the cumulative urinary and fecal excretion of nickel was 77.94 and 4.87%, respectively, during 6 d after i.p. exposure[149] and 89.8 and 3.0% during 4 d after i.v. exposure.[150] After i.m. administration of $^{63}Ni_3S_2$ to rats, the cumulative urinary and fecal excretion during 8 weeks was 67 and 7%, respectively.[151] After intratracheal instillation of $^{63}NiCl_2$ into rats, 96.5% of the dose was excreted in the urine after 21 d,[152] while the cumulative urinary and fecal elimination of nickel after 90 d was 64.3 and 6.4% in the study by English et al.[153]

Marzouk and Sunderman[101] followed biliary excretion of intravenously administered $^{63}NiCl_2$ in the rat by bile duct cannulation and bile collection. The cumulative biliary nickel excretion was calculated to be <0.5% during the first 24 hr postdosage. An increased biliary nickel concentration (62 μg/g) was observed at autopsy after an acute lethal nickel intoxication in a small girl who died 4 h after having swallowed about 15 g of nickel sulfate crystals.[154] Even if assuming the same high concentration in as much as 250 ml bile, nickel excretion via this route would only be about 0.1% of the dose. Thus, biliary excretion is likely to be of negligible importance as compared to renal excretion.

IV. DIAGNOSTIC TESTS FOR EXPOSURE

By measuring the nickel levels in urine and/or serum, an indication of the absorbed nickel dose rather than the exposure is obtained. Further, whole blood, nasal sinus biopsies, and various autopsy materials may be used for research purposes. Due to improved sampling, reduced contamination, and improved analytical technology, reference values for nickel in biological fluids of healthy nonexposed individuals have decreased steadily during recent years. Representative values are 0.34 ± 0.28 μg/l blood, 0.28 ± 0.24 μg/l serum,[155] and 2.0 ± 1.5 μg/g creatinine in urine (spot collection).[156]

Body fluid levels generally reflect recent exposures (the past few days) to soluble or relatively soluble nickel compounds. Half-times of plasma nickel in a group of nickel platers after exposure to soluble nickel were found to be between 20 and 34 h, and the half-times for urinary nickel were between 17 and 39 h in the same individuals.[157] Comparable urinary half-times were observed after exposure to less soluble forms of nickel in the glass industry[158] and during welding.[159] However, after pulmonary exposure to sparingly soluble compounds, nickel will be only slowly released from the lungs, and the serum or urine levels will then reflect the accumulated lung burden rather than recent exposure. Thus, even 3 to 4 years after cessation of exposure, nickel levels in plasma and urine were elevated in retired nickel workers exposed to sparingly soluble compounds in the roasting/smelting department.[160]

A number of studies[154] demonstrate a positive relation between air levels of nickel and serum/plasma levels of nickel after occupational exposure to various (insoluble, mixed, or soluble) forms of nickel. The air levels of nickel varied by three orders of magnitude, and serum/plasma nickel levels varied by about 30-fold within and between these studies. When occupational nickel exposure includes both easily and sparingly soluble compounds, urinary nickel levels will mainly reflect the exposure to the soluble compounds, since exposure to less soluble forms of nickel results in much smaller elevations at the same air levels of nickel. Even when looking solely at "pure" exposures to soluble nickel compounds, one does not find a strong correlation between air levels and urinary levels of nickel.

The use of urine samples for monitoring the exposure to mainly soluble nickel compounds offers certain advantages compared to serum or plasma, as the concentrations are usually much higher, most frequently 2 to 15 times higher.[161-166]

Normalization to adjust for variable urine dilution can be performed by expressing the results as milligrams nickel per gram of creatinine or by adjustment to a normal specific gravity.[157,161,167] However, these parameters may not necessarily reflect the changes in nickel levels with variations in urinary flow rates, and very dilute urine samples are particularly likely to cause erroneous results. Such samples should be discarded.

The best indicator of current exposure to soluble nickel compounds is a 24-h urine sample. Such sampling is, however, seldom applicable. The alternatives are venipuncture or sampling of spot urines in a uniform way.

The results of the biological monitoring data must be evaluated in relation to other medical information. Patients with prior heart or kidney disease may be more susceptible to increased circulating levels of nickel. On the other hand, hemodialysis as such may cause increased serum nickel values,[168] and some drugs, such as disulfiram which is metabolized to DDC, may result in high nickel concentrations in serum. The possible significance of other nickel exposures and possible release from, for example, orthopedic implants, must also be evaluated.

Provided that the above caveats are respected, biological monitoring may be very useful and should be made available as a voluntary test to occupationally exposed workers at intervals which depend on the exposure level and the nickel compounds in question.

V. PREVENTION

A. Preventive Strategies

The major adverse health effects due to nickel are nickel allergy, as reviewed in this volume, and cancer of the respiratory tract. Sensitization is due to cutaneous exposure to soluble nickel, although oral or respiratory exposure may play a role in provoking exacerbation of eczemas. However, nickel-associated cancer is due to inhalation exposure to sparingly soluble nickel compounds. Therefore, preventive efforts must focus on the particular exposure route. Further, the most efficient strategy is to control the hazard at the source. If the exposure cannot be abated, the releases must be kept at a sufficiently low level to minimize any adverse effects.

B. Cutaneous Exposures

The most important objective in relation to contact dermatitis is to prevent sensitization by limiting nickel release from items worn in close contact with the skin. Tolerance to metal objects can be obtained by nickel-sensitized patients only if nickel is absent. In nickel-sensitized guinea pigs, tolerance to skin contact with this metal can be obtained by using a 0.25-μm chromium plating.[169]

Based on a special test with a synthetic sweat solution, Danish authorities intend to enforce a limit of 0.05 $\mu g/cm^2$/week for metal objects which may be worn in long-term close contact with the skin, e.g., jewelry, buttons, and watches. This limit is one tenth of the no-effect level found by Menne et al.[13] From the data reviewed above, objects electrolytically plated with nickel or made from German silver would probably become illegal. In addition, the individual consumer would have the option of using the dimethylglyoxime test described above.

In the workplace, cutaneous exposures should also be prevented. If contact with soluble nickel compounds cannot be avoided, proper protective clothing must be worn.

C. Dietary Restrictions

As increased dietary nickel intake may provoke dishydrotic hand eczema in nickel-allergic patients, a low-nickel diet has been suggested in the treatment. Kaaber et al.[109] examined 17 patients and observed an improvement during a period of 6 weeks on a low-nickel diet. The dermatitis of seven of the nine patients worsened when a normal diet was resumed. Similar results were obtained by Veien et al.[170,171] The data available on nickel levels in food items have increased considerably during recent years (Table 2),[26,38] and this knowledge should be utilized when advising patients on possible dietary restrictions and in further elimination studies.

Disulfiram has been suggested for treatment of patients with severe hand eczema due to nickel allergy. Although hardly a preventive provision, employment of this drug may reduce possible adverse effects of nickel retained in the body. However, because disulfiram treatment may also increase the deposition of nickel in the central nervous system, chelation treatment should be instituted only when considered essential for the amelioration of severe cases, and then only for a limited time period.

D. Pulmonary Exposures

Exposure monitoring and engineering control constitute the mainstay of prevention in occupational and environmental health. Exposure limits are generally used in this connection, but no agreement on the level of such limits has been reached on an international scale.

Unfortunately, the criteria or basis used for the standard-setting in different countries are only available to a limited extent. In the Federal Republic of Germany, a summary is available.[172] According to this document, nickel metal, nickel sulfide or sulfide ores, nickel oxide, and nickel carbonate, as they occur during production and use, are considered carcinogenic. No exposure sure limit (MAK) has been determined; only a technical guideline has been provided. Due to the risk of sensitization caused by nickel metal and soluble compounds, nickel has received a "skin" denotation.

In the U.S., the American Conference of Governmental Industrial Hygienists (ACGIH)[173] has published the background material for the threshold limit values (TLV). Following a brief review of major studies considered, ACGIH concludes, "With the available knowledge, it is not felt that all forms of nickel are carcinogenic. Therefore, a TLC of 1.0 mg/m^3 is recommended for nickel metal and insoluble inorganic nickel compounds. A TLC of 0.1 mg/m^3 and a STEL of 0.3 mg/m^3 are recommended for soluble nickel compounds." Unfortunately, the basic criteria used when reaching this conclusion are not stated, although the limit of sparingly soluble nickel compounds is intended to protect against carcinogenic effects and the limit for soluble compounds against sensitization. NIOSH[61] proposed a limit of 15 μg/m^3.

A Dutch Working Group of Experts[174] concluded in a draft document that experimental evidence suggested a threshold for nonharmful effects below 0.2 to 0.3 mg/m^3 (6 h/d, 6 weeks) for soluble nickel compounds and below 0.1 mg/m^3 for nickel metal dust. They therefore recommended an exposure limit for nickel metal, inorganic compounds, and industrial nickel oxide (except primary nickel production) of 50 μg/m^3 as an 8-h limit and 100 μg/m^3 as a ceiling value for 15-min exposures. On the basis of carcinogenic effects, the experts followed the NIOSH recommendation and suggested limits of 10 μg/m^3 (8 h) and 20 μg/m^3 (15 min) for primary nickel production.

When examining exposure limits in various countries, the Norwegian limits are of particular interest. These limits apply to a large refinery, mines (at least in principle), and a range of nickel-using industries. The same exposure limit (at 0.1 mg/m^3) has been set for both soluble and insoluble nickel compounds, a decision which is less difficult to implement than several sets of standards. However, the need for prevention of cutaneous exposures, especially to soluble nickel compounds, must be emphasized.

VI. CONCLUSIONS

Nickel has many useful properties and a wide range of applications in alloys and a large number of compounds. Nickel exposures occur in a range of occupations, as a result of environmental dissemination and use of consumer products, and, to a lesser degree, due to natural levels in the environment. Although some food items and beverages are generally high in nickel, additional amounts may leach from containers and utensils.

Dermal absorption of solubilized nickel may result from prolonged skin contact with corrosive metals; sweat, particularly, may dissolve nickel when a high chloride concentration is present. A release of 0.5 μg/cm^2/week from a metal object appears to be the threshold for provocation of contact dermatitis. Gastrointestinal absorption of soluble nickel compounds is usually less than 5%, but may be more when the stomach is empty. Pulmonary absorption is of importance in occupational exposures and depends on the physicochemical properties of the particles and aerosols concerned. The significance of nondermal nickel exposures is still unclear, but increased oral intake of nickel appears to provoke an exacerbation of hand eczemas in some nickel-allergic patients. Nickel in the blood is mainly located in the serum where most is bound to albumin. Transfer to the kidneys and subsequent excretion occur relatively rapidly. Lesser amounts are transferred to the lungs and skin where, on the other hand, the retention time is longer.

Prevention should first aim at limiting prolonged skin contact with metal objects which may release significant amounts of nickel. In addition, avoidance of increased nickel absorption through food, beverages, or occupational exposures would be advisable in many nickel-allergic patients, especially in case of severe hand eczema.

ACKNOWLEDGMENT

This chapter is based in part on an extensive review carried out on behalf of the Ontario Ministry of Labour.

REFERENCES

1. **Anke, M., Groppel, B., Kronemann, H., and Grun, M.,** Nickel — an essential element, in *Nickel in the Human Environment,* Sunderman, F. W., Jr., Ed., IARC Scientific Publ. No. 53, International Agency for Research on Cancer, Lyon, 1984, 339.
2. **Sunderman, F. W., Jr.,** A review of the metabolism and toxicology of nickel, *Ann. Clin. Lab. Sci.,* 7, 377, 1977.
3. **Cronin, E., Di Michiel, A. D., and Brown, S. S.,** Oral challenge in nickel-sensitive women with hand eczema, in *Nickel Toxicology,* Brown, S. S. and Sunderman, F. W., Jr., Eds., Academic Press, London, 1980, 149.
4. **Malo, J.-L., Cartier, A., Doepner, M., Nieboer, E., Evans, S., and Dolovich, J.,** Occupational asthma caused by nickel sulfate, *J. Allergy Clin. Immunol.,* 69, 55, 1982.
5. **Grandjean, P., Andersen, O., and Nielsen, G. D.,** Carcinogenicity of occupational nickel exposure. An evaluation of the epidermiological evidence, *Am. J. Ind. Med.,* in press.
6. **Marcussen, P. V.,** Ecological considerations on nickel dermatitis, *Br. J. Ind. Med.,* 17, 65, 1960.
7. **Warner, J. S.,** Occupational exposure to airborne nickel in producing and using primary nickel products, in *Nickel in the Human Environment,* Sunderman, F. W., Jr., Ed., IARC Scientific Publ. No. 53, International Agency for Research on Cancer, Lyon, 1984, 419.
8. **Fisher, A. A.,** *Contact Dermatitis,* 2nd ed., Lea & Febiger, Philadelphia, 1973.
9. **Fischer, T., Fregert, S., Gruvberger, B., and Rystedt, I.,** Contact sensitivity to nickel in white gold, *Contact Dermatitis,* 10, 23, 1984.
10. **Brandrup, F. and Schultz Larsen, F.,** Nickel dermatitis provoked by buttons in blue jeans, *Contact Derm.,* 5, 148, 1979.
11. **Fisher, A. A.,** Metal dermatitis — some questions and answers, *Cutis,* 19, 156, 1977.
12. **Katz, S. A. and Samitz, M. H.,** Leaching of nickel from stainless steel consumer commodities, *Acta Derm. Venereol.,* 55, 113, 1975.
13. **Menne, T., Brandrup, F., Thestrup-Pedersen, K., Veien, N. K., Andersen, J. R., Yding, F., and Valeur, G.,** Patch test reactivity to nickel alloys, *Contact Derm.,* 16, 255, 1987.
14. **Wahlberg, J. E. and Skog, E.,** Nickel allergy and atopy, threshold of nickel sensitivity and immunoglobulin E determinations, *Br. J. Dermatol.,* 6, 414, 1971.
15. **Spruit, D., Bongaarts, P. J. M., and Malten, K. E.,** Dermatological effects of nickel, in *Nickel in the Environment,* Nriagu, J. O., Ed., John Wiley & Sons, New York, 1980, 601.
16. **Wahlberg, J. E.,** Nickel allergy and atopy in hairdressers, *Contact Derm.,* 1, 161, 1975.
17. **Marks, R. and Cronin, E.,** Hand eczema in hairdressers, *Aust. J. Dermatol.,* 18, 123, 1977.
18. **Wahlberg, J. E.,** Nickel allergy in hairdressers, *Contact Derm.,* 7, 358, 1981.
19. **Rystedt, I. and Fischer, T.,** Relationship between nickel and cobalt sensitization in hard metal workers, *Contact Derm.,* 9, 195, 1983.
20. **Wall, I. M. and Calnan, C. D.,** Occupational nickel dermatitis in the electroforming industry, *Contact Derm.,* 6, 414, 1980.
21. **Fregert, S.,** Occupational dermatitis in a 10 year material, *Contact Derm.,* 1, 96, 1975.
22. **Clemmensen, O. J., Menne, T., Kaaber, K., and Solgaard, P.,** Exposure of nickel and the relevance of nickel sensitivity among hospital cleaners, *Contact Derm.,* 7, 14, 1981.

23. **Acres Consulting Services, Ltd.**, Worker Exposure to Nickel and Its Compounds, Occupational Health and Safety Division, Standards and Programs Branch, Ontario Ministry of Labor, Ontario, Canada, 1982.
24. **Menne, T. and Bachmann, E.**, Permanent disability from hand dermatitis in females sensitive to nickel, chromium and cobalt, *Dermatosen,* 5, 129, 1979.
25. **Ellen, G., van den Bosch Tibbesma, G., and Douma, F. F.**, Nickel content of various Dutch foodstuffs, *Z. Lebensm, Unters. Forsch.*, 166, 145, 1978.
26. **Flyvholm, M., Nielsen, G. D., and Andersen, A.**, Nickel content of food and estimation of dietary intake, *Z. Lebensm. Unters. Forsch.*, 179, 427, 1984.
27. Nickel, *National Academy of Science, Washington, D.C.*, 1975.
28. **Bennett, B. G.**, Summary exposure assessments for mercury, nickel, tin, in *Exposure Commitment Assessments of Environmental Pollutants,* Vol. 1, No. 2, Monitoring and Assessment Research Centre, Chelsea College, University of London, 1981 (No. 25), 17.
29. **Hovmand, M. F.**, Nickel, in *Agricultural Use of Sewage Sludge,* (Danish), Vol. 2, Andersen, A., Jepsen, A., Larsen, K. E., Larsen, V., Hansen, I. A. S., Hovmand, M. F., and Tjell, J. C., Eds., Polyteknisk Forlag, Copenhagen, 1981, 121.
30. **Schroeder, H. A., Balassa, J. J., and Tipton, I. H.**, Abnormal trace metals in man — nickel, *J. Chronic Dis.*, 15, 51, 1962.
31. **Rondia, D.**, Sources, modes and levels of human exposure to chromium and nickel, in *Trace Metals: Exposure and Health Effects,* Di Ferrante E., Ed., Pergamon Press, Oxford, 1979, 117.
32. **Schlettwein-Gsell, D. and Mommsen-Straub, S.**, Trace elements in food (German), *Int. J. Vitam. Nutr. Res.*, Suppl. 13, 62, 1973.
33. **Rudzki, E. and Grzywa, Z.**, Exacerbation of nickel dermatitis by margarine, *Contact Derm.*, 3, 344, 1977.
34. **Myron, D. R., Zimmerman, T. J., Schuler, T. R., Klevay, L. M., Lee, D. E., and Nielsen, F. H.**, Intake of nickel and vanadium by humans. A survey of selected diets, *Am. J. Clin. Nutr.*, 31, 527, 1978.
35. **Nodiya, P. I.**, Cobalt and nickel balances in studies of an occupational technical school (Russian), *Gig. Sanit.*, 37, 108, 1972.
36. **Sidorenko, G. I., and Izkowa, A. I.**, Nickel as a factor of environmental pollution, *Z. Gesamte Hyg. Ihre Grenzgeb.*, 21, 733, 1975.
37. **Horak, E. and Sunderman, F. W., Jr.**, Fecal nickel excretion by healthy adults, *Clin. Chem.*, 19, 429, 1973.
38. **Nielsen, G. D. and Flyvholm, M.**, Risks of high nickel intake with diet, in *Nickel in the Human Environment,* Sunderman, F. W., Jr., Ed., IARC Scientific Publ. No. 53, International Agency for Research on Cancer, Lyon, 1984, 333.
39. **Snodgrass, W. J.**, Distribution and behavior of nickel in the aquatic environment, in *Nickel in the Environment,* Nriagu, J. O., Ed., John Wiley & Sons, New York, 1980, 203.
40. **Meranger, J. C., Subramanian, K. S., and Chalifoux, C.**, Survey for cadmium, cobalt, chromium, copper, nickel, lead, zinc, calcium, and magnesium in Canadian drinking water supplies, *J. Assoc. Off. Anal. Chem.*, 64, 44, 1981.
41. **McNeely, M. D., Nechay, M. W., and Sunderman, F. W., Jr.**, Measurements of nickel serum and urine as indices of environmental exposure to nickel, *Clin. Chem.*, 18, 992, 1972.
42. **Punsar, S., Erametsa, O., Karvonen, M. J., Ryhanen, A., Hilska, P., and Vornamo, H.**, Coronary heart disease and drinking water. A search in two Finnish male cohorts for epidemiologic evidence of a water factor, *J. Chron. Dis.*, 28, 259, 1975.
43. **Schuhmann, H.**, The nickel contamination of surface and potable water and some of its causal factors (German), in *3rd Spurenelement-Symposium Nickel,* Anke, M., Schneider, H.-J., and Bruckner, C., Eds., Karl-Marx-Universitat, Leipzig, and Friedrich-Schiller-Universitat, Jena, 1980, 155.
44. **Dominok, B., Geppert, H., and Kilz, C.**, Drinking-water poisoning with nickel salts (German), in *3rd Spurenelement-Symposium Nickel,* Anke, M., Schneider, H.-J., and Bruckner, C., Eds., Karl-Marx-Universitat, Leipzig, and Friedrich-Schiller-Universitat, Jena, 1980, 327.
45. **Anderseen, K. E., Nielsen, B. D., Flyvholm, M., Fregert, S., and Gruvberger, B.**, Nickel in tap water, *Contact Derm.*, 9, 140, 1983.
46. **Strain, W. H., Varnes, A. W., Davis, B. R., and Kark, E. C.**, Nickel in drinking and household water, in *3rd Spurenelement-Symposium Nickel, Anke M., Schneider, H.-J., and Bruckner, C., Eds., Karl-Marx-*Universitat, Leipzig, and Friedrich-Schiller-Universitat, Jena, 1980, 149.
47. **Boudene, C.**, Food contamination by metals, in *Trace Metals: Exposure and Health Effects,* Di Ferrante, E., Ed., Pergamon Press, Oxford, 1979, 163.
48. **Christensen, O. B. and Møller, H.**, Release of nickel from cooking utensils, *Contact Derm.*, 4, 343, 1978.
49. **Rasmussen, G.**, Release of Trace Elements (Arsenic, Cadmium, Chromium, Copper, Nickel, Lead, Antimony, Tin and Zinc) from Kitchen Utensils (Danish), National Food Institute, Søborg, 1983.
50. **Moffa, J. P.**, Biological effects of nickel-containing dental alloys, *J. Am. Dent. Assoc.*, 104, 501, 1982.

51. **Adamsson, E., Lind, B., Nielsen, B., and Piscator, M.,** Urinary and fecal elimination of nickel in relation to airborne nickel in a battery factory in *Nickel Toxicology,* Brown, S. S. and Sunderman, F. W., Jr., Eds., Academic Press, London, 1980, 103.
52. **Christensen, O. B. and Møller, H.,** External and internal exposure to the antigen in the hand eczema of nickel allergy, *Contact Derm.,* 1, 136, 1975.
53. **Jordan, W. P., Jr. and King, S. E.,** Nickel feeding in nickel-sensitive patients with hand eczema, *J. Am. Acad. Dermatol.,* 1, 506, 1979.
54. **Veien, N. K. and Kaaber, K.,** Nickel, Cobalt and chromium sensitivity in patients with pompholyx (dyshidtrotic eczema), *Contact Derm.,* 5, 371, 1979.
55. **Burrows, D., Creswell, S., and Merrett, J. D.,** Nickel, hands and hip prostheses, *Br. J. Dermatol.,* 105, 437, 1981.
56. **Veien, N. K., Svejgaard, E., and Menne, T.,** In vitro lymphocyte transformation to nickel: a study of nickel-sensitive patients before and after epicutaneous and oral challenge with nickel, *Acta Derm. Venereol. (Stockholm)* 59, 447, 1979.
57. **Schmidt, J. A. and Andren, A. W.,** The atmospheric chemistry of nickel, in *Nickel in the Environment,* Nriagu, J. O., Ed., John Wiley & Sons, New York, 1980, 93.
58. **Bennett, B. G.,** Environmental nickel pathways to man, in *Nickel in the Human Environment,* Sunderman, F. W. Jr., Ed., IARC Scientific Publ. No. 53, International Agency for Research on Cancer, Lyon, 1984, 487.
59. **Sunderman, F. W. and Sunderman, F. W., Jr.,** Nickel poisoning. XI. Implication of nickel as a pulmonary carcinogen in tobacco smoke, *Am. J. Clin. Pathol.,* 35, 203, 1961.
60. **Szadkowski, D., Schultze, H., Schaller, K.-H., and Lehnert, G.,** On the ecological significance of heavy metal contents of cigarettes (German), *Arch. Hyg.,* 153, 1, 1969.
61. Criteria for a Recommended Standard . . . Occupational Exposure to Inorganic Nickel, DHEW (NIOSH) Publ. No. 77-164, National Institute for Occupational Safety and Health, Cincinnati, 1977.
62. **Wilson, J. D., Stenzel, M. R., Lombardozzi, K. L., and Nichols, C. L.,** Monitoring personnel exposure to stainless steel welding fumes in confined spaces at a petrochemical plant, *Am. Ind. Hyg. Assoc. J.,* 42, 431, 1981.
63. **Jones, J. G. and Warner, C. G.,** Chronic exposure to iron oxide, chromium oxide, and nickel oxide fumes of metal dressers in a steelworks, *Br. J. Ind. Med.,* 29, 169, 1972.
64. **Stettler, L. E., Donaldson, H. M., and Grant, G. C.,** Chemical composition of coal and other mineral slags, *Am. Ind. Hyg. Assoc. J.,* 43, 235, 1982.
65. **Bernacki, E. J., Parsons, G. E., Roy, B. R., Mikac-Devic, M., Kennedy, C. D., and Sunderman, F. W., Jr.,** Urine nickel concentrations in nickel-exposed workers, *Ann. Clin. Lab. Sci.,* 8, 184, 1978.
66. **Grandjean, P., Selikoff, L. J., Shen, S. K., and Sunderman, F. W., Jr.,** Nickel concentrations in plasma and urine of shipyard workers, *Am. J. Ind. Med.,* 1, 181, 1980.
67. **Samitz, M. H. and Katz, S. A.,** Skin hazards from nickel and chromium salts in association with cutting oil operations, *Contact Dermatol.,* 1, 158, 1975.
68. **Clausen, J. and Rastogi, S. C.,** Heavy metal pollution among autoworkers. II. Cadmium, chromium, copper, manganese, and nickel, *Br. J. Ind. Med.,* 34, 216, 1977.
69. **Raithel, H. J. and Schaller, K. H.,** Toxicity and carcinogenicity of nickel and its compounds. A review of the current status, *Zentralbl. Bakteriol. Mikrobiol. Hyg. B,* 173, 63, 1981.
70. **Dewhurst, F.,** Laboratory usage of some suspect carcinogens, *Br. J. Cancer,* 44, 304, 1981.
71. **Brune, D., Beltesbrekke, J., and Melson, S.,** Dust in workroom air of dental laboratories, *Swed. Dent. J.,* 5, 247, 1981.
72. **Norseth, T.,** Urinary excretion of nickel as an index of nickel exposure in welders and nickel refinery workers, in Abstr. 18th Int. Congr. Occupational Health, Brighton, England, September 14 to 19, 1975, 327.
73. **Zober, A.,** Possible dangers to the respiratory tract from welding fumes (German), *Schweissen Schneiden,* 34, 77, 1982.
74. **Tandon, S. K., Mathur, A. K., and Gaur, J. S.,** Urinary excretion of chromium and nickel among electroplaters and pigment industry workers, *Int. Arch. Occup. Environ. Health,* 40, 71, 1977.
75. **Tola, S., Kilpiö, J., and Virtamo, M.,** Urinary and plasma concentrations of nickel as indicators of exposure to nickel in an electroplating shop, *J. Occup. Med.,* 21, 184, 1979.
76. **Mathur, A. K. and Tandon, S. K.,** Urinary excretion of nickel and chromium in occupational workers, *J. Environ. Biol.,* 2, 1, 1981.
77. **Samitz, M. H. and Katz, S. A.,** Nickel dermatitis hazards from prostheses, *Br. J. Dermatol.,* 92, 287, 1975.
78. **Lyell, A., Bain, W. H., and Thomsen, R. M.,** Repeated failure of nickel-containing prosthetic heart valves in a patient allergic to nickel, *Lancet,* 2, 657, 1978.
79. **Laing, P. G.,** Clinical experience with prosthetic materials, historical perspectives, current problems and future directions. Corrosion and degradation of implant materials, *ASTM,* STP 684, 119, 1979.

80. **Marek, M. and Treharne, R. H.,** An in vitro study of the release of nickel from two surgical implant alloys, *Clin. Orthop.,* 167, 291, 1982.
81. **Sunderman, F. W., Jr.,** Potential toxicity from nickel contamination of intravenous fluids, *Ann. Clin. Lab. Sci.,* 13, 1, 1983.
82. **Webster, J. D., Parker, T. F., Alfrey, A. C., Smythe, W. R., Kubo, H., Neal, G., and Hull, A. R.,** Acute nickel intoxication by dialysis, *Ann. Intern. Med.,* 92, 631, 1980.
83. **Olerud, J. E., Lee, M. Y., Uvelli, D. A., Goble, G. J., and Babb, A. L.,** Presumptive nickel dermatitis from hemodialysis, *Arch. Dermatol.,* 120, 1066, 1984.
84. **Evans, E. M., Freeman, M. A. R., Miller, A. J., and Vernon-Roberts, B.,** Metal sensitivity as a cause of bone necrosis and loosening of the prosthesis in total joint replacement, *J. Bone Jt. Surg.,* 56B, 626, 1974.
85. **Merritt, K. and Brown, S. A.,** Tissue reaction and metal sensitivity, *Acta Orthop. Scand.,* 52, 403, 1980.
86. **Pegum, J. S.,** Nickel allergy, *Lancet,* 1, 674, 1974.
87. **Tilsey, D. A. and Rotstein, M.,** Sensitivity caused by internal exposure to nickel, chrome and cobalt, *Contact Dermatitis,* 6, 175, 1980.
88. **Wells, G. C.,** Effects of nickel on the skin, *Br. J. Dermatol.,* 68, 237, 1956.
89. **Spruit, D., Mali, J. W. H., and De Groot, N.,** The interaction of nickel ions with human cadaverous dermis, *J. Invest. Dermatol.,* 44, 103, 1965.
90. **Kolpakov, F. I.,** Permeability of skin to nickel compounds, *Arkh. Patol.,* 25, 38, 1963.
91. **Lloyd, G. K.,** Dermal absorption and conjugation of nickel in relation to the induction of allergic contact dermatitis — preliminary results, in *Nickel Toxicology,* Brown, S. S. and Sunderman, F. W., Jr., Eds., Academic Press, London, 1980, 145.
92. **Samitz, M. H. and Pomerantz, H.,** Studies of the effects on the skin of nickel and chromium salt, *Arch. Ind. Health,* 18, 473, 1958.
93. **Jarisch, R., Ballczo, H., and Richter, W.,** Nickelallergie: Ausschaltung der Noxe durch Kationenaustauscher. I, *Z. Hautkr.,* 50, 33, 1975.
94. **Kurtin, A. and Orentreich, N.,** Chelation deactivation of nickel ion in allergic eczematous sensitivity, *J. Invest. Dermatol.,* 22, 441, 1954.
95. **Wahlberg, J. E.,** Immunoglobulin E, atopy, and nickel allergy, *Cutis,* 18, 715, 1976.
96. **Braathen, L. R.,** Studies on human epidermal Langerhans cells. III. Induction of T lymphocyte response to nickel sulphate in sensitized individuals, *Br. J. Dermatol.,* 103, 517, 1980.
97. **Braathen, L. R. and Thorsby, E.,** Studies on human epidermal Langerhans cells. Allo-activating and antigen-presenting capacity, *Scand. J. Immunol.,* 11, 401, 1980.
98. **Stingl, G., Katz, S. I., Clement, L., Green, I., and Shevach, E. M.,** Immunological functions of IA-bearing epidermal Langerhans cells, *J. Immunol.,* 121, 2005, 1978.
99. **Shelly, W. B. and Juhlin, L.,** Selective uptake of contact allergens by the Langerhans cells, *Arch. Dermatol.,* 113, 187, 1977.
100. **Silberberg-Sinakin, I. and Thorbecke, I.,** Contact hypersensitivity and Langerhans cells, *J. Invest. Dermatol.,* 75, 61, 1980.
101. **Marzouk, A. and Sunderman, F. W., Jr.,** Biliary excretion of nickel in rats, *Toxicol. Lett.,* 27, 65, 1985.
102. **Tedeschi, R. E. and Sunderman, F. W.,** Nickel poisoning. V. The metabolism of nickel under normal conditions and after exposure to nickel carbonyl, *Arch. Ind. Health,* 16, 486, 1957.
103. **Nomoto, S. and Sunderman, F. W., Jr.,** Atomic absorption spectrometry of nickel in serum, urine, and other biological materials, *Clin. Chem.,* 16, 477, 1970.
104. **Ho, W. and Furst, A.,** Nickel excretion by rats following a single treatment, *Proc. West. Pharmacol. Soc.,* 16, 245, 1973.
105. **Sunderman, F. W., Jr., Hopfer, S. M., Swift, T., Ziebka, L., Marcus, A. H., Most, B. M., and Creason, J.,** *Nickel Absorption and Elimination in Human Volunteers,* WHO Environmental Health Series, Vol. 20, World Health Organization, Copenhagen, 1987, 66.
106. **Christensen, O. B. and Lagesson, V.,** Nickel concentration of blood and urine after oral administration, *Ann. Clin. Lab. Sci.,* 11, 119, 1981.
107. **Solomons, N. W., Viteri, F., Shuler, T. R., and Nielsen, F. H.,** Bioavailability of nickel in man: effects of foods and chemically-defined dietary constituents on the absorption of inorganic nickel, *J. Nutr.,* 112, 39, 1982.
108. **Nielsen, G. D., Jørgensen, P. J., Keiding, K., and Grandjean, P.,** Urinary nickel excretion before and after loading with naturally occurring nickel, in Trace Elements in Human Health and Disease, 2nd Nordic Symp., Odense, Denmark, August 17 to 21, 1987.
109. **Kaaber, K., Veien, N. K., and Tjell, J. C.,** Low nickel diet in the treatment of patients with chronic nickel dermatitis, *Br. J. Dermatol.,* 98, 197, 1978.
110. **Spruit, D. and Bongaarts, P. J. M.,** Nickel content of plasma, urine and hair in contact dermatitis, *Dermatologica,* 154, 291, 1977.

111. **Lippmann, M., Yeates, D. B., and Albert, R. E.,** Deposition, retention, and clearance of inhaled particles, *Br. J. Ind. Med.*, 37, 337, 1980.
112. **Graham, J. A., Miller, F. J., Daniels, M. J., Payne, E. A., and Gardner, D. E.,** Influence of cadmium, nickel, and chromium on primary immunity in mice, *Environ. Res.*, 16, 77, 1978.
113. **Wehner, A. P. and Craig, D. K.,** Toxicology of inhaled NiO and CoO in Syrian golden hamsters, *Am. Ind. Hyg. Assoc. J.*, 33, 147, 1972.
114. **Kodama, Y., Ishimatsu, S., Matsuno, K., Tanaka, I., and Tsuchiya, K.,** Pulmonary deposition and clearance of a nickel oxide aerosol by inhalation, *Biol. Trace Element Res.*, 7, 1, 1985.
115. **Tanaka, I., Ishimatsu, S., Matsuno, K., Kodama, Y., and Tsuchiya, K.,** Biological half-time of deposited nickel oxide aerosol in rat lung by inhalation, *Biol. Trace Element Res.*, 8, 203, 1985.
116. **Merritt, K., Brown, S. A., and Sharkey, N. A.,** Blood distribution of nickel, cobalt, and chromium following intramuscular injection into hamsters, *J. Biomed. Mater. Res.*, 18, 991, 1984.
117. **Mohanakrishnan, P. and Chignell, C. F.,** Copper and nickel binding to canine serum albumin. A circular dichroism study, *Comp. Biochem. Physiol.*, 79C, 321, 1984.
118. **Mohanakrishnan, P., Chignell, C. F., and Cox, R. H.,** Chloride ion nuclear magnetic resonance spectroscopy probe studies of copper and nickel binding to serum albumins, *J. Pharm. Sci.*, 74, 61, 1985.
119. **Lucassen, M. and Sarkar, B.,** Nickel(II)-binding constituents of human blood serum, *J. Toxicol. Environ. Health*, 5, 897, 1979.
120. **Hendel, R. C. and Sunderman, F. W. Jr.,** Species variations in the proportions of ultrafilterable and protein-bound serum nickel, *Res. Commun. Chem. Pathol. Pharmacol.*, 4, 141, 1972.
121. **Callan, W. M. and Sunderman, F. W., Jr.,** Species variations in binding of ^{63}Ni(II) by serum albumin, *Res. Commun. Chem. Pathol. Pharmacol.*, 5, 459, 1973.
122. **Glennon, J. D. and Sarkar, B.,** Nickel(II) transport in human blood serum. Studies of nickel(II) binding to human albumin and to native-sequence peptide, and ternary-complex formation with L-histidine, *Biochem. J.*, 203, 15, 1982.
123. **Laussac, J.-P. and Sarkar, B.,** Nickel(II)-binding to the NH_2- terminal peptide segment of human serum albumin: ^{13}C- and ^{1}H-nuclear magnetic resonance investigation, *Can. J. Chem.*, 58, 2055, 1980.
124. **Laussac, J.-P. and Sarkar, B.,** Characterization of the copper(II)- and nickel(II)-transport site of human serum albumin. Studies of copper(II) and nickel(II) binding to peptide 1-24 of human serum albumin by ^{13}C and ^{1}H NMR spectroscopy, *Biochemistry*, 23, 2832, 1984.
125. **Nomoto, S., McNeely, M. D., and Sunderman, F. W., Jr.,** Isolation of a nickel a_2-macroglobulin from rabbit serum, *Biochemistry*, 10, 1647, 1971.
126. **Sakar, B.,** Nickel metabolism, in *Nickel in the Human Environment*, Sunderman, P. W., Jr., Ed., IARC Scientific Publ. No. 53, International Agency for Research on Cancer, Lyon, 1984, 367.
127. **Bonde, I., Beck, H.-I., Jørgensen, P. J., and Grandjean, P.,** Nickel levels in intercellular fluid from nickel-allergic patients and control, in Trace Elements in Human Health and Disease, 2nd Nordic symp., Odense, Denmark, August 17 to 21, 1987.
128. **Jasim, S. and Tjalve, H.,** Effect of thiuram sulphides on the uptake and distribution of nickel in pregnant and non-pregnant mice, *Toxicology*, 32, 297, 1984.
129. **Oskarsson, A. and Tjalve, H.,** Effects of diethyldithiocarbamate and penicillamine on the tissue distribution of $^{63}NiCl_2$ in mice, *Arch. Toxicol.*, 45, 45, 1980.
130. **Nielsen, G. D., Andersen, O., and Grandjean, P.,** *Effect of Diethyldithiocarbamate on Toxicokinetics of ^{57}Ni in Mice*, WHO Environmental Health Series, Vol. 20, World Health Organization, Copenhagen, 1987, 78.
131. **Hildebrand, H. F., Decaestecker, A. M., and Hetuin, D.,** *Binding of Nickel Sulfides to Lymphocyte Subcellular Structures*, WHO Environmental Health Series, Vol. 20, World Health Organization, Copenhagen, 1987, 82.
132. **Costa, M. and Mollenhauer, H. H.,** Carcinogenic activity of particulate nickel compounds is proportional to their cellular uptake, *Science*, 209, 515, 1980.
133. **Costa, M. and Mollenhauer, H. H.,** Phagocytosis of nickel subsulfide particles during the early stages of neoplastic transformation in tissue culture, *Cancer Res.*, 40, 2688, 1980.
134. **Haugen, A., Schafer, P. W., Lechner, J. F., Stoner, G. D., Trump, B. F., and Harris, C. C.,** Cellular ingestion, toxic effects, and lesions observed in human bronchial epithelial tissue and cells cultured with asbestos and glass fibers, *Int. J. Cancer*, 30, 265, 1982.
135. **Kuehn, K. and Sunderman, F. W., Jr.,** personal communication, 1985.
136. **Costa, M., Abbracchio, M. P., and Simmons-Hansen, J.,** Factors influencing the phagocytosis, neoplastic transformation, and cytotoxicity of particulate nickel compounds in tissue culture systems, *Toxicol. Appl. Pharmacol.*, 60, 313, 1981.
137. **Costa, M., Simmons-Hansen, J., Bedrossian, C. W. M., Bonura, J., and Caprioli, R. M.,** Phagocytosis, cellular distribution, and carcinogenic activity of particulate nickel compounds in tissue culture, *Cancer Res.*, 41, 2868, 1981.

138. **Herlant-Peers, M.-C., Hildebrand, H. F., and Kerckaert, J.-P.,** *In vitro* and *in vivo* incorporation of 63Ni(II) into lung and liver subcellular fractions of Balb/C mice, *Carcinogenesis,* 4, 387, 1983.
139. **Webb, M., Heath, J. C., and Hopkins, T.,** Intranuclear distribution of the inducing metal in primary rhabdomyosarcomata induced in the rat by nickel, cobalt and cadmium, *Br. J. Cancer,* 26, 274, 1972.
140. **Jasmin, G. and Riopelle, J. L.,** Renal carcinomas and erythrocytosis in rats following intrarenal injection of nickel subsulfide, *Lab. Invest.,* 35, 71, 1976.
141. **Abbracchio, M. P., Simmons-Hansen, J., and Costa, M.,** Cytoplasmic dissolution of phagocytized crystalline nickel sulfide particles: a prerequisite for nuclear uptake of nickel, *J. Toxicol. Environ. Health.,* 9, 663, 1982.
142. **Sen, P. and Costa, M.,** Induction of chromosomal damage in Chinese hamster ovary cells by soluble and particulate nickel compounds: preferential fragmentation of the heterochromatic long arm of the X chromosome by carcinogenic crystalline NiS particles, *Cancer Res.,* 45, 2320, 1985.
143. **Ciccarelli, R. B. and Wetterhahn, K. E.,** *In vitro* interaction of 63-nickel(II) with chromatin and DNA from rat kidney and liver nuclei, *Chem. Biol. Interact.,* 52, 347, 1985.
144. **Roychaudhury, S., Banerjee, A., and Bhattacharyya, B.,** Interaction of Ni^{2+} with the tubulin-microtubule system, *Biochem., Biophys. Acta,* 787, 46, 1982.
145. **Andersen, O.,** Evaluation of the spindle-inhibiting effect of Ni^{++} by quantitation of chromosomal supercontraction, *Res. Commun. Chem. Pathol. Pharmacol.,* 50, 379, 1985.
146. **Swierenga, S. H. H. and Basrur, P. K.,** Effect of nickel on cultured rat embryo muscle cells, *Lab. Invest.,* 19, 663, 1968.
147. **Hohnadel, D. C., Sunderman, F. W., Jr., Nechay, M. W., and McNeely, M. D.,** Atomic absorption spectrometry of nickel, copper, zinc, and lead in sweat collected from healthy subjects during sauna bathing, *Clin. Chem.,* 19, 1288, 1973.
148. **Christensen, O. B., Moller, H., Andrasko, L., and Lagesson, V.,** Nickel concentration of blood, urine and sweat after oral administration, *Contact Derm.,* 5, 312, 1979.
149. **Tandon, S. K.,** Disposition of nickel-63 in the rat, *Toxicol. Lett.,* 10, 71, 1982.
150. **Sunderman, F. W., Jr. and Selin, C. E.,** The metabolism of nickel-63 carbonyl, *Toxicol. Appl. Pharmacol.,* 12, 207, 1968.
151. **Sunderman, F. W., Jr., Kasprzak, K. S., Lau, T. J., et al.,** Effects of manganese on carcinogenicity and metabolism of nickel subsulfide, *Cancer Res.,* 36, 1790, 1976.
152. **Carvalho, S. M. M. and Ziemer, P. L.,** Distribution and clearance of ^{63}Ni administered as $^{63}NiCl_2$ in the rat: intratracheal study, *Arch. Environ. Contam. Toxicol.,* 11, 245, 1982.
153. **English, J. C., Parker, R. D. R., Sharma, R. P., and Oberg, S. G.,** Toxicokinetics of nickel in rats after intratracheal administration of a soluble and insoluble form, *Am. Ind. Hyg. Assoc. J.,* 42, 486, 1981.
154. **Daldrup, T., Haarhoff, K., and Szathmary, S. C.,** Toedliche Nickelsulfatintoxikation, *Ber. Gerichtlichen Med.,* 41, 141, 1983.
155. **Linden, J. V., Hopfer, S. M., Gossling, H. R., and Sunderman, F. W., Jr.,** Blood nickel concentrations in patients with stainless-steel hip prostheses, *Ann. Clin. Lab. Sci.,* 15, 459, 1985.
156. **Sunderman, F. W., Jr., Aitio, A., Morgan, L. G., and Norseth, T.,** Biological monitoring of nickel, *Toxicol. Ind. Health,* 2(1), 17, 1986.
157. **Tossavainen, A., Nurminen, M., Mutanen, P., and Tola, S.,** Application of mathematical modelling for assessing the biological half-times of chromium and nickel in field studies, *Br. J. Ind. Med.,* 37, 285, 1980.
158. **Raithel, H.-J., Schaller, K. H., Mohrmann, W., Mayer, P., and Henkels, U.,** Untersuchungen zur Ausscheidungskinetik von Nickel bei Beschaftigten in der Glas-und Galvanischen Industrie, *Jahrestagung Dtsch. Gesellschaft Arbeitsmedizin,* 22, 223, 1982.
159. **Zober, A., Weltle, D., and Schaller, K.-H.,** Study of the kinetics of chromium and nickel in biological material during a week of arc welding work using chromium-nickel-containing filler metals, *Weld. Cutting,* 10, E162, 1984.
160. **Boysen, M., Solberg, L. A., Torjussen, W., Poppe, S., and Hogetveit, A. C.,** Histological changes, rhinoscopical findings and nickel concentration in plasma and urine in retired nickel workers, *Acta Oto-Laryngol.,* 97, 105, 1984.
161. **Morgan, L. G. and Rouge, P. J. C.,** Biological monitoring in nickel refinery workers, in *Nickel in the Human Environment,* Sunderman, F. W., Jr., Ed., IARC Scientific Publ. No. 53, International Agency for Research on Cancer, Lyon, 507, 1984.
162. **Hogetveit, A. C., Barton, R. T., and Kostol, C. O.,** Plasma nickel as a primary index of exposure in nickel refining, *Ann. Occup. Hyg.,* 21, 113, 1978.
163. **Torjussen, W. and Andersen, I.,** Nickel concentrations in nasal mucosa, plasma, and urine in active and retired nickel workers, *Ann. Clin. Lab. Sci.,* 9, 289, 1979.
164. **Zwennis, W. C. M. and Franssen, A. G.,** Exposition to insoluble nickel compounds in a plant for nickel catalysts; relation between concentrations of nickel in urine and plasma, 2nd Int. Conf. Clin. Chem. Clin. Toxicol. Metals, Montreal, 1983, 128.

165. **Zober, A., Welte, D., Schaller, K.-H., and Valentin, H.,** Untersuchunger zur Kinetik von Chrom und Nickel in Biologischen Material Wahrend Einwochigen Lightbogenschweissens mit Chrom-Nickel-Haltigen Zusatzwerkstoffen, *Schweissen Schneiden,* 36, 461, 1984.
166. **Raithel, H.-J., Kress, W., Schaller, K.-H., Weltle, D., and Valentin, H.,** Toxicological and occupational medical investigations concerning nickel-exposure in different industrial areas in the FRG, in *Progress in Nickel Toxicology,* Brown, S. S. and Sunderman, F. W., Jr., Eds., Blackwell Scientific, Oxford, 1985, 219.
167. **Adams, D. B.,** The routine determination of nickel and creatinine in urine, in *Nickel Toxicology,* Brown, S. S. and Sunderman, F. W., Jr., Eds., Academic Press, London, 1980, 99.
168. **Hopfer, S. M., Linden, J. V., Cristostomo, M. C., Catalanatto, F. A., Galen, M., and Sunderman, F. W., Jr.,** Hypernickelemia in hemodialysis patients, *Trace Elements Med.,* 2, 68, 1985.
169. **Cavelier, C., Foussereau, J., and Massin, M.,** Nickel allergy: analysis of metal clothing objects and patch testing to metal samples, *Contact Derm.,* 12, 65, 1985.
170. **Veien, N. K., Hattel, T., Justesen, O., Nørholm, A.,** Oral challenge with metal salts. II. Various types of eczema, *Contact Derm.,* 9, 407, 1983.
171. **Veien, N. K., Hattel, T., Justesen, O., and Nørholm, A.,** Dietary treatment of nickel dermatitis, *Acta Derm. Venereol. (Stockhølm),* 65, 138, 1985.
172. **MAK-Werten,** Nickel und Nickelverbindungen, in *Toxikologisch-Arbeitsmedizinsche Begrundungen von MAK-Werten,* Weinheim/Bergstr, Verlag Chemie, 1971, 1.
173. Documentation of the Threshold Limit Values, American Conference of Governmental Industrial Hygienists, Cincinnati, 1984, 295.
174. Draft Report Relating to Threshold Limiting Value: Nickel and Nickel Compounds, Working Group of Experts of the National Maximum Workplace Concentration Committee, The Hague, 1985.

Chapter 3

CARCINOGENICITY

M. A. Cooke and R. A. Cooke

TABLE OF CONTENTS

I. Introduction 36

II. Nickel Refining 36

III. Nickel Plating 37

IV. Manufacture and Use of Nickel Alloys and Stainless Steel 37

V. Foundry Work 37

VI. Welding, Grinding, Polishing, and Other Metal Trades 37

VII. Nickel/Cadmium Battery Manufacture 38

VIII. Classification 38

IX. Mechanisms 38

X. Regulatory Requirements 40

XI. Future Research 41

References 41

I. INTRODUCTION

There are no reports of nickel or its compounds being skin carcinogens. Neither is skin absorption (see Chapter 2) likely to be significant in the context of malignancy at other sites, especially taking into account the much higher level of absorption likely from inhalation in the processes accepted as having a cancer risk. However, Boysen et al.[1] quote the opinion of the International Agency for Research on Cancer (IARC)[2] in 1976 that "it has not been established whether the inhalation of nickel compounds is the sole cause of respiratory tract carcinoma in nickel workers". In a consideration of nickel, certain nickel compounds, and nickel refining in 1982, IARC[3] stated that "it is still not possible to state with certainty which specific nickel compounds are human carcinogens, although metallic nickel seems less likely to be so than nickel subsulphide or nickel oxides".

In the event of nickel being carcinogenic, one should not ignore the possibility of secondary deposits occurring in the skin from primary lesions caused by nickel or its compounds in other organs.

The occupations which involve exposure to nickel are manifold, many being described elsewhere in this volume (Chapter 12). They range from those involving exposure to the more toxic compounds with possible carcinogenic potential, such as nickel carbonyl in nickel refining, through plating and work with nickel alloys to tasks involving minimal absorption and consequently minimal or no carcinogenic risk. The lesions in the etiology of which nickel may have a role are sinonasal cancer, laryngeal cancer, and lung cancer.

In a multicancer occupation exposure case-referent study of 12 tumor types (esophagus, stomach, colon, rectum, liver, pancreas, lung, melanoma of skin, prostate, bladder, kidney, lymphoid tissue) Gerin et al.[4] did not find any association except for nickel-lung, although they pointed out that exposure to chromium is highly correlated with nickel and any effects may be confounded with possible effects from chromium.

Sorahan et al.[5] studied 2689 nickel/chromium platers in which the chrome exposure was statistically related to cancers of stomach, liver, nose and nasal cavities, lung and bronchus, and "all cancers". A positive association was found for cancers of lung and bronchus, and it was concluded that nickel did not have a role in this.

The lung carcinomata are usually of epidermoid and anaplastic types.

The processes involved in nickel exposure include

- Nickel refining
- Nickel plating
- Manufacture and use of nickel alloys and stainless steels
- Foundry work
- Welding, grinding, polishing, and other metal trades
- Nickel/cadmium battery manufacture

II. NICKEL REFINING

Historically the first reported association of nickel and respiratory tract cancer was by Grenfell and Samuel[6] in 1932 in the nickel refining industry. The history of nickel mining, smelting, and refining has been described by Mastromatteo[7] in the Yant Memorial Lecture. Essentially this uses a process which combines nickel from impure ores with carbon monoxide to produce nickel carbonyl [$Ni(CO)_4$]. This is heated and decomposes to release carbon monoxide, which is recycled; nickel is deposited and formed into pellets.

Both the carbon monoxide and the nickel carbonyl are highly toxic and have been known to be dangerous from the inception of the current process at the Mond Nickel Company in 1900. The toxicity of the nickel arises from the decomposition of the nickel carbonyl with resultant deposition of finely divided, slightly soluble compounds of nickel over the respiratory mucous membrane.

In 1973 Pedersen et al.[8] reported a ratio of 40 observed per 1 expected case of nasal cancer in workers employed for more than 3 years in a Norwegian nickel refinery. Subsequently Andersen et al.[9] showed that this increased to 168/1 for workers with more than 25 years employment.

In addition to sinonasal cancer in exposed workers in nickel refineries, lesions have occurred in the lungs and possibly the larynx.

Elimination of the original methods of calcining and open sintering, improved ventilation, and respiratory protection by dust masks resulted in sharp decreases in respiratory cancer. The decrease in the period 1920 to 1930 was reported by Morgan[10] and Doll et al.[11]

III. NICKEL PLATING

The plating industry is difficult to assess by epidemiological surveys since there is a variety of types of exposure, degrees of exposure (often not quantified), and, perhaps most important of all, the presence of other materials such as chromium, arsenic, other dusts, and aerosols of soluble nickel and acids. It is very difficult to be specific concerning the relationship between any increase in mortality and a specific material known to be in the environment.

The report of the U.K. Registrar General[12] for 1970—72 had noted a significant increase in deaths due to cancer of the lung and intestine in platers and related workers; these could not be explained by social class differences.

Burges[13] in 1980 reported on 508 male nickel platers in the period 1945—77. The lung cancer rate was increased, but not to a statistically significant degree, for "all employees" (observed = 10, expected = 8.2), but there was no increase for those with over 1 year of exposure (observed = 1, expected = 1.7). There was a significant excess of deaths from stomach cancer in persons exposed for more than 1 year on data adjusted for social class and region (observed = 4, expected = 0.8, $0.005 < p < 0.025$). In a study in the metal polishing and plating industry in the U.S., Blair[14] found an increase of esophageal and liver cancer but no increase for lung and sinus cancer.

IV. MANUFACTURE AND USE OF NICKEL ALLOYS AND STAINLESS STEEL

In Chapter 4 Morgan and Flint discuss the release of nickel from alloys. Cox et al.[15] found no evidence of any occupational hazard in a mortality study of nickel alloy workers exposed to nickel and nickel oxides. These men were not exposed to nickel subsulfide, which was postulated as a causative agent by Ottolenghi et al.[16] in those cases observed in the obsolete processes of calcining and sintering and which most readily produces bronchial carcinoma in rats.

V. FOUNDRY WORK

Foundry workers may be exposed to dust and fumes containing nickel and its compounds. However, the levels in nickel foundries are generally low. Warner[17] has given 0.01 to 0.03 mg/m^3 as levels encountered. In a proportional mortality study in 26 U.S. foundries from 1968—79 among 851 foundry workers exposed to nickel chromium, Cornell and Landis[18] reported no increase in respiratory cancer rates.

VI. WELDING, GRINDING, POLISHING, AND OTHER METAL TRADES

This group, involved in miscellaneous employment of the above type, may be exposed not only to dust and fumes, but also to aerosols containing nickel and its compounds. Nickel exposures likely to have been encountered under maximal exposure in such work are unlikely to be significant compared with nickel refinery workers. In a case-control study of 42 nickel-exposed workers in miscellaneous employment such as welders, platers, sprayers, grinders, and polishers, Bernachi et al.[19] found no difference between nickel-exposed workers and controls. However, the observation time of this study was limited. Nevertheless, Godbold and Tompkins[20] found similar results in 814 workers exposed to powdered nickel catalyst. Work among other nickel-exposed workers, such as plumbers, steel workers, and copper smelter workers, has shown an increased cancer risk, but it is not possible to derive any indication from these whether or not nickel played a significant part in the findings. There is such a multiplicity of exposure that it is unlikely that epidemiological studies alone will provide an answer. Among other possible confounding factors are exposure to chromium, arsenic, and asbestos which are almost always present at some time in the working life of such workers.

In a proportional mortality study of metal polishers in nickel finishing, Blair[21] found no excess of lung or sinus cancer.

VII. NICKEL/CADMIUM BATTERY MANUFACTURE

These workers will also be exposed to dust, fumes, and aerosols of nickel and its salts, although at a relatively low level. Warner[17] has suggested typical levels of 0.04 mg/m^3. Nasal cancer has been reported in such workers and it seems unlikely that cadmium was the cause.

VIII. CLASSIFICATION

The U.S. Environmental Protection Agency (EPA)[22] categorized nickel and its compounds in respect to carcinogenicity as follows:

- Group A: Nickel refinery dust — from pyrometallurgical sulfide nickel matte refineries
- Group B: Nickel carbonyl

IARC[3] has classified nickel as follows:

- Group 1: Sufficient evidenct that nickel refining is carcinogenic to humans
- Group 2A: Limited evidence that nickel and certain nickel compounds are carcinogenic to humans; sufficient evidence of carcinogenicity in experimental animals; inadequate evidence of activity in short-term tests

The International Labour Office (ILO)[23] includes nickel carbonyl and the treatment of nickel ore, the latter being an "industrial process involving significant carcinogenic risk", in Group 3 of their list, which advises that "exposure should be kept to a minimum through the use of the most feasible and applicable controls".

IX. MECHANISMS

As with many carcinogens, the mechanisms by which nickel acts are the subject of much work, debate, and discussion.

Nishioka[24] has reported negative results with nickel compounds in microbial recombinational repair (Rec.) assays, but positive results in the Ames test have been reported by Hollstein and McCann[25] and by Lavelle and Witmer.[26]

Nishimura and Umeda[27] have demonstrated chromosomal aberrations in cultured mammalian cells. Cell transformation assays have also been reported as positive.

Sunderman[28] has described inhibition of DNA-dependent RNA synthesis by nickel carbonyl, and Ciccarelli et al.[29] have demonstrated DNA-protein cross links and DNA strand breaks in rat kidney. However, Gilman and Swierenga[30] report other work[31,32] which found no single-strand breaks in human blood cells and animal tissues.

Weisburger and Williams[33] have commented that of the inorganic carcinogens some, such as nickel and chromium, may be electrophiles, while others may induce changes in DNA by alterations of fidelity of DNA replication.

Lechner et al.[34] have studied the *in vitro* effects of nickel sulfate on growth and differentiation of normal human bronchial epithelial cells. They incubated cultures for 40 d with nickel sulfate and then in medium without nickel sulfate. Colonies of mitotic cells appeared in cultures exposed to 5 μg nickel sulfate per milliliter or more, but none appeared at lower concentration or in controls. Derived mass cultures demonstrated increased population doubling potential (>50 divisions). There were some aberrations in the squamous (terminal) differentiation process. Some lost the requirement for epidermal growth factors; aneuploidy and marker chromosomes were also noted. However, no nickel-altered cultures were anchorage independent nor did tumors result from injection into athymic nude mice.

In discussing the possible mechanism of cancer induction by nickel, Furst[35] emphasized that investigations have been limited mainly to interaction with nucleic acids, which is only one stage of the initiation process. Other considerations should include any possible role of nickel in the formation of ultimate carcinogens from a procarcinogen or inhibition of metabolism of active carcinogens to inactive conjugates. He also mentioned the possible indirect role of nickel as a stabilizer of free radicals formed by oxidation of some unsaturated molecules, such as carcinogenic aromatic hydrocarbons.

In 1987, IARC[36] was of the opinion that it has not been established whether the inhalation of nickel compounds is the sole cause of respiratory tract carcinoma in nickel workers.

In the U.S. Fourth Annual Report on Carcinogens,[37] there is a comment to the effect that "the toxicologic (and carcinogenic) risk from a metal is often a property less of the form of the metal administered than of the metal per se". It is pointed out that the anion or ligand of the metal can determine the physicochemical properties which may be involved in carcinogenesis, for example by affecting solubility, absorption, distribution, etc. It therefore concludes that all compounds containing nickel are potentially carcinogenic, as is the elemental metal, "given the property of most tissues of being able to solubilize most metals to some degree".

There is experimental animal evidence of carcinogenicity of nickel and several nickel compounds. Inhalation of nickel subsulfide by rats will produce lung cancer. Intramuscular injections of nickel powder, nickel oxide, hydroxide, carbonate and subsulfide, nickelocene, and nickel-iron sulfide matte have produced various tumors, including fibrosarcomas and rhabdomyosarcomas, in mice, rats, hamsters, and rabbits. Sunderman[38] has reviewed the carcinogenic effects of nickel compounds in animals including rats exposed to nickel carbonyl.[39,40]

Saknyn and Shabynina[41] found an eightfold increase in death rate from sarcoma, in addition to an increase in lung cancer, in a group of nickel refinery workers that included those involved with roasting and smelting.

Of particular interest is the animal model utilized by Albert et al.,[42] who injected nickel subsulfide into the vitreous cavity of the eyes of rats. Subsequent ocular tumors included malignant melanomas, retinoblastomas, and other neoplasms of the eye. There were no lesions in the untreated eyes, nor were there any primary extra-ocular tumors. The authors describe the merits of the test — short latent period, high tumor incidence, and ease of detection — but point out that "cells which undergo malignant transformation in ocular tissues may be relatively protected from contact by immunocompetent cells responsible for tumour rejection at other sites".

Boysen et al.[1] have correlated histological changes and rhinoscopical findings with nickel concentrations in plasma and urine in retired nickel workers. They found evidence of regression of mucosal swelling after retirement and the incidence of "hyperplastic/polypoid nasal mucosa was inversely related to the duration of retirement".

The presence of nickel in plasma and urine has been reported by Torjussen and Andersen.[43] Torjussen et al.[44] had previously demonstrated that dysplasia of nasal epithelium, probably precancerous, was related to age and duration of employment.

It is of interest that Leach and Sunderman[45] have reported nickel contamination of human albumin solutions from several different manufacturers. They presumed this came "from contact of the albumin solutions with stainless steel during processing". In solutions that contained 250 g of serum albumin per liter, nickel concentration averaged 300 times the mean nickel concentration in serum samples from 30 healthy adults in New England (0.28 μg/l — range <0.05 to 1.1 μg/l).

The retired nickel workers studied by Boysen et al.[1] had plasma nickel levels of 5.8 (±4.8) μg/l in 1976 and 2.7 (±1.7) in 1978—79. The urine levels were 48.2 (± 72.5) and 10.4 (±7.9) μg/l, respectively.

Allergic reactions to nickel in intravenous fluids or indwelling needles have been reported by Sunderman.[46] The relevance of nickel in plasma, and its excretion in urine, to systemic carcinogenesis is unknown. The relevance of skin absorption to systemic carcinogenesis is equally unknown, but such absorption is likely to be low or insignificant compared to inhalation in occupations with exposure to nickel dust or fume.

Sunderman[47] refers to the work of Yamane and Sakai,[48] who found that nickel acetate suppressed hepatoma induction by amino azodyes. This "was related to their capacity to stimulate hepatic aminoazo reductase activity and to reduce the hepatic concentration of protein-bound azodyes." He points out that the "inhibitory effects of metals are probably mediated by different mechanisms" in each experimental system.

In view of the known interaction of nickel compounds with DNA, it could be wrong to disregard skin and gastrointestinal absorption in its entirety, while accepting that absorption by these routes may be *de minimis* in respect to cancer of the respiratory tract.

X. REGULATORY REQUIREMENTS

The U.S. Occupational Safety and Health Administration (OSHA)[37] has adopted the following permissible levels:

Nickel carbonyl	0.007 mg/m^3 as an 8-h time-weighted average (TWA)
Nickel metal and soluble nickel compounds	1 mg/m^3 as an 8-h TWA

Cancer effects were not taken into account in defining these standards. The National Institute of Occupational Safety and Hygiene (NIOSH) had recommended in 1977 that exposure to nickel be limited to 15 μg/m^3 (10-h TWA) on account of observed carcinogenicity.

In the U.K.,[49] the recommended occupational exposure limits are

Nickel	1 mg/m^3 long-term exposure limit (8-h TWA)
Soluble nickel compounds (as Ni)	0.1 mg/m^3 long-term exposure limit (8-h TWA)
	0.3 mg/m^3 short-term exposure limit (10-min TWA)
Insoluble nickel compounds (as Ni)	1 mg/m^3 long-term exposure limit (8-h TWA)
	3 mg/m^3 short-term exposure limit (10-min TWA)

Nickel and its compounds are regulated under the Hazardous Waste Disposal rule of the Resource Conservation and Recovery Act. The EPA regulates nickel and nickel compounds under the Clean Water Act. (For full updating and details see current National Technical Information Service documents and appropriate copies of the *Federal Register*.)

The Food and Drug Administration (FDA) has no regulation of nickel as a food additive. Data available are not considered adequate to assess carcinogenicity through dietary exposure.

The Consumer Products Safety Commission (CPSC) made a preliminary determination in 1980 that nickel carbonyl was not present in consumer products under CPSC jurisdiction. No further action is anticipated until receipt of new information.

XI. FUTURE RESEARCH

Current studies in the U.S. National Toxicology Program[50] include

- Prechronic studies in progress (by inhalation in rats and mice):
  Nickel (II) oxide
  Nickel sulfate hexahydrate
- Prechronic tests completed — chemicals in review for further testing (by inhalation in rats and mice):
  Nickel subsulfide

There are 17 cancer epidemiology research programs specifically related to nickel and listed by IARC[51] and others may well include consideration of the role of nickel. It is often valuable to update and continue cohort studies, not only to increase their power, but also to determine trends with time, changing exposure, or other relevant factors. It would seem to be a general consensus that definitive and significant excesses of cancer deaths in those with industrial exposure to nickel and its compounds only occur in those substantially exposed to dust or fume, such as in nickel refineries with less stringent controls than some of the better plants currently in operation.

The role of nickel in the induction of cancer is still incompletely understood, and, consequently, its possible role in promotion or induction is not completely determined. Epidemiology, animal experiments of the bioassay type, and short-term screening tests are not likely to resolve such issues. The greatest hope would seem to be in studies of mechanisms and such possibilities as the role of nickel as a stabilizer of free radicals. The need for such studies has been discussed by Furst.[35]

REFERENCES

1. **Boysen, M., Solbery, L. A., Torjussen, W., Poppe, S., and Hogetveit, A. C.,** Histological changes, rhinoscopical findings and nickel concentration in plasma and urine in retired nickel workers, *Acta Oto Laryngol.*, 97, 105, 1984.
2. *Cadium, Nickel, Some Epoxides, Miscellaneous Industrial Chemicals and General Considerations on Volatile Anaesthetics,* International Agency for Research on Cancer, Lyon, France, 1976.
3. *Chemicals, Industrial Processes and Industries Associated with Cancer in Humans,* IARC Monogr. (Suppl. 4), International Agency for Research on Cancer, Lyon, France, 1982.
4. **Gerin, M., Siemiatycki, J., Richardson, L., Pellerin, J., Lakhani, R., and Dewer, R.,** Nickel and cancer associations from a multicancer occupation exposure case-referent study: preliminary findings, *IARC Sci. Publ.*, 53, 105, 1984.
5. **Sorahan, T., Burges, D. C., and Waterhouse, J. A. H.,** A mortality study of nickel/chromium platers, *Br. J. Ind. Med.*, 44, 250, 1987.
6. **Grenfell, J. P. W. and Samuel, H.,** Cancer among Welsh nickel workers, *Lancet,* 1, 375, 1932.
7. **Mastromatteo, E.,** Nickel, Yant Memorial Lecture, *Am. Ind. Hyg. Assoc.* 47(10), 589, 1986.
8. **Pedersen, E., Hogetveit, A. C., and Andersen, A.,** Cancer of respiratory organs among workers at a nickel refinery in Norway, *J. Natl. Cancer Inst.*, 12, 32, 1973.

9. **Andersen, A., Hogetveit, A. C., and Magnus, K.,** A follow-up study among Norwegian nickel workers, in *Nickel Toxicology,* Brown, S. S. and Sunderman, F. W., Jr., Eds., Academic Press, London, 1980, 31.
10. **Morgan, J. G.,** Some observations on the incidence of respiratory cancer in nickel workers, *Br. J. Ind. Med.,* 15, 224, 1958.
11. **Doll, R., Mathews, J. D., and Morgan, L. G.,** Cancers of the lung and nasal sinuses in nickel workers: a reassessment of the period of risk, *Br. J. Ind. Med.,* 34, 102, 1977.
12. Occupational mortality tables, *The Registrar-General's Decennial Supplement for England and Wales 1970—1972,* London, 1978.
13. **Burges, D. C. L.,** Mortality study of nickel platers, in *Nickel Toxicology,* Brown, S. S. and Sunderman, F. W., Jr., Eds., Academic Press, London, 1980.
14. **Blair, A.,** Mortality among workers in metal polishing and plating industry, *J. Occup. Med.,* 22, 158, 1980.
15. **Cox, J. E., Doll, R., Scott, W. A., and Smith, S.,** Mortality of nickel workers: experience of men working with metallic nickel, *Br. J. Ind. Med.,* 38, 235, 1981.
16. **Ottolenghi, A. D., Haseman, J. K., Payne, W. W., Falk, H. L., and Macfarland, H. N.,** Inhalation studies of nickel sulphide in pulmonary carcinogenesis of rats, *J. Natl. Cancer Inst.,* 54, 1165, 1975.
17. **Warner, J. S.,** Occupational exposure to airborne nickel in producing and using primary nickel products, in *Nickel in the Human Environment, IARC Sci. Publ.,* 53, 419, 1984.
18. **Cornell, R. G. and Landis, J. R.,** Mortality patterns among nickel/chromium alloy foundry workers, in *Nickel in the Human Environment, IARC Sci. Publ.,* 53, 87, 1984.
19. **Bernachi, E. J., Parsons, G. E., and Sunderman, J. W., Jr.,** Investigation into exposure to nickel and lung cancer mortality: case-control study at aircraft engine factory, *Ann. Clin. Lab. Sci.,* 8, 190, 1980.
20. **Godbold, J. H. and Tompkins, E. A.,** A long-term mortality study of workers occupationally exposed to metallic nickel at the Oak Ridge gaseous diffusion plant, *J. Occup. Med.,* 21, 799, 1979.
21. **Blair, A.,** Mortality among workers in metal polishing and plating industry, *J. Occup. Med.,* 22, 158, 1980.
22. Health Assessment Document for Nickel, Office of Research and Development, Environmental Criteria and Assessment Office, U.S. Environmental Protection Agency, Research Triangle Park, NC, 1986.
23. Occupational Cancer-Prevention and Control, Occupational Safety and Health Ser., No. 39, International Labour Office, Geneva, 1977.
24. **Nishioka, H.,** Mutagenic activities of metal compounds in bacteria, *Mutat. Res.,* 31, 185, 1975.
25. **Hollstein, M. and McCann, J.,** Short-term tests for carcinogens and mutagens, *Mutat. Res.,* 65, 133, 1979.
26. **Lavelle, J. M. and Witmer, C. M.,** *Environ. Mutagen.,* 3, 320, 1981.
27. **Nishimura, M. and Umeda, M.,** Induction of chromosomal aberrations in cultured mammalian cells by nickel compounds, *Mutat. Res.,* 68, 337, 1979.
28. **Sunderman, F. W., Jr.,** Metal carcinogenesis in experimental animals, *Food Cosmet. Toxicol.,* 9, 105, 1971.
29. **Ciccarelli, R. B., Hampton, T. H., and Jennette, K. W.,** Nickel carbonate induces DNA-protein crosslinks and DNA strand breaks in rat kidneys, *Cancer Lett.,* 12(4), 349, 1981.
30. **Gilman, J. P. W. and Swierenga, S. H. H.,** Inorganic carcinogenesis, in *Chemical Carcinogens,* Vol. 1, 2nd ed., Searle, C. E., Ed., American Chemical Society, Washington, D.C., 1984, 588.
31. **Casto, B. C., Pieczynski, W. J., Nelson, R. L., and DiPaolo, J. A.,** *Proc. Am. Assoc. Cancer Res.,* 17, 12, 1976.
32. **McLean, J. R., Williams, R. S., Kaplan, J. G., and Birnbo, H. C.,** in *Progress in Mutation Research,* Vol. 1, Elsevier/North Holland, Amsterdam, 1981, 828.
33. **Weisburger, J. H. and Williams, G. M.,** Bioassay of Carcinogens: in vitro and in vivo tests, in *Chemical Carcinogens,* Vol. 2, 2nd ed. Searle, C.E., Ed., American Chemical Society, Washington, D.C., 1984, 22.
34. **Lechner, J. F., Tokiwa, T., McClendon, I. A., and Haugen, A.,** Effects of nickel sulfate on growth and differentiation of normal human bronchial epithelial cells, *Carcinogenesis,* 5(12), 1697, 1984.
35. **Furst, A.,** Mechanism of action of nickel as a carcinogen: needed information, *IARC Sci. Publ.,* 53, 245, 1984.
36. Monographs on the evaluation of carcinogenic risk of chemicals to man, in *Nickel and Inorganic Nickel Compounds,* Vol. 2, International Agency for Research on Cancer, Lyon, France, 1973, 126.
37. Fourth Annual Report on Carcinogens, National Technical Information Service, U.S. Department of Commerce, 1985, 315.
38. **Sunderman, F. W., Jr.,** Carcinogenicity of nickel compounds in animals, *IARC Sci. Publ.,* 53, 127, 1984.
39. **Sunderman, F. W., Donnelly, A. J., West, B., and Kincuid, J. F.,** Nickel poisoning. IX. Carcinogenesis in rats exposed to nickel carbonyl, *AMA Arch. Ind. Health,* 20, 36, 1959.

40. **Sunderman, F. W. and Donnelly, A. J.,** Studies of nickel carcinogenesis. Metastasizing pulmonary tumours in rats induced by the inhalation of nickel carbonyl, *Am. J. Pathol.,* 46, 1027, 1965.
41. **Saknyn, A. C. and Shabynina, N. K.,** Some statistical materials on carcinogenic hazards in the production of nickel on an ore oxide base, *Gig. Tr. Prof. Zabol.,* 14, 10, 1970.
42. **Albert, D. M., Gonder, J. R., Papale, J., Craft, J. L., Dohlman, H. G., Reid, M. C., and Sunderman, F. W., Jr.,** Induction of ocular neoplasms in Fischer rats by intraocular injection of nickel subsulfide, *Invest. Ophthalmol. Vis. Sci.,* 22, 768, 1982.
43. **Torjussen, W. and Andersen, I.,** Nickel concentration in nasal mucosa, plasma, and urine in active and retired nickel workers, *Ann. Clin. La. Sci.,* 9, 289, 1979.
44. **Tourjussen, W., Solberg, L. A., and Hogetveit, A. C.,** Histopathological changes of the nasal mucosa in nickel workers. A pilot study, *Cancer (Brussels),* 44, 963, 1979.
45. **Leach, C. N., Jr. and Sunderman, F. W., Jr.,** Nickel contamination of human serum albumin solutions, correspondence, *N. Engl. J. Med.,* 313(19), 1232, 1985.
46. **Sunderman, F. W., Jr.,** Potential toxicity from nickel contamination of intravenous fluids, *Ann. Clin. Lab. Sci.,* 13, 1, 1983.
47. **Sunderman, F. W.,** Carcinogenicity and anticarcinogenicity of metal compounds, in *Environmental Carcinogenesis* Emmelot, P. and Kriek, E., Eds., Elsevier, Amsterdam, 1979.
48. **Yamane, Y. and Sakai, K.,** Suppressive effect of concurrent administration of metal salts on carcinogenesis by 3-methyl-4-(dimethyl amino) azo benzene, and of the effect of these metals on amino azo dye metabolism during carcinogenesis, *Gann,* 64, 563, 1973.
49. Occupational Exposure Limits, Guidance Note EH 40/87, Health and Safety Executive, Her Majesty's Stationery Office, London, 1987.
50. *Management Studies Report,* U.S. National Toxicology Program, July 11, 1987.
51. Directory of On-Going Research in Cancer Epidemiology, IARC Sci. Publ. No. 80, Lyon, France, 1986.

Chapter 4

NICKEL ALLOYS AND COATINGS: RELEASE OF NICKEL

L. G. Morgan and G. N. Flint

TABLE OF CONTENTS

I. Introduction 46

II. The Corrosive Environment 46

III. Solubility of Nickel Alloys and Coatings 46
- A. Copper/Nickel and Related Alloys 48
- B. Other Non-Chromium-Containing Nickel Alloys 49
- C. Alloys Containing Chromium: Passive Condition 49
- D. Loss of Passivity 50

IV. Metal Prostheses 51

V. Nickel and Nickel-Alloy Coatings 52

References 53

I. INTRODUCTION

The virtues of nickel as an alloying element were recognized by various civilizations long before its identification as a separate metal in 1751. Today, nickel alloys provide an essential and frequently unique constituent of many components on which modern industrial and domestic life is dependent for its satisfactory operation. The invention of nickel plating in about 1840 resulted in a major improvement in the function, life, and decorative appeal of many steel, zinc alloy, and brass components and articles.

Exposure of humans to nickel alloys and coatings arises from handling of many everyday objects or from the use of nickel alloys in prostheses. In domestic and commercial life the nickel-containing alloys in common use are cupronickels used for coinage, nickel silver in cutlery and architectural features, and stainless steel in kitchenware, cutlery, architectural trim, watchcases, and household and industrial components of many kinds.

Common nickel-plated articles include keys, fasteners, handles, tools, toys, kitchenware, paper clips, trinkets and jewelry. Frequently a topcoat, usually of chromium, is applied to the nickel coating. However, where the surfaces are kept clean and bright by the nature of use and for nondecorative applications, it may not be applied.

In clinical practice, nickel alloyed with chromium and other metals is used in surgical implants and dental prostheses.

Occupational exposure to nickel alloys is extensive, as would be expected from their evident importance to industry. This is illustrated by the applications shown in Figure 1. The extent of that exposure will vary greatly according to the nature of the application, but those workers exposed to nickel metal, alloys, and compounds in powder form and to spray from solutions of compounds are at greatest risk.

II. THE CORROSIVE ENVIRONMENT

Assuming that the first step in the development of nickel dermatitis is the formation of soluble nickel compounds in contact with skin, the corrosiveness of sweat, saliva, and other body fluids to nickel and nickel alloys is of primary importance. The composition of sweat varies from person to person and in the same person, according to physiological requirements. Individuals known as "rusters", with high concentrations of chloride in their sweat, are known to have increased potential for causing corrosion of metals.[1] Evaporation will increase salt concentration, and possible reaction with the atmosphere may also affect the constituents present.

Regarding corrosion by sweat, the two constituents of greatest significance are oxygen and chloride. Oxygen, essential for the electrochemical process by which corrosion of nickel alloys in neutral or near-neutral solutions occurs, is normally present at saturation levels in thin films of sweat, but occasions arise where oxygen levels may vary. For example, sweat present in a crevice formed between metal or alloy surface and skin, as in watchcases, buttons, clips, or earrings, is likely to be low in oxygen content. The differential aeration cell set up with oxygen-rich areas outside the crevice will promote corrosion in the crevice area. Additionally, in chromium-containing alloys, the low oxygen level will lead to breakdown of the oxide film in the crevice area and loss of passivity (see Section III.D).

The chloride ion is particularly effective in promoting corrosion and has the ability to cause breakdown of protective oxide films. The presence of lactic acid, amino acids, and ammonium compounds in sweat is also of importance because of their effect on pH value and ability to complex the nickel ion.

III. SOLUBILITY OF NICKEL ALLOYS AND COATINGS

The resistance of nickel alloys to corrosion by sweat, to a large extent, governs their ability to cause contact dermatitis. In this connection it is appropriate to distinguish between:

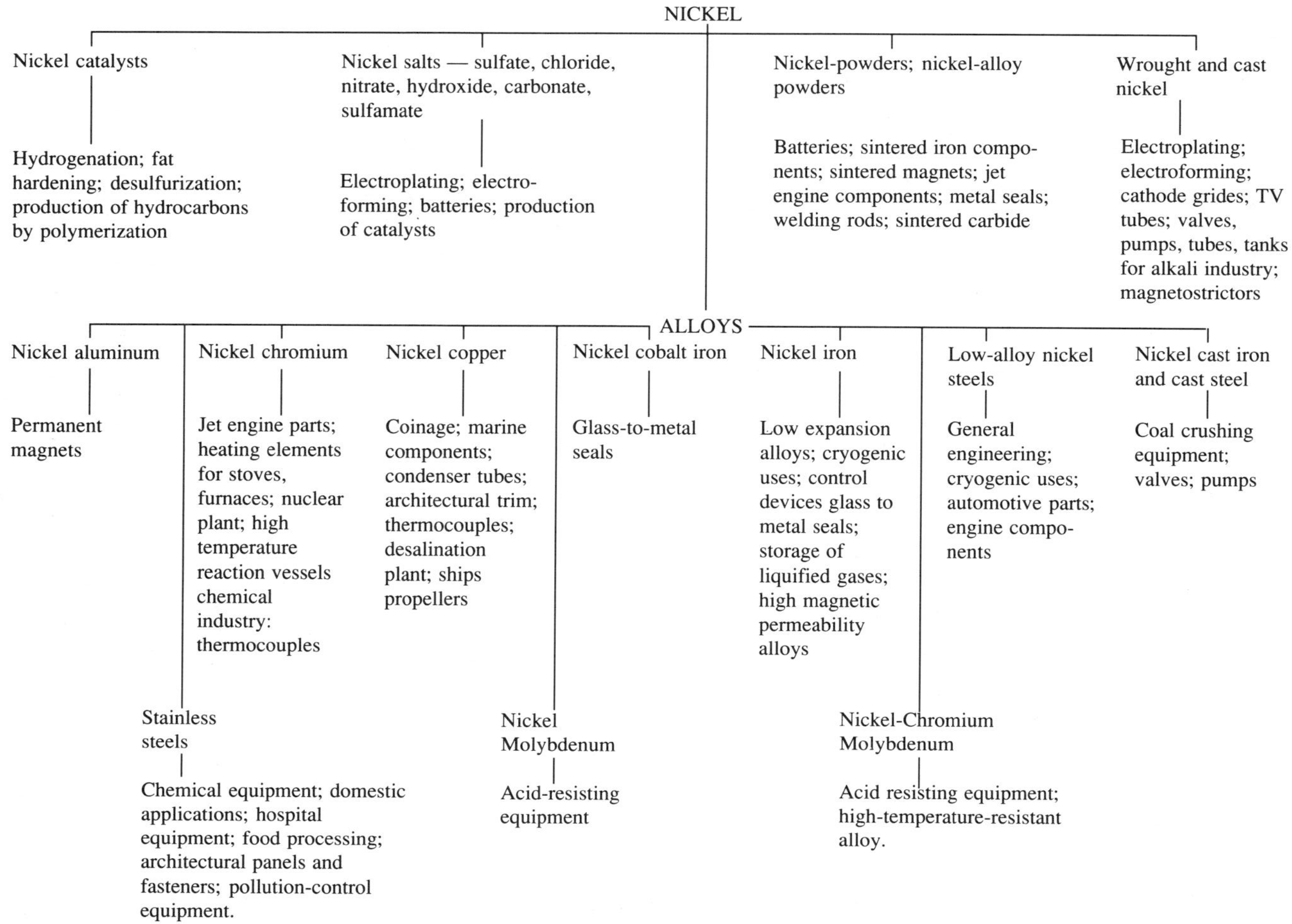

FIGURE 1. Usage of nickel in industry.

Table 1
WEIGHT LOSS (g/dm^2) AFTER EXPOSURE FOR 1 YEAR IN THE PANAMA CANAL ZONE[2]

	Immersion in seawater	Intermittent tidal immersion	Marine atmosphere
Nickel	5.40	0.78	0.02
67 Ni/30 Cu alloy (Monel)	3.67	0.23	0.10
30 Ni/69 Cu alloy	1.15	0.19	0.08
18 Ni/64 Cu/17 Zn (nickel silver)	—	—	0.09

1. Nickel-containing alloys such as copper/nickel, copper/zinc/nickel, and/or nickel/iron (and nickel itself) that do not form effectively protective, oxide films
2. Nickel/chromium-containing alloys such as nickel/chromium itself, iron/nickel/chromium, iron/nickel/chromium/molybdenum alloys that form oxide films consisting mainly of chromium oxide (Cr_2O_3) with iron oxide (Fe_2O_3), and/or nickel oxide (NiO) which protect the alloys to a very large extent from corrosive attack — this group contains the stainless steels and the special alloys developed for use in surgical and dental implants

In regard to electrodeposited coatings, it is necessary to distinguish between:

1. Those in which nickel forms the outer layer
2. Those in which the nickel layer is covered by a more noble topcoating such as chromium, gold, or rhodium, since the mode and rate of corrosion are affected by the top coat
3. Those composed of a nickel alloy such as tin/nickel or zinc/nickel

A. Copper/Nickel and Related Alloys

The addition of copper to nickel increases its nobility and improves resistance to chloride solutions. The results of corrosion tests in tropical seawater showing the influence of additions of copper to nickel are given in Table 1.[2] The low corrosion rate of copper-nickel alloys is responsible for their extensive use in industrial applications requiring resistance to corrosion by seawater and brines. However, some attack does occur, thus releasing copper and nickel corrosion products into the environment. A rather higher rate of attack would be expected in the sweat present on the skin, bearing in mind the high degree of aeration and the lower pH value and higher temperature in comparison with seawater.

Hypersensitivity resulting from exposure to cupronickel coinage is not unexpected and has been reported in the literature.[3,4] The nickel content of coinage can vary greatly. The composition of various coins is given in Table 2. The release of nickel by the action of sweat on cupronickel coinage has been demonstrated by Pedersen (Table 3).[5] Even higher rates of release were observed by Katz, but the experimental conditions may not have been comparable.[6] The accelerating effect of increase in temperature was established by Menne and is also shown in Table 3.[7]

In view of the experimental results and the extensive handling of coins, it is perhaps surprising that a higher incidence of dermatitis from contact with cupronickel coins is not observed in practice. Factors of significance in this connection are the generally short duration of contact and the protection provided by the thick horny layer of the palm.[8] Persons who, from habit or from the nature of their occupation, have prolonged contact with coins show more susceptibility.[3]

Table 2
COMPOSITION OF COINAGE

	Al	Cr	Mo	Fe	Ni	Cu	Zn
Belgium							
1 franc 1973					25.5	74.0	0.5
5 francs 1971					26.6	72.7	0.7
5 francs 1986					2.7	96.4	0.9
20 francs 1980					6.9	92.3	0.8
Canada							
1 cent 1978						97.9	2.1
10 cents 1986				0.2	99.8		
Finland							
5 pennia 1981					0.4	96.6	3.0
10 pennia 1975					2.7	97.3	
20 pennia 1981	5.5		0.05		2.4	92.1	
1 markka 1978			0.4		21.2	78.4	
France							
5 cent 1966	9.2				2.5	88.3	
20 cents 1984	8.1			0.15	2.5	89.2	
1 franc 1973				0.15	99.9		
2 francs 1980				0.2	99.8		
German							
10 pfennig 1971						76.8	23.2
Greek							
5 drachmas 1980					8.3	75.4	16.3
Italy							
100 lira 1977		19.5		80.4	0.1		
200 liras 1978	6.9				2.6	90.5	
500 liras 1984	8.8				2.6	88.6	
U.K.							
2 pence 1979				0.18		94.2	3.3
5 pence 1971				0.3	18.6	81.1	
2 shillings 10 pence 1965/76				0.2	25.2	74.6	
50 pence 1986				0.3	18.7	81.0	
1 pound 1983				0.14	7.1	71.9	20.8
U.S.							
1 cent 1969					0.4	95.5	4.1

B. Other Non-Chromium-Containing Nickel Alloys

The nickel-iron and nickel-steels are the most common alloys in this class, e.g., low alloy nickel-steels used in engineering and/or low expansion alloys used in control devices and other applications (Figure 1), but the extent of persistent contact of such alloys with the skin is minimal.

The nickel-containing alloys used in jewelry are of much more significance in this connection. Thus some white gold alloys contain nickel, often in combination with silver, copper, palladium, and/or zinc. The range of composition is very wide, but a common alloy in the U.S., for example, contains approximately 75 Au/17 Ni/2 Cu/5 Zn. Surprisingly, in view of their nobility and consequent excellent resistance to corrosion, the white gold alloys have been reported to cause a dermatitic reaction in nickel-sensitive women subjected to a patch test.[9] The test alloys contained 2.5 to 15% nickel and, curiously, a positive result was obtained even when the alloy surfaces were covered by a rhodium topcoat.

C. Alloys Containing Chromium: Passive Condition

Alloys of nickel and/or iron containing more than about 13% chromium are protected from corrosion in many media by the presence of a surface film consisting mainly of

Table 3
NICKEL RELEASE FROM COINAGE AND STAINLESS STEEL

Material	Medium	Temp. (°C)	Duration	Volume (ml)	Nickel release (μg)	Ref.
Swedish 1-krona coin	Distilled water	Ambient	24 h	10	49	5
(5% Ni)	Synthetic sweat	Ambient	24 h	10	96	5
Swedish 1-krona coin	Distilled water	Ambient	24 h	10	15	5
(25% Ni)	Synthetic sweat	Ambient	24 h	10	137	5
Type 302 stainless steel (18Cr/9Ni)	Synthetic sweat	20—24	7 d	5	1.80	6
Type 303 stainless steel (18Cr/9Ni/0.15 Min. S)	Synthetic sweat	20—24	7 d	5	51.1	6
U.S. 5-cent coin (1964)	Synthetic sweat	20—24	7d	5	5637	6
U.S. 10-cent coin (1972)	Synthetic sweat	20—24	7 d	5	3348	6
Danish 1-krone coin (25% Ni)	Synthetic sweat	20	24 h	10	409	7
Danish 1-krone coin (25% Ni)	Synthetic sweat	35	24 h	10	691	7

Note: Synthetic sweat — 0.5% sodium chloride, 0.1% lactic acid, 0.1% urea; pH 6.5.

chromium oxide Cr_2O_3. The surface film is not wholly impermeable and, even in this protected (passive) condition, a minute passage of metal ions can occur across the metal/solution interface as a result of metal dissolution at flaws in the film and healing of such flaws by film growth. For stainless steels containing 18 Cr/8 Ni in chloride solutions there is evidence to indicate that in the passive condition the amount of nickel entering solution would be less than 0.2 $\mu g/cm^2/day$, depending on solution composition, temperature, and other factors such as steel composition.

Common experience shows that stainless steels and similar chromium-containing alloys, when in the fully passive condition, do not give rise to contact dermitis.[8] However, the passive condition may be difficult to maintain and a number of cases have recently been reported of eczematous and other dermatitic symptoms arising from close contact with stainless-steel articles. Thus, Fischer et al.[10] report dermatitis in nickel-sensitive women from use of stainless-steel studs, clasps, and earrings. Oakley et al.[11] describe incidences of dermatitis in female patients from the use of stainless-steel skin clips following surgery. In both situations a narrow crevice existed between skin and article, creating a region where ultimate breakdown of protective oxide film on 18 Cr/8 Ni stainless steel was probable (see below).

The Fischer report correlated incidence of dermatitis with results from separate tests that determined the release of nickel from stainless-steel articles freely exposed without crevices to synthetic sweat solutions. It was indicated that dermatitis could occur in sensitized women from use of earrings that released as little as 0.05 μg nickel in a test involving 1-week exposure to synthetic sweat solution. However, with earrings, the release of nickel in the crevice could have been much greater. Similarly, in regard to the ocurrence of dermatitis reported by Oakley, it is not known whether the clips used were made from stainless steel of a type resistant to crevice corrosion, such as is used for implants. It remains doubtful whether stainless steel, when in the fully passive condition, can release nickel at a rate sufficient to elicit a dermatitic reaction, even in sensitized persons.

D. Loss of Passivity

When exposed to many nonoxidizing acid solutions, the oxide film is unstable and the nickel-iron-chromium alloys may lose their passivity and suffer corrosion at a rate similar

Table 4
SOLUBILIZATION OF NICKEL FROM STAINLESS STEEL[15]

Material	Nickel found in sweat (ppm)	Nickel found in physiological saline solution (PSS) (ppm)
Stainless steel		
Type 302 18 Cr/9 Ni	0.3	N.D.
Type 303 18 Cr/9 Ni/0.15 (min) S	10.2	1.1
Type 316 L 17 Cr/12 Ni/$2^1/_2$ mo	N.D	N.D.

Note: Specimens had surface areas varying from 2 to 5 cm^2. Volume of sweat or PSS was "no more than 10 ml". Duration of exposure — 1 week. Temperature — 20—22°C. N.D. = none detected, limit of detection about 1.0 μg.

to that of the chromium-free nickel and/or iron. Long exposure to neutral or near-neutral chloride solutions may result in partial loss of passivity due to the ability of the chloride ion to penetrate the oxide film and create conditions of local acidity. Localized corrosion in the form of pitting then occurs. A survey of industrial and domestic experience indicated that 18 Cr/8 Ni stainless steel is liable to suffer localized corrosion in waters containing more than about 200 ppm chloride.[12] The molybdenum-containing variety, such as Type 316, would be expected to be resistant at chloride levels up to at least 1000 ppm.

The region most likely to suffer localized corrosion is in crevices where oxygen can be completely consumed followed by development of local acidity. Conditions governing onset and rate of corrosion in crevices on stainless-steel surfaces have been analyzed in detail and the relationship established between crevice geometry, electrolyte composition, alloy composition, and duration of exposure.[13,14] It was shown that as the development of acidity in the solution within the crevice progressed, a critical pH value, dependent on alloy composition, was reached at which the oxide film broke down and a high rate of attack then ensued.

Alloys containing nickel, iron, and chromium that are to be used for critical applications where crevices occur should also contain molybdenum, which greatly improves the resistance to localized corrosion by chloride-containing solutions. The influence of molybdenum is indicated in the results shown in Table 4.[15] Table 4 also shows the adverse effect of a high sulfur content which is used to impart free-machining properties to stainless steels.

IV. METAL PROSTHESES

Metal implants made from nickel-containing alloys are currently used for a wide variety of clinical purposes including orthopedic pins, cardiac pacemakers, dental prostheses, and skin clips. The nickel-containing alloys currently specified for the purpose are (1) austenitic stainless steels containing a minimum of 17% chromium, generally 13 to 16% nickel, and 2.35 to 3.5% molybdenum and (2) wrought cobalt-chromium-tungsten-nickel alloy containing 9 to 11% nickel. The cast cobalt-chromium-molybdenum alloy may also contain up to 2.5% nickel.

There is evidence that products derived from chromium-containing prostheses can sensitize the patient to the metals present in the alloys.[16] Cobalt products are particularly likely to cause sensitization, but nickel can also be implicated. Such products can be obtained by wear or by corrosion, the latter possibility arising from fretting corrosion, corrosion fatigue, stress corrosion, or loss of passivity. Fretting corrosion may be considered a form of wear while corrosion fatigue and stress corrosion, although possibly resulting in failure of the prosthesis, are unlikely to generate sufficient loss of metal to cause sensitization. Loss of passivity in crevices is one of the major causes of the small proportion of failures that have occurred in stainless-steel prostheses and can give rise to significant amounts of corrosion products.[17,18] Thus, concerning sensitization of the patient, detritus from wear and products from corrosion in crevices are the most probable sources of metal derived from prostheses.

Evidence of correlation of sensitization with wear characteristics of alloy prostheses was found by Elves et al.[16] The incidence of metal sensitivity was greatest in joints with a high wear rate. In this connection it was noted that metal-on-plastic joint replacements generate considerably smaller amounts of metallic wear products than those with metal-on-metal joints and are now more generally used (Chapter 16).

Clearly, alloys used for prostheses require resistance to crevice corrosion since, if the fully passive condition can be maintained, it is estimated that the amount of nickel released into the body would be less than $^1/_{100}$ of that ingested in the diet. It has been shown that the blood nickel concentration in patients with stainless-steel prostheses was not dissimilar to that in healthy controls.[19]

Although the body environment is not particularly aggressive, the crevice gaps formed by prostheses are likely to be very narrow and the depth sufficiently extensive to create a critical condition for the less highly alloyed types of stainless steel.[14] A crevice formed by dissimilar alloys is especially dangerous.[18] The importance of molybdenum content in improving the resistance of stainless steels to crevice corrosion is well established.[20] A large proportion of stainless-steel implants that have undergone corrosion in service has been associated with a deficiency of molybdenum.[21]

Nickel-containing prostheses are also used in some countries in dentistry, either as dental caps to cover teeth or as plate material. The environment in the mouth can be aggressive and crevices inevitably exist at the site of the prosthesis. The significance of chromium content in dental alloys has been reported.[22] In this connection it should be noted that dental alloys frequently have a high carbon content to aid their castability, and in consequence may contain chromium carbide. The formation of chromium carbide and other chromium compounds in the alloy removes chromium from the matrix and may thus hinder the development of a satisfactorily protective oxide surface film.

It should be stressed that, while nickel-containing prosthetic devices may cause problems in sensitized individuals, hundreds of thousands are used every year worldwide. If the complications of their use were widespread, their application would have ceased long since. It is nevertheless a wise precaution to perform patch tests on persons undergoing elective surgery and to inquire into the history of metal sensitivity (nickel, chromium, and cobalt) in persons undergoing emergency treatment (Chapter 16). It might be considered more appropriate to conduct the patch tests using the alloys involved in the implant in preference to the salts of the metals contained in the alloys, since in practice the latter are unlikely to be released in a quantity sufficient to cause reaction in other than highly sensitized persons.

V. NICKEL AND NICKEL-ALLOY COATINGS

The literature abounds with references to occurrences of dermatitis arising from close rubbing contact of the skin with nickel-plated objects such as buttons, clips, costume jewelry, and earrings.[8] Nickel exposed to chloride-containing solutions such as sweat suffers an irregular, overall attack. The bright nickel deposits commonly used for decorative coatings on consumer articles are more active and less corrosion resistant than the dull nickel electrodeposits because of the high sulfur content derived from the organic brighteners used in the electrodeposition process.

When the nickel coating is covered by a topcoat of a more noble metal such as chromium, gold, or rhodium, the release of nickel is very much reduced, but may not be entirely eliminated. The topcoats generally are not more than a few micrometers thick and consequently are porous or, as with chromium, are cracked or become cracked as a result of stresses in service. The nickel undercoat is subject to corrosion at these discontinuities in the topcoat and may provide sufficient corrosion product to cause a dermatitic reaction. Thus the amount of nickel released from a nickel undercoat covered with a more noble topcoat depends greatly on the continuity of that topcoat. Once the corrosive attack has penetrated the nickel coating to a more active substrate such as steel, zinc alloy, or brass, corrosion of the nickel is very much reduced, if not entirely halted, by the attack being diverted to the substrate.[23]

If exposure to the composite coating is not continuous, e.g., earrings are periodically removed and the surface wiped clean of corrosion product, the likelihood of dermatitis occurring is much reduced.

Nickel alloy coatings in commercial use comprise the zinc-nickel deposit used as an undercoat for paint systems and tin-nickel. It is unlikely that the former will ever by used in a situation that will result in persistent contact with the skin. The latter has decorative qualities and has been used in applications involving handling. The coating is composed of an intermetallic compound, NiSn, that is inert and reported to have good resistance to corrosion by chloride solutions. The coating is brittle and may be porous. It is therefore normally applied over an undercoat of copper or nickel. Tests conducted in synthetic sweat showed tin-nickel to release more nickel than 18 Cr/9 Ni stainless steel but less than that from a 75 Cu/25 Ni alloy.[24]

In conclusion, it is clear that nickel release from alloys depends both on biological and physicochemical factors. Alloys can be designed for specific purposes, and with proper care and attention to their selection most contact dermatitis can be avoided.

REFERENCES

1. **Buckley, W. R. and Lewis, C. E.,** The "ruster" in industry, *J. Occup. Med.*, 2, 23, 1960.
2. **Southwell, C. R. and Alexander, A. L.,** Corrosion of metals in tropical environment, *Mater. Prot.*, 8(3), 39, 1969.
3. **Husain, S. L.,** Nickel coin dermatitis, *Br. Med. J.*, October 1977, p. 998.
4. **Rothman, S.,** *JAMA*, 97, 336, 1931.
5. **Pedersen, N. B., Fregert, S., Brodelius, P., and Gruvberger, B.,** Release of nickel from silver coins, *Acta Derm. Venereol. (Stockholm)*, 54, 231, 1975.
6. **Katz, S. A. and Samitz, M. H.,** Leaching of nickel from stainless steel consumer commodities, *Acta Derm. Venereol. (Stockholm)*, 55, 113, 1975.
7. **Menne, T. and Solgaard, P.,** Temperature-dependent nickel release from nickel alloys, *Contact Dermatitis*, 5, 82, 1979.
8. **Fisher, A. A.,** *Contact Dermatitis*, 3rd ed., Lea & Febiger, Philadelphia, 1986.
9. **Fischer, T., Fregert, S., Gruvberger, B., and Rystedt, I.,** Contact sensitivity to nickel in white gold, *Contact Dermatitis*, 10, 23, 1984.
10. **Fischer, T., Fregert, S., Gruvberger, B., and Rystedt I.,** Nickel release from ear piercing kits and earrings, *Contact Dermatitis*, 10, 39, 1984.
11. **Oakley, A. M. M., Ive, F. A., and Carr, M. M.,** Skin clips are contraindicated when there is nickel allergy, *J. R. Soc. Med.*, 80, 290, 1987.
12. **Flint, G. N.,** Resistance of stainless steels to corrosion in naturally occurring waters, *Corros. y Protecc.*, 8, 23, 1977.
13. **Oldfield, J. W. and Sutton, W. H.,** Crevice corrosion of stainless steels, 6th Eur. Congr. on Corrosion, London, September 1987.
14. **Oldfield, J. W.,** Crevice corrosion of stainless steels — the importance of crevice geometry and alloy composition, 19th Conf. Journee des Aciers Speciaux, Saint Eteinne, May 1980.
15. **Samitz, M. H. and Katz, S. A.,** Nickel dermatitis hazards from prostheses, *Br. J. Dermatol.*, 92, 287, 1975.

16. **Elves, M. W., Wilson, J. W., Scales, J. T., and Kept, H. B. S.,** Incidence of metal sensitivity in patients with total joint replacements, *Br. Med. J.*, p. 376, 1975.
17. **Scales, J. T., Winter, G. D., and Shirley, H. T.,** *J. Bone Jt. Surg.*, 41-B, 810, 1959.
18. **Levine, D. L. and Staehle, R. W.,** Crevice corrosion in orthopaedic implant metals, *J. Biomed. Mater. Res.*, 11, 553, 1977.
19. **Linden, J. V., Hopfer, S. M., Gossling, H. R., and Sunderman, F. W.,** Blood nickel concentrations in patients with stainless steel prostheses, *Ann. Clin. Lab. Sci.*, 15(b), 459, 1985.
20. **Oldfield, J. W. and Sutton, W. H.,** A new technique for predicting the performance of stainless steels in seawater and other chloride containing environments, *Br. Corros. J.*, 15(1), 31, 1980.
21. **Harris, B.,** Corrosion of stainless steel surgical implants, *J. Med. Eng. Technol.*, 3, 117, 1979.
22. **Espevik, S.,** Corrosion of base metal alloys in vitro, *Acta Ondotol. Scand.* 36, 113, 1978.
23. **Flint, G. N. and Melbourne, S. H.,** The corrosion of nickel + chromium coatings. A metallographic and potential study, *J. Inst. Met. Fin.*, 38, 35, 1961.
24. **Menné, T., Brandrup, F., Thestrup-Pedersen, K., Velen, N. K., Andersen, J. R., Yding, F., and Valeur, G.,** Patch test reactivity to nickel alloys, *Contact Dermatitis,* 16, 255, 1987.

Chapter 5

RECENT ADVANCES IN IMMUNOLOGY OF NICKEL SENSITIZATION

Rik J. Scheper, Mary von Blomberg, Karel J. J. Vreeburg, and Ingrid M. W. van Hoogstraten

TABLE OF CONTENTS

I. Allergic Contact Dermatitis (ACD): A Short History 56

II. Current View on the Development of ACD 56

III. Nickel as a Hapten in ACD 57

IV. Concluding Remarks 60

Acknowledgments 61

References 61

I. ALLERGIC CONTACT DERMATITIS (ACD): A SHORT HISTORY

In 1896, J. Jadassohn[1] demonstrated in humans that characteristic skin inflammations caused by certain chemicals were due to a specific increase in skin sensitivity. This hypersensitivity was acquired by repeated contact with the incriminated substance and could be demonstrated by applying the chemical to a normal skin site of the patient in a non-skin-irritating concentration. The present diagnosis of ACD is still based on this method.

Experimental contact sensitization in guinea pigs was established in 1930 by W. Jadassohn[2] with phenylhydrazine and in the same year by Bloch and Steiner-Wourlisch[3] with a primula extract. These investigators showed that the skin reactions thus obtained very closely resembled allergic eczema in man both macroscopically and histologically.

In 1935, Landsteiner and Jacobs[4] noted a relationship between allergenicity and chemical structure. They showed that small chemical compounds (haptens), such as 2,4-dinitrochlorobenzene (DNCB) and picrylchloride, are allergens because of their capacity to conjugate with macromolecular carriers. Landsteiner and Chase[5] subsequently demonstrated that contact sensitivity to picrylchloride could be transferred to naive recipients with living lymphoid cells, but not with serum.

In the last two decades our understanding of the basic mechanisms underlying induction, expression, and regulation of ACD has increased considerably. Highlights in this period include the identification of specialized regions in the lymphoid tissue related to cellular immune functions,[6] the discovery of the thymus as the cradle for those lymphocytes (from then on called T-lymphocytes) mediating contact hypersensitivity,[7] the development of reagents (monoclonal antibodies) that allowed the further dissection of T-cells in functionally active subsets (helper and suppressor/cytotoxic)[8] and, very recently, the identification of the allergen-specific T-cell receptor.[9,10]

II. CURRENT VIEW ON THE DEVELOPMENT OF ACD

It is convenient to distinguish afferent and efferent limbs of ACD.[11,12] The afferent limb includes the events following a first contact with the allergen and is complete when the individual is sensitized and capable of giving a positive elicitation reaction. The efferent limb is activated during the elicitation (challenge) phase. The entire process of the afferent limb requires from at least 3 d to several weeks, whereas full development of the elicitation phase only requires 24 to 48 h.

The main events which occur in the afferent limb are consecutively:

1. Conjugation of hapten with skin components
2. Recognition of these hapten-modified components by specific T-cells
3. Proliferation of specific T-cells in draining lymph nodes
4. Propagation (dissemination) of specific T-cell progeny over the body

Upon primary contact, an allergen penetrating the skin associates with class II molecules, encoded by the major histocompatibility complex. These molecules, coded in humans by HLA-D-region genes, are abundantly present in the epidermis on dendritic cells, the Langerhans cells (LCs). This association may be covalent, as in the case of DNCB, or noncovalent, as with nickelcations. Allergen-carrying LCs travel via the afferent lymphatic to the regional lymph nodes, where they settle in the paracortical areas.[13,14]

Here, conditions are optimal for allergen-carrying LCs (allergen-presenting cells [APC]) to encounter T-cells that specifically recognize the allergen-class II molecule complex. Notably, in nonsensitized individuals the frequency of T-cells with a given specificity would be far below one per thousand. The dendritic morphology of the LC strongly facilitates multiple cell contacts, ultimately leading to binding and activation of sufficient numbers of allergen-specific T-cells.

Supported by interleukin 1, a mediator released by the allergen-presenting cells, activated T-cells start producing several other mediators called lymphokines, including interleukin 2 (IL-2). Following activation, the density of IL-2 receptors on T-cells increases transiently and results in a vigorous blast formation within a few days.[15]

The expanded progeny (effector T-cells) is subsequently released into the circulation to join the pool of recirculating lymphocytes and may enter the peripheral tissues, including the skin.[11]

The efferent limb of the cellular immune response is based on the increased frequency of T-cells with a given specificity throughout the body of a sensitized individual. Whereas at primary allergenic skin contact too few specific T-cells are locally available to allow for macroscopically detectable skin reactivity, when challenging a sensitized individual a local inflammatory reaction will follow. Local interaction will be possible between allergen-presenting cells and specific T-cells and lead to local lymphokine production within the skin. The release of these mediators, many of which have a pro-inflammatory action, causes the arrival of more T-cells, including specific T-cells, that amplify the local mediator release. This leads to a gradually developing eczematous reaction which reaches a maximum after 18 to 48 h. The lymphokines that play a major role in the development of contact allergic reaction are those with stimulatory effects on other lymphocytes (IL-2, interferon-gamma), on mononuclear phagocytes (chemotactic factor, migration inhibition factor, interferon-gamma), and on mast cells and vasculature (skin reactive factor, interferon-gamma).[16-18]

III. NICKEL AS A HAPTEN IN ACD

Many important contributions to our present knowledge on ACD result from studies with experimental animals, particularly mice and guinea pigs. Remarkably, despite nickel being the most frequent sensitizer in man, reproducible sensitization of guinea pigs to this allergen has met with some difficulties,[19,20] whereas no convincing sensitization of mice has been reported so far[21] (Chapter 6). Most recent developments in the field of nickel allergy, therefore, stem from clinical *in vivo* and *in vitro* studies. In this section we will focus on some recently developed views regarding induction and expression of nickel hypersensitivity, along the lines set out in the previous section.

Nickel cations, potentially generated upon oxidation of metallic nickel, may readily penetrate the epidermis and associate with immunologically relevant carriers.[22] For comparison, a different situation exists with another important metal allergen, i.e., chromium, where the active hapten (Cr^{3+}) poorly penetrates the epidermis, but may be locally generated by reduction of readily penetrating chromate anions.[23]

As to the formation of immunogenic nickel complexes at the surface of antigen-presenting cells, three possibilities exist:

1. Nickel binds directly to class II molecules.
2. Nickel binds to other cell membrane-associated molecules located in the close vicinity of class II molecules.
3. Nickel binds to soluble carrier molecules that, following uptake and, eventually, degradation by APC, are exposed at the cell surface close to class II molecules.

Recent data favor the view that direct nickel modification of class II molecules is of major relevance. Nickel-specific T-cell clones, derived from nickel-hypersensitive individuals, can be fully activated to proliferate and produce lymphokines *in vitro* in the absence of skin constituents, provided class II molecule-bearing cells are added.[24] Moreover, such *in vitro*

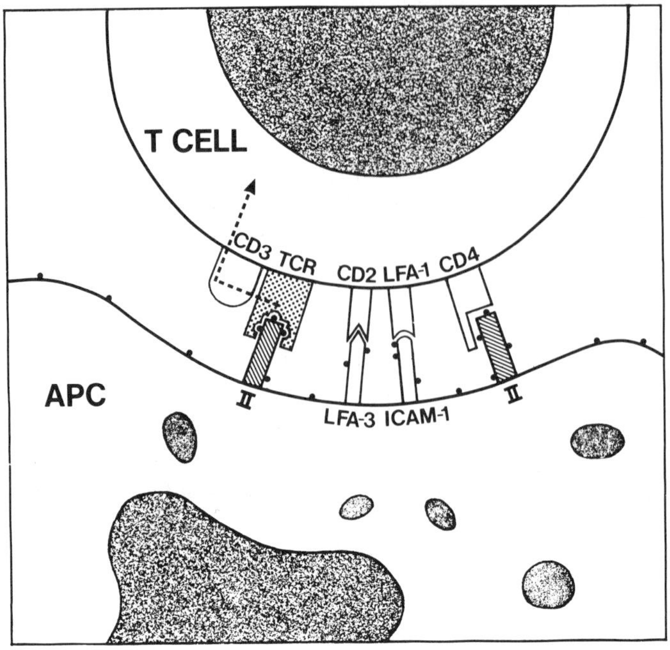

FIGURE 1. Proposed intermolecular interactions during cell-cell contact between an APC bearing nickel (small black dots) and a nickel-specific T-helper cell. The T-cell receptor recognizes a nickel-modified class II molecule, thereby allowing a receptor-associated CD3 molecule to provide a signal for cellular activation (---▶). Additional binding and signals result from CD2/LFA-3, LFA-1/ICAM-1, and CD4/class II interactions.

stimulation could be obtained with serum-free culture media and chloroquin-treated APC, in the absence of intracellular uptake and degration of metal-protein complexes.[25] In line with this view, Silvennoinen-Kassinen[26] recently suggested that one of the two major nickel-binding proteins isolated from mononuclear cells is the beta chain of the class II molecule. Probably the many nickel cations that, following skin contact, associate with other carriers either remain immunologically silent (null event) or possibly induce suppressor cell activation (tolerance induction).

Although it cannot yet be excluded that hapten-specific T-cells with the cytotoxic phenotype may contribute to contact hypersensitivity (i.e., by killing hapten-binding epidermal cells), most effector T-cells belong to the helper subclass. As for this concept in man, oviously no direct proof can be obtained from passive transfer experiments. The major evidence comes from *in vitro* studies analyzing peripheral blood lymphocytes (PBL)[47] or skin infiltrates obtained from nickel-allergic patients. Using culture conditions for *in vitro* cloning of PBL both with helper and cytotoxic function, all the nickel-specific clones we could isolate were CD3+;. CD4+, CD8− (T-helper phenotype).[24] This was also found for clones derived from nickel-induced skin lesions.[27]

How would these cells at first have been activated *in vivo* to proliferate during primary sensitization? Based on recent data, activation of nickel-specific T-helper cells by nickel-bearing APC can be depicted as shown in Figure 1. The T-cell receptor accounts for allergen specificity through the polymorphic nature of its alpha and beta chains. Nickel-specific T-cell receptors tend to bind to nickel-modified class II molecules, thereby allowing receptor-associated CD3 (a trimolecular complex) to provide an intracellular signal for activation. An additional signal is provided by interaction between CD2 molecules (formerly known as the sheep erythrocyte receptor) and the lymphocyte function-associated antigen 3 (LFA-3).[28]

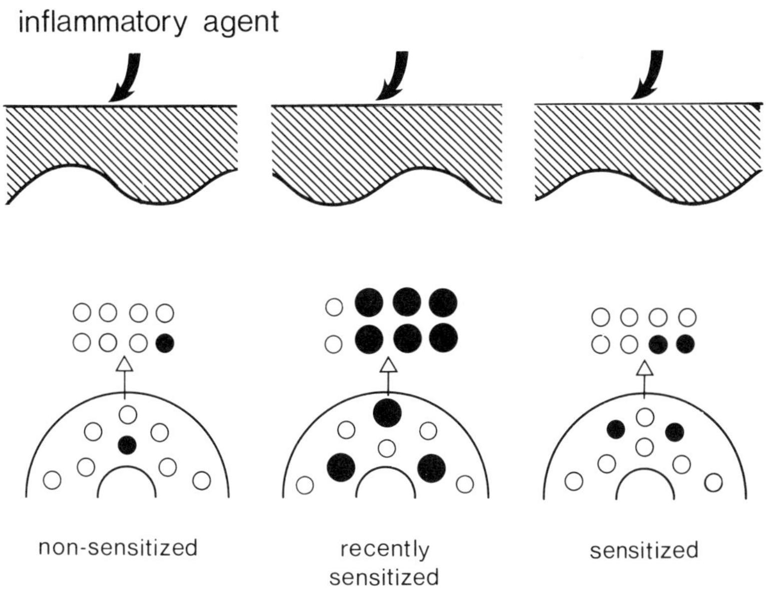

FIGURE 2. Local accumulation of specific T-cells by allergen-independent infiltration (—▷). After sensitization the frequency of allergen-specific T-cells (closed circles) shows a transient peak. Moreover, recently activated T-cells possess a strong capacity to migrate into any inflammatory lesion.

The chance of these activation steps taking place may be further increased by other interaction molecules and their counterstructures present on T-helper cells and APC: LFA-1/intercellular adhesion molecule (ICAM) 1[29] and CD4/class II, respectively. Apparently, these interactions are not readily inhibited by surface-bound allergens at a concentration that provides sufficient class II modification for T-cell receptor triggering.

Despite the strong specificity of T-cells imposed by their T-cell receptors, cross-reactivity may be observed. A recent study using nickel-reactive T-cell clones showed donor-dependent cross-reactivity with either copper or palladium (at the right hand and immediately below nickel, respectively, in the periodic table of the elements). Cross-reactivity at the clonal level was never observed with other metals like cobalt and chromium (immediately and further left from nickel, respectively).[30] Interestingly, copper- and palladium-contact allergic reactions have only been reported to occur in nickel-allergic patients.[31,32]

During the sensitization process, proliferation of nickel-specific T-cells is assumed to take place within the paracortical areas of those lymph nodes that drain the skin contact site(s).These conditions may be mimicked *in vitro* in the lymphocyte transformation test with PBL.[33]

Systemic dissemination of the resulting large numbers of nickel-specific T-cells forms the basis for the development of nickel hypersensitivity. Skin hyperreactivity to nickel may therefore be observed at all skin sites, provided that sufficient LCs are locally available.

Marked preferential entry of nickel-specific T-cells may occur at nickel-challenged skin sites. After sensitization, the frequency of nickel-specific T-cells in the circulation is persistently increased. In addition, this frequency will show transient peaks shortly after the first and every further contact with the allergen (Figure 2, middle). An inflammatory reaction, therefore, when elicited within a few weeks after sensitization or boosting, will contain an increased frequency of nickel-specific T-cells. Moreover, recently activated T-cells possess a uniquely high migratory capacity.[34,35]

At the skin level, selective accumulation of nickel-specific T-cells at nickel-challenged skin sites may be further enhanced by two other mechanisms. First, vascular endothelial

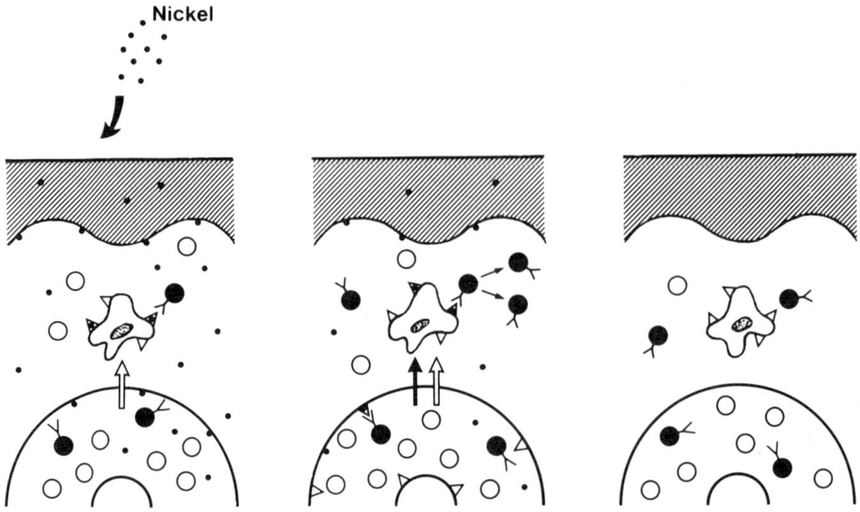

FIGURE 3. Local accumulation of specific T-cells by allergen-dependent infiltration (⟶) and proliferation, in addition to entry of nonspecific T cells (⟹). Preferential accumulation of nickel-specific T-lymphocytes (closed circles) at specific contact reaction site through nickel (●)-class II (△) mediated local trapping and proliferation (middle), leading to a highly increased frequency of residual specific T-lymphocytes (right).

cells contain low levels of relevant interaction molecules (class II, ICAM), the expression of which is increased by locally released interferon-gamma.[36,37] Upon local mediator release, therefore, nickel binding to these surfaces may contribute to local adhesion and extravasation of nickel-specific cells. In support of this view, Kapsenberg et al.[27] demonstrated that approximately 10% of the T-cells present at 48 h in nickel-induced contact skin lesions were nickel-specific. In this study, inflammatory T-cells were cloned with a 100% cloning efficiency, thereby providing a full reflection of the T-cell repertoire within the lesional infiltrates.

As a corollary of the activation of nickel-specific T-cells at a contact reaction skin site, these cells also start to proliferate. This further contributes to a strongly increased frequency of nickel-specific T-cells at a nickel skin reaction site and accounts for strong and protracted eczematous reactivity as long as sufficient allergen is locally available. It should be emphasized that the local frequency of specific T-cells may still increase in a subsiding skin reaction despite a large fall in absolute numbers of residual T-cells (Figure 3). The recurrence of nickel hypersensitivity reactions in patients, i.e., flare-up reactions following systemic supply of nickel,[38,39] can now be well understood in the light of the above-mentioned mechanisms.

IV. CONCLUDING REMARKS

The more we have learned about nickel presentation and T-effector cell function,the less we seem to know about mechanisms controlling development and expression of nickel ACD. No obvious association with HLA antigens could be detected[40] (Chapter 10), and development of nickel allergy may therefore primarily depend on frequency and intensity of skin contacts with this allergen. Ear piercing and use of cheap jewelry certainly represent important risk factors.[41,42]

From animal studies on ACD, many data have been obtained concerning the nature and role of suppressor cells in preventing development (afferent action) and expression (efferent action) of ACD.[11,43] In man, relevant data are still virtually absent. In nickel allergy, no suppressor function of T-cells with the suppressor/cytotoxic phenotype could be detected.[44] Nevertheless, evidence suggesting the existence in man of afferently acting suppressor cells in nickel allergy, i.e., preventing sensitization, was recently obtained in an epidemiological study. The frequency of nickel hypersensitivity was remarkably low in a group of junior nurses who had oral contacts with this allergen through orthodontic dental treatment.[41] Earlier, in guinea pigs, we had succeeded in demonstrating the induction of immunological tolerance to nickel by oral administration before attempted sensitization.[45] In mouse models with other allergens, intravenously and orally induced tolerance have been demonstrated to depend on suppressor cell action.[43,46,48] Further studies into the possibility of exploiting this principle in man are warranted. Obviously, reducing the risk of nickel sensitization through the introduction of low nickel-releasing alloys[49] would represent a primary step in the prevention of nickel ACD in man.

ACKNOWLEDGMENTS

We thank Dr. H. Bril for critically reviewing the manuscript, and Miss J. Ambagtsheer for skillful secretarial help. This work was supported by a grant from the Dutch Praeventiefonds.

REFERENCES

1. **Jadassohn, J.,** Zur Kenntnis der Medicamentösen Dermatosen, in Verhandlungen der Deutschen Dermatologischen Gesellschaft, Jarisch and Neisser, Eds., Fünfter Congress, Graz, Austria, 1895, 103.
2. **Jadassohn, W.,** Sensibilisierung der Haut des Meerschweinchens auf Phenylhydrazin, *Klin. Wochenschr.,* 9, 551, 1930.
3. **Bloch, B. and Steiner-Wourlisch, A.,** Die Sensibilisierung des Meerschweinchens gegen Primeln, *Arch. Dermatol. Syphilol.,* 162, 349, 1930.
4. **Landsteiner, K. and Jacobs, J.,** Studies on the sensitization of animals with simple chemical compounds, *J. Exp. Med.,* 61, 643, 1935.
5. **Landsteiner, K. and Chase, M. W.,** Experiments on transfer of cutaneous sensitivity to simple compounds, *Proc. Soc. Exp. Biol. Med.,* 49, 688, 1942.
6. **Oort, J. and Turk, J. L.,** A histological study of lymph nodes during the development of contact sensitivity in guinea-pigs, *Br. J. Exp. Pathol.,* 46, 147, 1965.
7. **Miller, J. F. A. P. and Osaba, D.,** Current concepts of the immunological function of the thymus, *Physiol. Rev.,* 47, 437, 1967.
8. **Reinhertz, E. L., Kung, P. C., Goldstein, G., and Schlossman, S. F.,** Separation of functional subsets of human T cells by a monoclonal antibody, *Proc. Natl. Acad. Sci. U.S.A.,* 76, 4061, 1979.
9. **Meuer, S. C., Cooper, D. A., Hodgdon, J. C., et al.,** Identification of the receptor for antigen and major histocompatibility complex on human inducer T lymphocytes, *Science,* 222, 1239, 1983.
10. **Royer, H. D. and Reinherz, E. L.,** T lymphocytes: ontogeny, function and relevance to clinical disorders, *N. Engl. J. Med.,* 317, 1136, 1987.
11. **Turk, J. L.,** *Delayed Hypersensitivity,* 3rd ed., North-Holland, Amsterdam, 1980.
12. **Polak, L.,** Immunological aspects of contact sensitivity, in *Monographs in Allergy,* Vol. 15, S. Karger, Basel, 1980.
13. **Shelley, W. B. and Juhlin, L.,** Langerhans cells form a reticuloepithelial trap for external contact allergens, *Nature (London),* 261, 46, 1976.
14. **Drexhage, H. A., Mullink, H., de Groot, J., Clarke, J., and Balfour, B. M.,** A study of cells present in peripheral lymph of pigs with special reference to a type of cell resembling Langerhans' cells, *Cell Tissue Res.,* 202, 407, 1979.
15. **Isakov, N. and Altman, A.,** Lymphocyte activation and immune regulation, *Immunol. Today,* 7, 155, 1986.

16. **Geczy, C. L.,** The role of lymphokines in delayed-type hypersensitivity reactions, *Springer Semin. Immunopathol.*, 7, 321, 1984.
17. **Trinchieri, G. and Perussia, B.,** Immune interferon: a pleotropic lymphokine with multiple effects, *Immunol. Today*, 6, 131, 1985.
18. **Dinarello, C. A. and Mier, J. W.,** Lymphokines, *N. Engl. J. Med.*, 317, 940, 1987.
19. **Samitz, M. H., Katz, S. A., Scheiner, D. M., and Lewis, J. E.,** Attempts to induce sensitization in guinea pigs with nickel complexes, *Acta Derm. Venereol. (Stockholm)*, 55, 475, 1975.
20. **Andersen, K. E. and Maibach, H. I.,** Guinea pig sensitization assays, in *Current Problems in Dermatology*, Vol. 14, Andersen, K. E. and Maibach, K. I., Eds., S. Karger, Basel, 1985, 263.
21. **Möller, H.,** Attempts to induce contact allergy to nickel in the mouse, *Contact Dermatitis*, 10, 65, 1984.
22. **Sunderman, F. W.,** A review of the metabolism and toxicology of nickel, *Ann. Clin. Lab. Sci.*, 7, 377, 1977.
23. **Polak, L.,** Immunology of chromium, in *Chromium: Metabolism and Toxicity*, Burrows, D., Ed., CRC Press, Boca Raton, Fl, 1983, 82.
24. **Sinigaglia, F., Scheidegger, D., Garotta, G., Scheper, R., Pletscher, M., and Lanzavecchia, A.,** Isolation and characterization of Ni-specific T cell clones from patients with Ni-contact dermatitis, *J. Immunol.*, 135, 3929, 1985.
25. **Kapsenberg, M. L., van der Pouw-Kraan, T., Stiekema, F. E. M., Schootemeijer, A., and Bos, J. D.,** Direct and indirect nickel-specific stimulation of T lymphocytes from patients with allergic contact dermatitis to nickel, *Eur. J. Immunol.*, 18, 977, 1988.
26. **Silvennoinen-Kassinen, S., Jakkula, H., and Karvonen, J.,** Binding of nickel to the mononuclear cells and in the serum of nickel-sensitive and healthy subjects, *Clin. Exp. Dermatol.*, 12, 265, 1987.
27. **Kapsenberg, M. L., Res, P., Bos, J. D., Teunissen, M. B. M., and van Schooten, W.,** Nickel-specific T lymphocyte clones derived from nickel-contact dermatitis lesions in man. I. Heterogeneity based on requirement of dendritic cell subsets, *Eur. J. Immunol.*, in press.
28. **Breitmeyer, J. B.,** How T cells communicate, *Nature (London)*, 329, 760, 1987.
29. **Rothlein, R., Dustin, M. L., Marlin, S. D., and Springer, T. A.,** A human intercellular adhesion molecule (ICAM-1) distinct from LFA-1[1], *J. Immunol.*, 137, 1270, 1986.
30. **Scheper, R. J., von Blomberg, M., Bruynzeel, D. P., Vreeburg, K. J. J., Kapsenberg, M., and Sinigaglia, F.,** Studies on T cell fine specificity using nickel-reactive T cell clones, *J. Invest. Dermatol.*, 87, 165, 1986.
31. **Walton, S.,** Investigation into patch testing with copper sulphate, *Contact Dermatitis*, 9, 89, 1982.
32. **van Loon, L. A. J., van Elsas, P. W., van Joost, T. H., and Davidson, D. L.,** Contact stomatitis and dermatitis to nickel and palladium, *Contact Dermatitis*, 11, 294, 1984.
33. **von Blomberg-van der Flier, M., van der Burg, C. K. H., Pos, O., van de Plassche-Boers, E. M., Bruynzeel, D. P., Garotta, G., and Scheper, R. J.,** In vitro studies in nickel allergy: diagnostic value of a dual parameter analysis, *J. Invest. Dermatol.*, 88, 362, 1987.
34. **Asherson, G. L. and Allwood, G. G.,** Inflammatory lymphoid cells, *Immunology*, 22, 493, 1972.
35. **van Dinther-Janssen, A. C. H. M., van Maarsseveen, A. C. M. T., de Groot, J., and Scheper, R. J.,** Comparative migration of T and B lymphocyte subpopulations into skin inflammatory sites, *Immunology*, 48, 519, 1983.
36. **Wagner, C. R., Vetto, R. M., and Burger, D. R.,** Expression of I-region associated antigen (Ia) and interleukin 1 by subcultured human endothelial cells, *Cell. Immunol.*, 93, 91, 1985.
37. **Haskard, D. O., Cavender, M. R. C. P., Fleck, R. M., Sontheimer, R., and Ziff, M.,** Human dermal microvascular endothelial cells behave like umbilical vein endothelial cells in T cell adhesion studies, *J. Invest. Dermatol.*, 88, 340, 1987.
38. **Christensen, O. B., Lindström, C., Löfberg, H., and Möller, H.,** Micromorphology and specificity of orally induced flare-up reactions in nickel-sensitive patients, *Acta Derm. Venereol. (Stockholm)*, 61, 505, 1981.
39. **Menné, T. and Hjorth, N.,** Reactions from systemic exposure to contact allergens, *Semin. Dermatol.*, 1, 15, 1982.
40. **Braathen, L. R., Haaveisrud, O., and Thorsby, E.,** HLA-antigens in patients with allergic contact sensitivity to nickel, *Arch. Dermatol. Res.*, 275, 355, 1983.
41. **van der Burg, C. K. H., Bruynzeel, D. P., Vreeburg, K. J. J., von Blomberg, B. M. E., and Scheper, R. J.,** Hand eczema in hairdressers and nurses: a prospective study, *Contact Dermatitis*, 14, 275, 1986.
42. **Larsson-Stymne, B. and Widström, L.,** Ear-piercing — a cause of nickel allergy in schoolgirls?, *Contact Dermatitis*, 13, 289, 1985.
43. **Gautam, S. C. and Battisto, J. R.,** Orally induced tolerance generates an efferently acting suppressor T cell and an acceptor T cell that together downregulate contact sensitivity, *J. Immunol.*, 135, 2975, 1985.
44. **Silvennoinen-Kassinen, S., Ilonen, J., Tiilikainen, A., and Karvonen, J.,** Inhibition of in vitro nickel sulphate reaction by anti-HLA-DDR antisera. Lack of demonstrable suppressor cells in peripheral blood in nickel unresponsiveness in man, *J. Invest. Dermatol.*, 77, 417, 1981.

45. **Vreeburg, K. J. J., de Groot, K., and Scheper, R. J.,** Induction of immunological tolerance by oral administration of nickel and chromium, *J. Dent. Res.,* 63, 124, 1984.
46. **Knop, J., Reichman, R., and Macher, E.,** Modulation of suppressor mechanisms in allergic contact dermatitis, *J. Invest. Dermatol.,* 77, 469, 1981.
47. **Silvennoinen-Kassinen, S., Jakkula, H., and Karvonen, J.,** Helper cells (Leu-3a+) carry the specificity of nickel sensitivity reaction in vitro in humans, *J. Invest. Dermatol.,* 86, 18, 1986.
48. **Asherson, G. L., Colizzi, V.,and James, B. M.,** The control of the contact sensitivity skin reaction. T-suppressor afferent cellblocks. The production of antigen-specific T-helper factor, *Immunology,* 54, 521, 1985.
49. **Menné, T., Branderup, F., Thestrup-Pedersen, K., Veien, N. K., Andersen, J. R., Yding, F., and Valeur, G.,** Patch test reactivity to nickel alloys, *Contact Dermatitis,* 16, 255, 1987.

Chapter 6

NICKEL: ANIMAL SENSITIZATION ASSAYS

J. E. Wahlberg

TABLE OF CONTENTS

I. Introduction 66

II. Methods for the Induction of Nickel Sensitivity 66
- A. Epicutaneous Methods 66
- B. Intradermal Methods 66
- C. Guinea Pig Maximization Test (GPMT) 66
- D. Other Methods Using Combinations of Epicutaneous, Intradermal, and Intramuscular Administration 67

III. Comparison of Various Nickel Salts 69

IV. Toxicity of Nickel 69

V. Comparison of Different Species (Guinea Pigs, Rabbits, Mini-Pigs, and Mice) and Strains of Guinea Pig 69

VI. Interlaboratory and Intralaboratory Comparisons 69

VII. Recommendation for the Choice of Sensitization Method 71

VIII. Conclusions 72

References 72

I. INTRODUCTION

Contact sensitivity to nickel is common in humans. It is thus natural that much effort has been spent developing animal models for study of different steps of the induction process, effects of treatment, and preventive means. As will be shown in this review, however, it has been surprisingly difficult to induce contact allergy in experimental animals, which implies that nickel is not the potent contact allergen we have imagined.

II. METHODS FOR THE INDUCTION OF NICKEL SENSITIVITY

The methods that have been used for inducing nickel sensitivity involve epicutaneous, intradermal, or intramuscular administration of various nickel salts, with or without stimulation with Freund's complete adjuvant (FCA), and combinations of these routes. For technical descriptions of the methods, refer to the original papers and to monographs on predictive testing.[1-5] Studies reported up to 1949 have been reviewed by Hunziker.[1]

A. Epicutaneous Methods (Table 1)

Nickel salts are applied topically with or without occlusion on normal skin or on skin pretreated with irritants in guinea pigs or mice.[6-13] To increase the possibility of inducing contact sensitivity, various tensides were added to the nickel solutions, and a marginal effect was observed for some anionic detergents.[1,6,7]

Maurer et al.[9] found a sensitization rate of 40% for skin pretreated with sodium lauryl sulfate (SLS), and Popov and Penev[10] succeeded in sensitizing 70% of their guinea pig subjects by painting 20% $NiSO_4$ on skin previously treated with "supersound".

A breakthrough for the epicutaneous method (without pretreatment) came with the findings of Lammintausta et al.,[12,13] where the skin of guinea pigs was painted with 25% $NiSO_4$ 5 d/week for 4 weeks, a procedure adapted from the open epicutaneous test of Klecak et al.[3] Lammintausta et al.[12,13] obtained a sensitization rate of 11/22 and 4/7 in a Hartley strain of guinea pigs. These favorable results need confirming from other laboratories using other guinea pig strains.

B. Intradermal Methods (Table 2)

Nickel salts, salts mixed with FCA, or FCA administered by separate injections have been applied in different protocols.[1,9,11,14-18] The optimization test[9] gave a high sensitization rate — 20/20 guinea pigs at intradermal challenge compared to 7/20 (occluded) and 12/20 (open) at epicutaneous challenge. Sensitization in the rabbit following the injection of $NiCl_2$ mixed with FCA was recorded using intradermal and in vitro tests.[17] Regarding other methods involving FCA, Turk and Parker[15] succeeded in sensitizing 3/5 guinea pigs, but Hicks et al.[16] failed with the split adjuvant technique. However, the majority of the intradermal methods have failed to induce sensitization to nickel.

C. Guinea Pig Maximization Test (GPMT) (Table 3)

Even where the prescribed protocol is followed, there is surprisingly great variation concerning the concentrations used for induction and challenge.[2,9,12,13,16,17-23] The concentration used for epicutaneous induction, whether there has been pretreatment with SLS or whether control groups[19] have been used, is rarely stated. With the exception of Zissu et al.,[23] all authors succeeded in sensitizing their subjects with a rate varying between 10 and 100%. My own study[20] was aimed at sensitizing guinea pigs in order to perform passive transfer experiments with nickel. For technical reasons it was therefore necessary to use intradermal challenge.

Table 1
EPICUTANEOUS METHOD FOR THE INDUCTION OF NICKEL SENSITIVITY

Year	Nickel salt	Method	Species	Results	Ref.
1955	$NiSO_4$	Skin painting	Guinea pigs	0/5	6
	$NiSO_4$ + SLS	Skin painting	Guinea pigs	1/5	
1960	$NiSO_4$ + tensides	Occlusive pads	Guinea pigs	Eczematous reactions to nickel mixed with 2 anionics	7
1965	$NiSO_4$ $Ni(NO_3)_2$	Closed patch	Guinea pigs	No sensitization	8
1969	$NiSO_4$ + SLS, Triton X 100	Pretreatment with irritants, skin painting	Guinea pigs	No sensitization	1
1979	$NiSO_4$	Pretreatment with SLS, skin painting	Guinea pigs	Open challenge 8/20 Closed challenge 0/20	9
1982	$NiSO_4$	Pretreatment with supersound, skin painting	Guinea pigs	7/10	10
1984	$NiSO_4$	Skin painting + SLS, stripping	Mice	Sensitization	11
	$NiSO_4$, $NiCl_2$		Mice	No sensitization	
1985-86	$NiSO_4$	Skin painting	Guinea pigs	11/22, 4/7	12, 13

Table 2
INTRADERMAL METHODS FOR THE INDUCTION OF NICKEL SENSITIVITY

Year	Nickel salt	Method	Species	Results	Ref.
1969	$NiSO_4$	Injections, FCA	Guinea pigs	No sensitization	1
1975	$NiSO_4$ Ni complexes	Injections, no FCA	Guinea pigs Mini-pigs	No sensitization	14
1975	$NiSO_4$	Optimization test, FCA	Guinea pigs	i.d. chall. 20/20 e.c. occl. 7/20 e.c. open 12/20	9
1977	$NiSO_4$	Split adjuvant, FCA	Guinea pigs	3/5	15
		Modified maxim, FCA	Guinea pigs	1/6	
		Polak (FCA)	Guinea pigs	0/6	
1979	$NiCl_2$	Split adjuvant technique, FCA	Guinea pigs	No sensitization	16
1980	$NiCl_2$	Injections, FCA	Rabbits	Sensitization	17
1981	$NiCl_2$, $NiSO_4$	Modified Draize, no FCA	Guinea pigs	i.d., e.c. open chall. 0/10	18
1981	$NiCl_2$, $NiSO_4$	Single injection adjuvant test, FCA	Guinea pigs	chall. occl. 6 h $NiCl_2$: 0/10 $NiSO_4$: 2/10	18
1984	$NiCl_2$, $NiSO_4$	Injections, FCA	Mice	No sensitization	11

Note: i.d. = intradermal, e.c. = epicutaneous, and chall. = challenge.

D. Other Methods Using Combinations of Epicutaneous, Intradermal, and Intramuscular Administration (Table 4)

In some other methods, intramuscular administration (with FCA), and intradermal and epicutaneous administration (with Triton X 100, SLS, etc.) have been used. Sensitization to nickel was obtained in most cases.[16,20,25,26] Compared to the GPMT, the sensitization rate

Table 3
GPMT FOR THE INDUCTION OF NICKEL SENSITIVITY

Year	Nickel salt	Induction conc	Challenge conc	Results	Ref.
1970	$NiSO_4$	i.d. 1.0% e.c. 5.0%	0.5%	11/20	2
1976	$NiSO_4$	i.d. 1.0% e.c. 5.0%	i.d. 0.5—0.0625%	Sensitization	20
1979	$NiSO_4$	i.d. 2.0% e.c. 2.0%		5/10	21
1979	$NiCl_2$	i.d. 1.0%	0.25% + stripping	8/12	16
1979	$NiSO_4$	i.d. 0.1% e.c. 5.0%	0.5%	7/20	9
1981	$NiCl_2$	i.d. 0.25% e.c. 2.0%	0.1%	1/10	18
1981	$NiSO_4$	i.d. 0.25% e.c. 0.7%	0.15%	1/10	18
1985-86	$NiSO_4$	i.d. 5.0% e.c. 25%	5.0%	7/31, 7/7	12, 13
1986	$NiSO_4$	i.d. 2.0% e.c. 5.0%	5.0%	8/9	22
1987	$NiSO_4$	i.d. 1.0%	2.0%	0/30	23

Note: i.d. = intradermal, e.c. = epicutaneous.

Table 4
METHODS — BESIDES THE GPMT — USING EPICUTANEOUS, INTRADERMAL, AND SOMETIMES INTRAMUSCULAR ADMINISTRATION FOR THE INDUCTION OF NICKEL SENSITIVITY

Year	Nickel salt	Method	Species	Results	Ref.
1965	$NiSO_4$	FCA, potassium alum injections	Guinea pigs	13/14	24
1976	$NiSO_4$	Polak & Turk (FCA)	Guinea pigs	Sensitization at i.d. chall.	20
1979	$NiCl_2$	Polak et al. (FCA)	Guinea pigs	i.d. chall. 4/8 e.c. chall. 2/8	16
1979	$NiCl_2$	Gross et al. (FCA)	Guinea pigs	No sensitization	16
1979	$NiSO_4$	Brulos (FCA)	Guinea pigs	9/14	21
1984	$NiSO_4$	FCA, cyclophosphamide	Mice	No sensitization	11
1985	$NiSO_4$	Open adm. + FCA	Guinea pigs	4/13	12
1985	$NiSO_4$	TINA test (FCA), SLS, DMSO	Guinea pigs	5/20, 1/19, 4/19, 8/33	25
1987	$NiSO_4$	TINA test	Guinea pigs	8/20	26
1987	$NiSO_4$	Open adm. + SLS + potassium alum injections	Guinea pigs	21/30, 27/30	23

Note: i.d. = intradermal, e.c. = epicutaneous, chall. = challenge.

was somewhat lower, and in spite of using FCA there were two failures in inducing nickel sensitivity.[11,16]

Bezian et al.[24] and Zissu et al.[23] obtained a sensitization rate of 63 to 93% when the epicutaneous administration of $NiSO_4$ was combined with injections and with potassium alum as adjuvant. In my study[20] a high sensitization rate was obtained, especially when the administration (epucutaneous and intradermal) was repeated.

III. COMPARISON OF VARIOUS NICKEL SALTS

Most studies have been carried out with nickel sulfate — the compound used in patch testing — and some authors have also tried nickel chloride, nickel nitrate, and nickel complexes. Goodwin et al.,[18] using the GPMT, found the same sensitization rate (10%) when comparing nickel chloride and nickel sulfate. However, no comparative studies with different nickel salts administered in equimolar concentrations have been published. Whether nickel chloride is a more potent contact allergen than nickel sulfate or is more efficient at detecting nickel-contact allergy has thus not been elucidated in animal experiments.

IV. TOXICITY OF NICKEL

Few authors have commented on the toxicity of nickel. Möller[11] observed that nickel chloride was more toxic than nickel sulfate in mice. Wahlberg[20] found that 3/25 guinea pigs died using the Polak-Turk method after receiving repeated injections and topical treatment exceeding the prescribed protocol. Zissu et al.[23] obtained a mortality rate of 10% using the GPMT.

V. COMPARISON OF DIFFERENT SPECIES (GUINEA PIGS, RABBITS, MINI-PIGS, AND MICE) AND STRAINS OF GUINEA PIG

Most sensitization attempts have been performed in the guinea pig and only in a few cases have other species been tried. Merritt and Brown[17] succeeded in sensitizing rabbits and Möller[11] obtained a weak sensitivity to nickel in painting experiments in mice; other methods of administration failed. Samitz et al.[14] compared guinea pigs and mini-pigs, but failed to induce nickel sensitivity in both species.

Using the optimization test, Maurer et al.[27] observed no difference in sensitization rate among three strains of guinea pig (Pirbright white, Dunkin Hartley, and Himalyan spotted). In the TINA test, Ziegler and Süss[25] observed a sensitization rate of 5/20 in an outbred and 1/19 in an inbred guinea pig strain.

VI. INTERLABORATORY AND INTRALABORATORY COMPARISONS

Comparisons between sensitization methods developed in one laboratory and those carried out at other laboratories with other methods are of limited value. On the other hand, intralaboratory comparative studies are of greater value, since it can be anticipated that animal facilities, strains, chemicals, challenge concentrations, technicians, etc. are the same.

In Table 5, the results from some GPMTs are compared with other sensitization assays carried out in the same laboratory. At epicutaneous challenge the GPMT and the optimization test were equivalent, while at intradermal challenge the optimization test gave a sensitization rate of 100% (Table 2). When Lammintausta et al.[12] compared the GPMT with their own method, the latter was more efficient (7/31 vs. 11/22) in inducing nickel sensitivity. However, in later studies, these authors[13] obtained a sensitization rate of 93 to 100% with the GPMT.

Wahlberg[20] compared the GPMT and the Polak-Turk methods, finding no difference in

Table 5
RESULTS OF GPMT AND OTHER SENSITIZATION ASSAYS CARRIED OUT BY THE SAME LABORATORY — $NiSO_4$, GUINEA PIGS

GPMT	Landsteiner-Draize	Brulos	Optimization	Epidermal test	SIAT	Modified Draize	Open e.c. sens.	Open e.c. sens. + FCA	Open adm. + potassium alum	Ref.
11/20	0/20	—	–	–	–	–	–	–	–	2
5/10	—	9/14	–	–	–	–	–	–	–	21
7/20	—	–	7—12/20	0—8/20	–	–	–	–	–	9
1/10	—	–	. –	–	2/10	0/10	–	–	–	18
7/31	—	–	–	–	–	–	11/22	4/13	–	12, 13
7/7	—	–	–	–	–	–	4/7	–	–	
0/30	—	–	–	–	–	–	–	–	21—27/30	23

Note: e.c. = epicutaneous.

Table 6
QUESTIONS TO CONSIDER WHEN PLANNING A STUDY ON NICKEL SENSITIVE ANIMALS

1. Is a high sensitization rate essential?
2. Should epicutaneous or intradermal challenge be used?
3. What is the time required?
4. What is the cost of experimental animals?
5. Are adequate animal facilities and technicians (considering staff intensiveness) available?

effectiveness provided that the guinea pigs had not been tested before sensitization started. Pretested animals were more difficult to sensitize, and this was interpreted as a tolerance.

The only obvious deviation is shown in a comparative study carried out by Zissu et al.[23] They failed to sensitize guinea pigs with the GPMT method, but obtained a sensitization rate of 21 to 27/30 with their own method (open administration plus injections of potassium alum).

Other authors who have carried out mutual comparisons of different sensitization assays are Hicks et al.,[16] Turk and Parker,[15] and Möller.[11] However, since the sensitization rate was low or the experiments failed entirely, the original publications are recommended for further reading.

VII. RECOMMENDATIONS FOR THE CHOICE OF SENSITIZATION METHOD

More than 25 different methods have been used in attempts to induce contact sensitivity to nickel in experimental animals, and the sensitization rate has varied greatly (Tables 1 to 4). If the GPMT is regarded as a reference method, at least nine studies have been published that show a sensitization rate of 0 to 100% (Table 3). Only one experiment failed to induce nickel sensitivity.

The optimization test[9] gave 100% sensitization at intradermal challenge, but a lower rate at epicutaneous challenge (Table 2). The favorable results with this test method have not been confirmed by other laboratories (Table 5).

Among the epicutaneous methods, repeated painting of nickel solutions was most efficient, and a sensitization rate of 50 to 70% has been reported[10,12,13] (Table 1). However, the method is time consuming since the painting has to be carried out 5 d/week for 4 weeks. It does, however, demonstrate that nickel sensitivity can be induced without the addition of FCA injections.

The methods employing combinations of epicutaneous, intradermal and intramuscular administration (Table 4) also gave a high sensitization rate with nickel sulfate,[16,20,25,26] but this must be balanced by the time they require and the availability of technicians (cost-effectiveness).

When planning a study involving animals sensitive to nickel, some questions may be raised (Table 6). Based on my own experience and the present review of the literature, I would recommend starting with a GPMT. The chance of getting at least some nickel-sensitive animals is high, and the procedure is rapid (3 weeks) and less laborious for the technicians.

If the GPMT fails, there are several alternatives, but since no comparative studies have been carried out it is difficult to recommend one method in favor of another. The optimization test[9] gives a high sensitization rate, but a drawback might be that it is based on intradermal challenge. The Polak-Turk[16,20] and the TINA test[25,26] methods give high sensitization rates but are laborious for the technicians, and nickel toxicity was found to be a limiting factor in the Polak-Turk method.[20]

Repeated epidermal painting[10,12,13] gives high sensitization rates and does not seem to be toxic to the animals, but the method is also laborious. Skin painting combined with injections of potassium alum[23,24] gives a high sensitization rate. This favorable finding should be confirmed by other laboratories.

VIII. CONCLUSION

The basic question of why nickel is such a common cause of contact allergy in the female population, but not a potent contact allergen in experimental animals, can probably be explained by exposure conditions. The inherent allergenic potential of nickel is comparatively low as shown in this review, but other risk factors for humans, e.g., concentration and penetration of nickel, duration, occlusion, and exposure of irritated or damaged skin, will aggravate this.

By using some of the recommended sensitization assays we hope to be able to clarify the relative importance of these risk factors and different steps of the induction process, to evaluate various methods of treatment and prevention, to elucidate the relationship between nickel allergy and the loosening of implants in nickel allergic patients,[17,28] and the induction of tolerance.[29]

REFERENCES

1. **Hunziker, N.,** *Experimental Studies on Guinea Pig's Eczema,* Springer-Verlag, Heidelberg, 1969.
2. **Magnusson, B. and Kligman, A. M.,** Allergic contact dermatitis in the guinea pig, in *Identifications of Contact Allergens,* Charles C Thomas, Springfield, IL, 1970.
3. **Marzulli, F. N. and Maibach, H. I., Eds.,** Dermatotoxicology and pharmacology, in *Advances in Modern Toxicology,* Vol. 4, Hemisphere, Washington, D.C., 1977.
4. **Maurer, T.,** *Contact and photocontact allergens:* A Manual of Predictive Test Methods, Marcel Dekker, New York, 1983.
5. **Andersen, K. E. and Maibach, H. I.,** Contact allergy. Predictive tests in guinea pigs, in *Current Problems in Dermatology,* Vol. 14, Andersen, K. E. and Maibach, K. I., Eds., S. Karger, Basel, 1985.
6. **Nilzén, Å. and Wikström, K.,** The influence of lauryl sulphate on the sensitization of guinea pigs to chrome and nickel, *Acta Derm. Venereol.,* 35, 292, 1955.
7. **Vinson, L. J. and Choman, B. R.,** Percutaneous absorption and surface active agents, *J. Soc. Cosmet. Chem.,* 11, 127, 1960.
8. **Buehler, E. V.,** Delayed contact hypersensitivity in the guinea pig, *Arch. Dermatol.,* 91, 171, 1965.
9. **Maurer, T., Thomann, P., Weirich, E. G., and Hess, R.,** Predictive evaluation in animals of the contact allergenic potential of medically important substances. II. Comparison of different methods of cutaneous sensitization with "weak" allergens, *Contact Dermatitis,* 5, 1, 1979.
10. **Popov, I. and Penev, Z.,** Experimental contact allergic dermatitis to nickel and cobalt in guinea pigs, *Khig. Zdraveopaz. (Sofia),* 25, 36, 1982.
11. **Möller, H.,** Attempts to induce contact allergy to nickel in the mouse, *Contact Dermatitis,* 10, 65, 1984.
12. **Lammintausta, K., Kalimo, K., and Jansén, C. T.,** Experimental nickel sensitization in the guinea pig: comparison of different protocols., *Contact Dermatitis,* 12, 258, 1985.
13. **Lammintausta, K., Korhonen, K., and Jansén, C. T.,** Method of sensitization determines if UVB irradiation inhibits the development of delayed type hypersensitivity to nickel in guinea pigs, *Photodermatology,* 3, 102, 1986.
14. **Samitz, M. H., Katz, S. A., Sheiner, D. M., and Lewis, J. E.,** Attempts to induce sensitization in guinea pigs with nickel complexes, *Acta Derm. Venereol. (Stockholm),* 55, 475, 1975.
15. **Turk, J. L. and Parker, D.,** Sensitization with Cr, Ni, and Zr salts and allergic type granuloma formation in the guinea pig, *J. Invest.Dermatol.,* 68, 341, 1977.
16. **Hicks, R., Hewitt, P. J., and Lam, H. F.,** An investigation of the experimental induction of hypersensitivity in the guinea pig by material containing chromium, nickel and cobalt from arc welding fumes, *Int. Arch. Allergy Appl. Immunol.,* 59, 265, 1979.
17. **Merritt, K. and Brown, S. A.,** Tissue reaction and metal sensitivity. An animal study, *Acta Orthop. Scand.,* 51, 403, 1980.
18. **Goodwin, B. F. J., Crevel, R. W. R., and Johnson, A. W.,** A comparison of three guinea-pig sensitization procedures for the detection of 19 reported human contact sensitizers, *Contact Dermatitis,* 7, 248, 1981.

19. **Wahlberg, J. E. and Boman, A.,** Guinea pig maximization test, in *Current Problems in Dermatology,* Vol. 14, Andersen, K. E. and Maibach, K. I., Eds., S. Karger, Basel, 1985, 59.
20. **Wahlberg, J. E.,** Sensitization and testing of guinea pigs with nickel sulfate, *Dermatologica,* 152, 321, 1976.
21. **Bourrinet, P., Puchault, P., Sarrazin, G., and Berkovic, A.,** Étude comparée de quelques substances allergéniques chez l'homme et chez le cobaye, *J. Pharm. Belg.,* 34, 21, 1979.
22. **Lammintausta, K., Jansén, C. T., and Kalimo, K.,** Langerhans' cell population in topical patch-test and systemic flare-up sites of nickel-sensitive guinea pigs, *Clin. Exp. Dermatol.,* 11, 246, 1986.
23. **Zissu, D., Cavelier, C., and De Ceaurriz, J.,** Experimental sensitization of guinea-pigs to nickel and patch testing with metal samples, *Food Chem. Toxicol.,* 25, 83, 1987.
24. **Bezian, J. H., Dervillée, E., Tardivel, M., and Vervialle, Y.,** Étude expérimentale de la sensibilisation au nickel, *Arch. Mal. Prof. Med. Trav. Secur. Soc.,* 26, 229, 1965.
25. **Ziegler, V. and Süss, E.,** The TINA test, in *Current Problems in Dermatology,* Vol. 14, Andersen, K. E. and Maibach, K. I., 1985, 172.
26. **Ziegler, V., Ritter V., and Stiller, A.,** Ergebnisse und 10 jährige Erfahrungen mit dem TINA-Test sowie Empfehlungen zu seiner Modifikation, *Dermatol. Monatsschr.,* 173, 20, 1987.
27. **Maurer, T., Weirich, E. G., and Hess, R.,** Predictive contact allergenicity influence of the animal strain used, *Toxicology,* 31, 217, 1984.
28. **Lewin, J., Lindgren, U., and Wahlberg, J. E.,** Screw fixation in bone of guinea pigs sensitized to nickel and cobalt, *Acta Orthop. Scand.,* 53, 675, 1982.
29. **Vreeburg, K. J. J., de Groot, K., von Blomberg, M., and Scheper, R. J.,** Induction of immunological tolerance by oral administration of nickel and chromium, *J. Dent. Res.,* 63, 124, 1984.

Chapter 7

NICKEL: HUMAN SENSITIZATION ASSAYS

Ruixi Yang and Howard I. Maibach

TABLE OF CONTENTS

I. Introduction 76

II. Early Human Studies 76

III. The Vandenberg and Epstein Study 76

IV. A. M. Kligman Studies 76

V. Comments 77

References 78

I. INTRODUCTION

As attested to by several chapters in this text, e.g., Epidemiology (Chapter 11), nickel remains a common allergen in most industrialized nations, in spite of our awareness of the problem and several metal and plastic alternatives that could be employed to lessen the sensitization rate. With approximately 10% of normal (nondermatitis) volunteers reactive, nickel presents an ideal antigen for utilization as a standard benchmark for ascertaining clinical relevance in guinea pig and mouse sensitization assays.

The animal sensitization data relating to this clinical experience are summarized in Chapter 6. In this brief chapter, the sparse literature on predictive assays with nickel in man is evaluated.

II. EARLY HUMAN STUDIES

Early studies preclude scientific interpretation, as the challenge patch test was 10% $NiSO_4$. Since even 5% $NiSO_4$ may be irritating, we are not comfortable in accepting the earlier literature utilizing the higher concentrations. Furthermore, the volunteers were not pretested to rule out preexisting sensitization.[1] Nevertheless, Haxthausen[2] and Burckhardt[3] each believed that they had sensitized subjects. It is possible that they made the correct judgment on the basis of morphology and time course of response, but we cannot ascertain this from the available information.

III. THE VANDENBERG AND EPSTEIN STUDY[1]

In 1963, Vandenberg reported a 9% nickel experimental sensitizing rate. A 3-cm circle of skin on a subject's upper arm was frozen for 3 s with dichlorodifluoromethane (Freon 12) to enhance penetration, then immediately a Lintine disc saturated with 25% NiCl and 0.1% sodium lauryl sulfate (SLS) solution was applied to the frozen area and occluded firmly with plastic tape for 2 d. The procedure was repeated three times at 5-d intervals. The challenge was performed on the opposite forearm 10 d after the third freeze with 1, 2, 5, and 10% NiCl in water. Among 172 subjects, 16 (9%) had been sensitized. After 4 months, 20 of the negative reactors previously exposed to nickel by the triple freeze technique were reexposed by the same method. One developed sensitivity in the interim, and 5 of 19 subjects were sensitized by the second series. This suggests that prolonged exposure raised the frequency of sensitization to nickel.

They retested eight reactive individuals for cross-sensitization with cobalt chloride (2 + 1%), copper sulfate (0.5%), and potassium dichromate (0.5 + 0.1%). No relevant positives were observed. The most intriguing fact of their study was that no experimentally sensitized subject demonstrated clinical sensitivity to nickel on daily contact.They inferred that genetics or some factor other than hypersensitivity was required to develop a clinical eruption.

IV. A. M. KLIGMAN STUDIES[4]

A. M. Kligman performed extensive studies attempting to establish an improved method for identifying contact allergens. Having assessed some predictive tests, namely complete and incomplete Schwarze-Peck tests, Draize Repeat Insult Test, and Shelanski-Shelanski Test by a 200-subject group, he regarded these procedures as insensitive.

He also studied several factors which might influence the experimental contact dermatitis in man. These factors were physical and chemical insults to enhance penetration, pretreatment with SLS vs. combinations of SLS with the allergen, area of sensitizing exposure and challenge, number of exposures, fixation of exposures, intervals between exposures, duration of exposures, concentration, vehicles, racial differences, etc. In this unique research, Dr. Kligman investigated nickel contact sensitization from different aspects.

Concentration — In inducing stage, the concentration ranged from 1 to 10%; 5 to 10% was preferable. For challenging, 5, 1, 0.1, and 0.01% were compared, and 2.5% was the threshold concentration allowing the highest sensitizing rate with no irritant response.

Vehicle — Petrolatum, hydrophilic ointment, water, lanolin, acquaphor-water and Carbowax ointment were tested. The results were that the first three vehicles possessed the same ability to sensitize (human). However, the latter three were not as efficient as the others.

Fixation of exposure area — Multiple exposures were compared to one extremity vs. single exposures to different extremities. It was found that 6 of 20 were sensitized by fixing exposure to one extremity, while only 1 of 23 was sensitized by exposure to four extremities.

Racial differences — To answer the question of whether the Negro's immunological capacity for contact sensitization was less than that of the Caucasian, Kligman tested sensitization rates of 2 groups of 25 subjects each, 1 white and 1 black. He found 3 of 24 black subjects were sensitized (by nickel) and 8 of 25 white subjects were sensitized. He suggested that the lower sensitization rate of the black people may be more apparent than real, and the difference not immunologic but one of lesser reactivity to an inflammatory stimulus.

Kligman established the maximization test in identifying contact allergen by human assay. Fundamentally, the procedure does not directly assess safety in use (except when negative). It classifies substances according to the sensitization capabilities exhibited under an arbitrarily defined set of experimental circumstances. The specific points of the procedure were

1. A panel of 25 healthy adult subjects is used.
2. Sensitizing patches are applied to an extremity, either the forearm or the lower leg in the calf region.
3. Concentration of induction is the greatest one to produce a moderate erythema within 48 h. For nickel, it is 10%.
4. Using occlusive patch, pretreat the induction area with 1 ml 5% aqueous SLS for 24 h.
5. Apply to the same site a 48-h occlusive patch with the test material.
6. Repeat this sequence of alternating 24-h irritant and 48-h allergen patches for a total of five exposures with 1 d rest between each exposure.
7. The challenge concentration is 2.5%.
8. The challenge reaction is read immediately after removal of the 48-h patch, and again in another 2 d.
9. A control patch is necessary.
10. The highest rate obtained was 12 of 25 subjects, utilizing an induction concentration of 10% and eliciting concentration of 2.5%.

V. COMMENTS

How does one place this information into clinical perspective?

It is clear that nickel presents a simple model for inducing sensitization in man. The two relevant studies, Vandenberg-Epstein and Kligman, sensitized 9 and 48% of subjects, respectively, and in a short period of time. The second challenge of Vandenberg increased their rate from 9 to approximately 26%.

Why is it that interest in this model decreased? First, investigators may have feared that they would produce not only patch test reactivity but also clinical disease. Neither study, in fact, noted this. Why they failed to induce clinical disease presents a critical and highly relevant area of research. Is it that daily exposure presents too low a dose level to elicit, or do clinically sensitized subjects represent a special, unique (perhaps a special genetic) class?

We find this issue to be the most important and challenging to result from the human studies. We suspect that the answer will reside in the extensive research currently underway in the immunologic forces which control sensitization and induce tolerance. It is provocative to note that these patch test-sensitized subjects are in fact able to tolerate daily nickel exposure.

REFERENCES

1. **Vandenberg, J. and Epstein, W.,** Experimental skin contact sensitization in man, *J. Invest. Dermatol.*, 41, 413, 1963.
2. **Haxthausen, H.,** Verwands-chraftreaktionen' bei Nickel und Kobalt-Allergic der Haut, *Arch. Dermatol. Syphilol.*, 174, 17, 1936.
3. **Burckhardt, W.,** Beitrage zur Ekzemfrage. III. Die Rolle der Alkalischädigung Gegen Nickel, *Arch. Dermatol. Syphilol.*, 173, 262, 1935.
4. **Kligman, A. M.,** Identification of contact allergies by human assay. III. Maximazation test: a procedure for screening and rating contact sensitizers, *J. Invest. Dermatol.*, 47; 393, 1966.

Chapter 8

IN VITRO TESTING IN NICKEL CONTACT HYPERSENSITIVITY

C. McMillan and D. Burrows

TABLE OF CONTENTS

I. Introduction 80

II. Immunology 80

III. Migration Inhibition 81

IV. Lymphocyte Transformation Test 82

V. Leukocyte Procoagulant Activity 84

VI. Interferon 84

VII. Discussion 85

References 86

I. INTRODUCTION

The concept of patch testing was introduced by Jadassohn[1] and involved reproducing allergic eczema, following application of antigen to a small area of unbroken skin in a sensitized individual. Almost a century later the technique, despite its empirical basis, has with refining and standardization developed into a useful tool in the investigation of contact dermatitis. Patch testing, however, gives rise to numerous problems. Interpretation is subjective and relies on observer experience. False-positive and false-negative reactions can occur.[2,3] Many patients who are patch test positive give no supporting history of nickel allergy. Does this indicate a true allergic response? Burrows et al.[2] noted that almost 50% of patients with a history of reacting to metal jewelry produced a negative result on patch testing. Are these individuals really not allergic to nickel? In addition, the procedure is not without possible risks. In particular, patch testing may cause active sensitization of previously nonallergic patients. Despite standardization, universal agreement on the patch testing procedures for the investigation of contact allergy to nickel has not yet been achieved.[4] The solutions to problems involving optimum patch test concentration and patch test-negative nickel-hypersensitive patients remain controversial.

There is, therefore, a requirement for a rapid, reproducible *in vitro* test. *In vitro* testing procedures have provided useful investigative tools for use in research, but have not gained widespread acceptance in the clinical field. This is explained, at least in part, by the fact that these techniques all lack simplicity, are time-consuming, and are not readily reproducible. While patch testing would continue to provide a useful screening technique, the development of a simple reliable *in vitro* procedure would have the advantages of safety, it could be carried out irrespective of the level of the patients' eczema (''Angry Back Syndrome''),[5] and it could be used to investigate false-positive and false-negative results.

II. IMMUNOLOGY

Increased understanding of the immunological mechanisms governing contact allergic dermatitis has enabled the development of *in vitro* test systems.

Acute contact dermatitis is well established as a delayed cell-mediated hypersensitivity.[6] Most allergens have low molecular weight and penetrate the skin where they act as haptens by covalently binding to carrier proteins in the skin to become immunogenic. This antigenic complex is believed to be processed by Langerhans cells (LC), dendritic cells found both in the skin and in lymph nodes. They possess the ability to move from the epidermis to the dermis and via the lymphatics to regional lymph nodes. They act as antigen-presenting cells and their significance was first demonstrated by Silberberg et al.[7] showing apposition of LC and lymphocytes at the site of positive patch test reactions. The LC possesses similar surface markers to the macrophage such as Ia antigen, complement, and Fc fragment of IgG. Antigenic information is transferred to adjacent T-lymphocytes which are complimentary to the contact antigen and receptors for the HLA-DR antigen on the membrane of the LC. The small number of T-cells thus activated are retained in the paracortical area of the draining lymph node where they differentiate into specific memory and effector cells.

Following epicutaneous challenge of hypersensitive patients, specific effector cells, present at or attracted to the site of application, are stimulated by the antigenic complex formed in the skin to release lymphokines.[8] These are soluble nonantibody products of lymphocytic activation by specific antigen, which modulate the inflammatory reaction, recruiting nonspecific cells enabling amplification of the inflammation-inducing effect of a relatively small number of T-effector cells. LC release interleukin 1 which triggers activated T-lymphocytes to release interleukin 2 which in turn promotes activated T-cell proliferation and release of immune interferon. The latter promotes expression of HLA-DR antigens on the LC membrane and T-cell cytotoxic activity. Lymphokines with a variety of biologic activities have been demonstrated in contact sensitivity. Mitogenic factor[9] induces mitotic division of lymphocytes. Migration inhibitor factor[10,11] causes macrophages to remain in the reaction area. Others include cytotoxic factor,[12] chemotactic factor,[13] and angiogenic factor.[14] Many lymphokine activities have now been described, but as these substances become fully characterized it is possible that different activities may be ascribed to the same molecule.

The detection of lymphokines provides a means of assessing cell-mediated immunity. Techniques that are of use in the *in vitro* investigation of contact allergy to nickel include migration inhibition, lymphocyte transformation, procoagulant activity, and viral plaque inhibition. However, most attention has been directed to measurement of lymphocyte proliferative response (transformation) and cell migration inhibition.

III. MIGRATION INHIBITION

Rich and Lewis[15] initiated *in vitro* studies on delayed hypersensitivity by demonstrating that antigen inhibited the migration of cells from tissue explants taken from sensitized animals. George and Vaughan[16] demonstrated inhibition of the outward circular migrations from capillary tubes of peritoneal exudate cells from sensitized guinea pigs on exposure to the specific antigen. Using this method, Bloom and Bennett[17] and David[18] showed that the inhibition was due to a soluble factor released by sensitized lymphocytes following stimulation. This factor was called "migration inhibition factor" (MIF). A migration assay demonstrating the presence of this factor is a useful test in the evaluation of delayed hypersensitivity in man.[19] Two major lymphokines influencing cell migration have been described. Leukocyte inhibiton factor[20] (LIF) inhibits the migration of polymorphonuclear leukocytes, but not monocytes or macrophages. MIF inhibits the migration of monocytes and macrophages, but not polymorphonuclear leukocytes.

In order to measure cell migration inhibition, two different types of assay, namely the direct and indirect methods, are used. In the direct migration system, leukocytes from hypersensitive patients are filled into capillaries. After centrifugation, the capillaries are cut at the liquid cell interface and the part containing cells is placed in chambers filled with medium, with or without specific antigen. In positive tests, the migration of cells is inhibited by antigen-containing media compared to controls and is expressed as the migration index (MI)

$$\text{MI} = \frac{\text{Average of migration areas with antigen}}{\text{Average of migration areas without antigen}} \times 100$$

Direct tests are quick and simple to perform, but negative results fail to differentiate between lack of target cell response and lack of lymphokine production. It has also been suggested that inhibition may be mediated by mechanisms other than lymphokine release.[21]

In the indirect test lymphocytes from hypersensitive patients are incubated in the presence or absence of specific antigen, and the supernatants from these cultures are placed in chambers where capillaries filled with cells from nonsensitized guinea pigs are placed. Only media containing LIF or MIF inhibit the migration of the appropriate normal cells. This method, although more time consuming, provides stronger evidence that substances produced by the lymphocytes are indeed the source of inhibition.

Investigations of nickel-hypersensitive patients using these assays initially failed to show significant cell migration inhibition. Grosfeld et al.,[22] using skin explants from patients allergic to nickel, did not find any differences in outgrowths compared to controls. Following improvements in methodology, Thulin and Zachariae[23] demonstrated specific leukocyte migration inhibition in chromate-hypersensitive patients. Shortly afterwards, Mirza et al.[24,25]

found significant inhibition in patients with contact allergy to nickel. This was followed by similar reports from Thulin[26] and Jordan and Dvorak.[27] Using the direct assay technique, these various investigators demonstrated significant differences in LIF production in affected patients compared to controls. Although toxic concentrations of allergens caused migration inhibition of both controls and hypersensitive patients, lower concentrations demonstrated a dose-dependent relationship with optimum inhibition at just below toxic levels.

A modified technique in which inhibition is assessed by measuring the linear migration of leukocytes along sealed capillaries was described by Jones.[28] Results with purified protein derivative (PPD)-induced migration inhibition of tuberculin-sensitive leukocytes suggested a good correlation to that of cut capillaries.[28] This sealed capillary method offers the advantages of being simpler, more rapid, and requiring less antigen.[29] Investigation of nickel-allergic subjects by Nordlind and Sandberg using this technique, however, failed to discriminate hypersensitive subjects from controls.[29]

In the absence of a standardized method, the various investigators used different nickel salts as antigen and found differing optimal antigenic concentrations. The need to complex the nickel salt to a carrier protein was confirmed by some workers,[25,26] but found unnecessary by others.[27] Overlap of the migration indices between controls and hypersensitive patients[25,26] led Thulin et al. to conclude that while being a useful test to investigate groups of patients, migration indices taken in isolation could not be used to discriminate whether an individual was hypersensitive to nickel. In contrast, Jordan and Dvorak[27] showed the two populations to be significantly different and without overlap between the groups. Interestingly, LIF responses were found in some patch test-negative patients who had a past history suggestive of nickel contact allergy. This could be interpreted as increased sensitivity of the LIF method compared to patch testing, but awaits further investigation.

IV. LYMPHOCYTE TRANSFORMATION TEST

In 1960, Nowell[30] first reported that lymphocytes, cultured in the presence of phytohaemagglutinin (PHA), enlarge and change into blast cells. This process, known as transformation, can be defined as the reversible redifferentiation of small lymphocytes to larger lymphoblasts, culminating in mitosis.[31] Such changes are associated with increased synthesis of phospholipids,[32] lymphokines,[17,33] RNA,[34] and DNA.[35] Blast transformation may be assessed morphologically using various techniques to stain for RNA, DNA, and nucleoli or by evaluation of relative DNA content in the microdensitometric assay.[36] The finding that uptake of radiolabeled thymidine by lymphocyte nucleic acids in culture correlates well with transformation,[37] however, has led to this being used as a more objective and accurate method of assessment. PHA causes nonspecific lymphocyte transformation. Pokeweed[38] and Concanavalin A[39] also cause similar changes. In addition to these nonspecific mitogens, anitigens also possess the capability of transforming lymphocytes.[40,41] Mills[42] demonstrated that this transformation was specific for the antigen to which the organism had become sensitized and that it correlated strongly with delayed hypersensitivity. In the mixed lymphocyte response (MLR),[43] specific transformation also occurs when two genetically different T-cell populations are mixed and react to each other's histocompatibility antigens. Aspegren and Rorsman,[44] following the report by Nowell,[30] investigated the ability of nickel to transform lymphocytes from nickel-allergic subjects. At high nickel concentrations they found suppression of mitosis, while at lower concentrations they were unable to demonstrate any differences in transformation compared to controls. Pappas et al.[45] also found nonspecific transformation in response to nickel acetate, similar to previous reports of the nonspecific mitogenic effects of mercury salts.[46,47] In contrast, Macleod et al.[48] found no significant thymidine uptake in controls, but significant uptake occurred in 7 out of 12 nickel-sensitive patients. The same authors, using nickel sulfate and nickel acetate as antigens, further reported specific transformation in nickel-sensitive patients with neither salt acting in a nonstimulatory capacity.[49] Following this, there have been reports of similar findings,[50,51] but several authors,[52-56] in addition to demonstrating transformation in response to nickel challenge of lymphocytes from nickel-allergic subjects, found a weak nonspecific mitogenic effect in control patients. Furthermore, cord blood lymphocytes showed the capacity to transform following incubation with nickel,[55] being more pronounced at higher nickel concentrations. Comparison of results by the various investigators using the lymphocyte transformation test (LTT) as an *in vitro* means of assessing nickel hypersensitivity is difficult because of numerous technical difficulties encountered in the test and the many different techniques used.[57] Methods of white cell separation vary among authors even though the techniques used may affect the overall levels of transformation. These methods include gravity sedimentation ± dextran and density sedimentation using Ficoll-Paque or Percoll. Gravity sedimentation results in less pure lymphocyte cultures, but may lead to higher transformational values following stimulation with nickel[58] and PHA.[49] In contrast, Nordlind[56] compared separation techniques in the *in vitro* study of nickel dermatitis using the LTT and found highest lymphocyte thymidine uptake in Percoll-separated cells, with less uptake in those separated using Ficoll-Plaque and lowest in gravity-sedimented lymphocytes.

While there is universal agreement among investigators that the concentration of nickel used to stimulate lymphocytes is critical, the optimum stimulatory concentration varies from laboratory to laboratory. Too high a concentration causes toxic effects, while too low a concentration fails to exhibit a mitogenic effect. Amounts of antigen vary from 23,770 μg with positive results, down to 0.01 μg with negative results.[50] Certain authors[51] suggest testing at more than one concentration because of "high dilution responders" and "low dilution responders". Nickel sulfate is the most commonly used source of nickel and corresponds with that used in patch testing, but nickel acetate and chloride are also reported.

Radiolabeled thymidine uptake, despite being the most widely used method of assessing lymphocyte transformation, is not standardized. Different radiolabeled isotopes may be added to cell culture from 1 to 24 h prior to assessment of uptake. Duration of cell culture varies from 3 to 7 d, with maximum response usually reported from 5 to 7 d. Following culture, the preparation of cells prior to measurement of thymidine uptake also varies.

The timing of the LTT in relation to epicutaneous testing might be important. Powell et al.[59] noted increased transformation following epicutaneous challenge of human volunteers with experimentally induced contact allergy to dinitrochlorobenzene. This problem was considered by Veien et al.[60] in the context of nickel hypersensitivity. They found no difference in transformation before and after epicutaneous challenge, although they did report increased transformation following oral challenge.

In spite of the inability to compare results between different laboratories, nonetheless several groups of investigators[53-56,60] report significant differences of lymphocyte transformation in nickel-hypersensitive patients from controls. Unfortunately, there is often a degree of overlap between affected subjects and controls,[53,54,56] which makes it unreliable to diagnose nickel allergy on the basis of the LTT alone.

Interestingly, the LTT was reported as positive in a nickel-hypersensitive patient whose patch testing had reverted to negative.[52] Similar findings were reported when testing *in vitro* using the migration inhibition assay.[25] It would, however, be premature to draw the conclusion that *in vitro* testing may, in certain circumstances, be more sensitive than testing *in vivo*.

Further points of interest resulting from *in vitro* investigation of contact allergy to nickel using the LTT include the finding of significant lymphocyte transformation in an atopic who developed a pustular patch test reaction to nickel,[52] suggesting that such a reaction is not necessarily nonspecific. Also reported is the case of a nickel-sensitive patient who was tested while taking 40 mg Prednisone orally.[50] The LTT was positive to nickel challenge, but PHA stimulation, used as a control, was depressed. On repeat testing after discontinuation of steroid, the nickel transformation remained unchanged, but the PHA stimulation reverted to normal.

V. LEUKOCYTE PROCOAGULANT ACTIVITY

Blood clotting mechanisms play an important role in cell-mediated hypersensitivty.[61-63] The cutaneous manifestation of delayed hypersensitivity in experimental animals is modified by anticoagulant therapy.[62,63] In man, fibrin deposition is a consistent feature of allergic contact dermatitis.[61] Normal leukocytes can be stimulated *in vitro* to generate a procoagulant that activates the extrinsic cascade of blood clotting and is identified as tissue factor.[64] Human mononuclear cells produce significant levels of procoagulant activity following antigenic stimulation[65] and it is suggested that lymphokine-induced macrophage procoagulant activity may explain the inhibition of macrophage migration due to the cross-linking of fibrinogen on the cell surface.[66] Geczy and Meyer[67] assessed procoagulant activity in man, using mumps antigen and tuberculin (PPD) as antigenic stimuli, and found it to be an *in vitro* correlate of delayed hypersensitivity. Comparison with blast transformation and migration inhibition suggested it to be a more sensitive assay.

In brief, this method involves leukocyte separation (Ficoll) followed by incubation in the presence or absence of antigen for 20 h. Procoagulant activity is assessed (in minutes) using a recalcification assay. This may be carried out visually or by automatic coagulocytometer. The activity is calculated from the clotting time of plasma incubated with cells in the absence of antigen (t1) and clotting time of plasma incubated with cells in the presence of antigen (t2)

$$\text{Activity} = \frac{t1 - t2}{t1} \times 100$$

Investigation of nickel hypersensitivity using this method was carried out by Aldridge et al.[68] These authors noted a marked increase in procoagulant activity in nickel-allergic subjects with increasing stimulatory concentrations of nickel. At concentrations of 5 and 10 μg/ml, there was a statistically significant difference ($p < 0.01$) from controls. This method appears to have the advantages of sensitivity and rapid determination of results over other established *in vitro* techniques, but awaits confirmation.

VI. INTERFERON

The ability of interferon (IFN) to inhibit viruses led to its discovery by Isaacs and Lindenmann[69] in 1957. More recently, however, attention has focused on the role of IFN as an immune modulator. Three types of human interferon (HuIFN) are recognized at present:

1. HuIFN α or leukocyte IFN (type 1)
2. HuIFN β or fibroblast IFN (type 1)
3. HuIFN γ or immune IFN (type 2)

IFN acts primarily as an immune modulator with some antiviral potential. In contrast, IFN α and IFNβ are primarily antiviral, with some potential for immune modulation.[70] Modulatory effects on immune function include activation of macrophages, promotion of antibody formation, and natural killer cell function and increased expression of HLA-DR antigen, important in transfer of antigenic information to T-lymphocytes.

The ability of T-lymphocytes to produce IFN (IFNγ) was first demonstrated by Wheelock[33] using PHA mitogen. Green et al.[71] showed that sensitized T-lymphocytes, following exposure to specific antigen, released a lymphokine which activated macrophages. Originally known as macrophage activating factor, this is now known to be IFNγ.

McKimm-Breschkin et al.[72] investigated the ability of antigen to induce IFN production in two continuous mouse T-cell lines which were shown to exhibit contact hypersensitivity to the sensitizing agent oxazalone. IFN was measured by means of its ability to inhibit growth of viral plaques, expressing titers as the reciprocal of the dilution causing a 50% reduction in the number of plaques compared to controls. These authors demonstrated antiviral activity which they characterized as IFNγ because it was species specific, stable at 50°C for 30 min, and inactivated by treatment at pH 2 (IFNα and IFNβ are acid stable). Using an inhibitor of protein synthesis, they also demonstrated inhibition of IFN, only when T-lymphocytes were treated.

Using a similar viral plaque inhibition technique, McMillan et al.[73] investigated IFN production in nickel-contact hypersensitivity. Following incubation of leukocytes with nickel sulfate, these investigators demonstrated significant ($p = 0.002$) antiviral activity produced by cells taken from nickel-hypersensitive subjects compared to controls. This activity was found to be heat stable and acid labile, similar to findings of McKimm-Breschkin et al., suggesting that this was caused by IFNγ. Further characterization, however, is required to confirm that this activity is produced by a T-cell factor which is indeed IFN.

VII. DISCUSSION

In the investigation of contact allergy to nickel, several *in vitro* tests are capable of distinguishing nickel-sensitive subjects from controls. Results from most of these studies, however, reveal overlap between affected individuals and those who are nonallergic. These tests, therefore, while providing a useful means of looking at specific groups of individuals, cannot be relied upon to make the diagnosis of nickel allergy when looking at an individual result taken in isolation.

Almost a century after its introduction, patch testing has, by refinement and standardization, become the single most important test in assessing contact allergy. In contrast, even in the more established *in vitro* tests such as the MIF and LTT assays, there is complete lack of standardization between laboratories. No attempt has been made to reach agreement even on fundamental questions of which nickel salt to use and what is the optimal concentration of nickel for antigenic challenge. This lack of standardization makes difficult the comparison of data from different laboratories using the same test and meaningful comparison of data from different laboratories using different tests is almost impossible. Von Blomberg-van der Flier et al.[74] evaluated and compared the LTT and MIF tests as *in vitro* correlates of nickel allergy. They found the MIF test to be less sensitive (26% vs. 7% false negatives), but more specific (13% vs. 39% false positives) than the LTT for predicting skin reactivity to nickel. These authors also found that although LTT and MIF data appeared to be only weakly correlated in the individuals tested, using a dual parameter analysis, there was excellent correlation between skin test and *in vitro* reactivity for individuals with matching *in vitro* results (60% of all individuals tested). Silvennoinen-Kassinen et al.,[54] using the LTT to assess *in vitro* response to nickel, reported a positive reaction in approximately 90% of patients with clinical allergy and a positive patch test. False-positive *in vitro* responses occurred in 10% of clinically healthy subjects. Al-Tawil et al.,[75] also using the LTT, investigated the correlation between quantitative *in vivo* and *in vitro* response in nickel-allergic patients. By testing *in vivo*, with a range of dilutions of nickel sulfate in water, they determined the lowest concentration eliciting a cutaneous response. This was then compared with the *in vitro* lympocyte response. The best correlation coefficients obtained were 0.42 (linear) and 0.46 (logarithmic). In addition, there were several patients who showed a weak response *in vivo*, but had high lymphocyte reactivity *in vitro* and vice versa. Hutchinson et al.,[49] comparing the clinical state of nickel-allergic subjects with their lymphocyte transformation responses, were unable to find any correlation.

The series of events involving cellular interactions and release of many immunopharmacological mediators in contact dermatitis is extremely complex. Patch testing, through its production of a "mini" patch of eczema, encompasses the entire reaction. It is almost certainly simplistic to consider that one *in vitro* assay, such as measurement of a lymphokine, will provide a definitive test that is a direct correlate of the entire *in vivo* response. A single lymphokine may exert several different effects and may be released in inflammatory responses other than contact allergic hypersensitivity. Nevertheless, current investigative techniques enable differentiation of groups of allergic individuals from normal controls. Refinement and standardization of procedure may enable *in vitro* assessment of an individual's susceptibility to a particular contact allergen such as nickel. Such a technique must be simple, quick, and reproducible. Up to now the techniques used to assess lymphokines have tended to be indirect. Rather than measure the effects of these mediators, advances using developments in fields of monoclonal antibodies, radioimmunoassays, and enzyme-linked immunosorbent assay[76] should enable direct measurement of specific lymphokines.

The diagnosis of contact allergy will continue to rely on the skills of the clinician, combined with *in vivo* testing.[77] The development of a suitable *in vitro* test, using a standardized technique, could provide complimentary information to enable the correct diagnosis to be established, particularly in cases where patch testing is contraindicated or the results inconclusive.

REFERENCES

1. **Jadassohn, J.,** Zur Kenntnis der medicamentosen Dermatosen, in *Verhand der Deutschen Dermatologischen Gesellschaft,* 5th Congress, Jarish & Neisser, Graz, Austria, 1895, 103.
2. **Burrows, D., Creswell, S. N., and Merrett, J. D.,** Nickel, hands and hip prostheses, *Br. J. Dermatol.,* 105, 437, 1981.
3. **Moller, H. and Svensson, A.,** Metal sensitivity: positive history but negative test indicates atopy, *Contact Dermatitis,* 14, 57, 1986.
4. **Fisher, A. A.,** Problems attending patch testing for nickel sensitivity, *Cutis,* 29, 148, 1982.
5. **Mitchell, J. C.,** Angry back syndrome, *Contact Dermatitis,* 7, 359, 1981.
6. **Chase, M. W.,** Hypersensitivity to simple chemicals, *Harvey Lect.,* 61, 169, 1967.
7. **Silberberg, I., Baer, R. L., and Rosenthal, S. A.,** The role of Langerhans cells in contact allergy, *Acta Derm. Venereol.,* 54, 321. 1974.
8. **Dumonde, D. C., Wolstencroft, R. A., Panayi, G. S., Matthew, M., Morley, J., and Howsen, W. J.,** Lymphokines: non-antibody mediators of cellular immunity generated by lymphocyte activation, *Nature (London),* 224, 38, 1969.
9. **Maini, R. N., Bryceson, A. D. M., Wolstencroft, R. A., and Dumonde, D. C.,** Lymphocyte mitogenic factor in man, *Nature (London),* 224, 43, 1969.
10. **Nishioka, K. and Amos, H. E.,** Contact sensitivity in vitro: the production of macrophage inhibition factors from DNCB sensitized lymphocytes by subcellular organelles obtained from DNCB epidermal tissue, *Immunology,* 25, 423, 1973.
11. **Nishioka, K.,** Detection of human contact sensitivity to dinitrochlorobenzene by the migration inhibition test, *J. Invest. Dermatol.,* 66, 351, 1976.
12. **Delescluse, J. and Turk, J. C.,** Lymphocyte cytotoxicity: a possible in vitro test for contact dermatitis, *Lancet,* 1, 75, 1970.
13. **Cohen, S., Ward, P. A., Yoshida, T., et al.,** Biologic activity of extracts of delayed hypersensitivity skin reaction sites, *Cell. Immunol.,* 9, 363, 1973.
14. **Nishioka, K. and Katayama, I.,** Angiogenic activity in culture supernatant of antigen-stimulated lymph node cells, *J. Pathol.,* 126, 63, 1978.
15. **Rich, A. R. and Lewis, M. R.,** The nature of allergy in tuberculosis as revealed by tissue culture studies, *Bull. Johns Hopkins Hosp.,* 50, 115, 1932.

16. **George, M. and Vaughan, J. H.,** In vitro cell migration as a model for delayed hypersensitivity, *Proc. Soc. Exp. Biol. Med.,* 111, 514, 1962.
17. **Bloom, B. R. and Bennett, B.,** Mechanism of a reaction in vitro associated with delayed hypersensitivity, *Science,* 153, 80, 1966.
18. **David, J. R.,** Delayed hypersensitivity in vitro: its mediation by cell-free substances formed by lymphoid cell-antigen interaction, *Proc. Natl. Acad. Sci. U.S.A.,* 56, 72, 1966.
19. **David, J. R.,** Lymphocyte mediators and cellular hypersensitivity, *N. Engl. J. Med.,* 288, 143, 1973.
20. **Rocklin, R. E.,** Products of activated lymphocytes; leukocyte inhibitory factor (LIF) distinct from migration inhibitory factor (MIF), *J. Immunol.,* 112, 1461, 1974.
21. **Hamblin, A. S. and Maini, R. N.,** An evaluation of lymphokine measurement in man, in *Recent Advances in Clinical Immunology,* 2nd ed., Thompson, R. A., Ed., Churchill Livingstone, Edinburgh, 1980, 243.
22. **Grosfeld, J. C. M., Penders, A. J. M., de Grood, R., and Verwilghen, L.,** In vitro investigations of chromium- and nickel-hypersensitivity with culture of skin and peripheral lymphocytes, *Dermatologica,* 132, 189, 1966.
23. **Thulin, H. and Zachariae, H.,** The leucocyte migration test in chromium hypersensitivity, *J. Invest. Dermatol.,* 58, 55, 1972.
24. **Mirza, A. M., Perera, M. G., and Bernstein, I. L.,** Leukocyte migration inhibition in nickel dermatitis, *Fed. Proc.,* 33, 728, 1974.
25. **Mirza, A. M., Perera, M. G., Maccia, C. A., Dziubynskyj, O. G., and Bernstein, I. L.,** Leukocyte migration inhibition in nickel dermatitis, *Int. Arch. Allergy Appl. Immunol,* 49, 782, 1975.
26. **Thulin, H.,** The leukocyte migration test in nickel contact dermatitis, *Acta Derm. Venereol.,* 56, 377, 1976.
27. **Jordan, W. P. and Dvorak, J.,** Leukocyte migration inhibition assay (LIF) in nickel contact dermatitis, *Arch. Dermatol.,* 112, 1741, 1976.
28. **Jones, B. M.,** Sealed capillary leucocyte migration test, *Med. Lab. Technol.,* 30, 245, 1973.
29. **Nordlind, K. and Sandberg, G.,** Leukocytes from patients allergic to chromium and nickel examined by the sealed capillary migration technique, *Int. Arch. Allergy Appl. Immunol.,* 70, 30, 1983.
30. **Nowell, P. C.,** Phytohemagglutinin, an initiator of mitosis in cultures of normal human leucocytes, *Cancer Res.,* 20, 462, 1960.
31. **Oppenheim, J. J.,** Immunological relevance of antigen and antigen antibody complex induced lymphocyte transformation, *Ann. Allergy,* 27, 305, 1969.
32. **Fisher, D. B. and Mueller, G. C.,** An early alteration in the phospholipid metabolism of lymphocytes by phytohemagglutinin, *Proc. Natl. Acad. Sci. U.S.A.,* 60, 1396, 1968.
33. **Wheelock, E. F.,** Interferon-like virus inhibitor induced in human leukocytes by phytohemagglutinin, *Science,* 149, 310, 1965.
34. **Cooper, H. L. and Rubin, A. D.,** RNA metabolism in lymphocytes stimulated by phytohemagglutinin: initial responses to phytohemagglutinin, *Blood,* 25, 1014, 1965.
35. **Mackinney, A. A., Stohlman, F.,and Brecher, G.,** The kinetics of cell proliferation in cultures of human peripheral blood, *Blood,* 19, 349, 1962.
36. **Cooper, E. H., Barkhan, P., and Hale, A. J.,** Observations on the proliferation of human leucocytes cultured with phytohaemagglutinin, *Br. J. Haematol.,* 9, 101, 1963.
37. **Caron, G. A., Sarkany, I., Williams, H. S., and Todd, A. P.,** Radioactive method for the measurement of lymphocyte transformation in vitro, *Lancet* 2, 1266, 1965.
38. **Farnes, P., Barker, B. E., Brownhill, L. E., and Fanger, H.,** Mitogenic activity in *Phytolacca americana* (pokeweed), *Lancet,* 2, 1100, 1964.
39. **Powell, A. E. and Leon, M. A.,** Reversible interaction of human lymphocytes with the mitogen concanavalin A, *Exp. Cell Res.,* 62, 315, 1970.
40. **Marshall, W. H. and Roberts, K. B.,** Tuberculin-induced mitosis in peripheral blood leucocytes, *Lancet,* 1, 773, 1963.
41. **Pearmain, G., Lycette, R. R., and Fitzgerald, P. H.,** Tuberculin-induced mitosis in peripheral blood leucocytes, *Lancet,* 1, 637, 1963.
42. **Mills, J. A.,** The immunologic significance of antigen induced lymphocyte transformation in vitro, *J. Immunol.,* 97, 239, 1966.
43. **Bach, F. and Hirschhorn, K.,** Lymphocyte interaction: a potential histocompatibility test in vitro, *Science,* 143, 813, 1964.
44. **Aspegren, N. and Rorsman, H.,** Short-term culture of leucocytes in nickel hypersensitivity, *Acta Derm. Venereol.,* 42, 412, 1962.
45. **Pappas, A., Orfanos, C. E., and Bertram, R.,** Non-specific lymphocyte transformation in vitro by nickel acetate, a possible source of errors in lymphocyte transformation test (LTT), *J. Invest. Dermatol.,* 55, 198, 1970.

46. **Schopf, E., Schulz, K. H., and Isensee, I.,** Untersuchungen uber den Lymphocytentransformations-Test bei Quecksilber-Allergie, unspezifische Transformation durch Hg-Verbindungen, *Arch. Klin. Exp. Dermatol.,* 234, 420, 1969.
47. **Schopf, E., Schulz, K. H., and Gromm, M.,** Transformation und Mitosen von Lymphocyten in vitro durch Quecksilber II-Chlorid, *Naturwissenschaften,* 5, 568, 1967.
48. **Macleod, T. M., Hutchinson, F., and Raffle, E. J.,** The uptake of labelled thymidine by leucocytes of nickel sensitive patients, *Br. J. Dermatol.,* 82, 487, 1970.
49. **Hutchinson, F., Raffle, E. J., and Macleod, T. M.,** The specificity of lymphocyte transformation in vitro by nickel salts in nickel sensitive subjects, *J. Invest. Dermatol.,* 58, 362, 1972.
50. **Millikan, L.E., Conway, F., and Foote, J. E.,** In vitro studies of contact hypersensitivity: lymphocyte transformation in nickel sensitivity, *J. Invest. Dermatol.,* 60, 88, 1973.
51. **Gimenez-Camarasa, J. M., Garcia-Calderon, P., Asensio, J., and DeMaragas, J. M.,** Lymphocyte transformation test in allergic contact nickel dermatitis, *Br. J. Dermatol.,* 92, 9, 1975.
52. **Kim, C. W. and Schopf, E.,** A comparative study of nickel hypersensitivity by the lymphocyte transformation test in atopic and non-atopic dermatitis, *Arch. Dermatol. Res.,* 257, 57, 1976.
53. **Svejgaard, E., Morling, N., Svejgaard, A., and Veien, N. K.,** Lymphocyte transformation induced by nickel sulphate: an in vitro study of subjects with and without a positive nickel patch test, *Acta Derm. Venereol.,* 58, 245, 1978.
54. **Silvennoinen-Kassinen, S.,** The specificity of a nickel sulphate reaction in vitro: a familial study and a study of chromium-allergic subjects, *Scand, J. Immunol.,* 13, 231, 1981.
55. **Al-Tawil, N. G., Marcusson, J. A., and Moller, E.,** Lymphocyte transformation test in patients with nickel sensitivity: an aid to diagnosis, *Acta Derm. Venereol.,* 61, 511, 1981.
56. **Nordlind, K.,** Lymphocyte transformation test in diagnosis of nickel allergy, *Int. Arch. Allergy Appl. Immunol.,* 73, 151, 1984.
57. **Hutchinson, F., Macleod, T. M.,and Raffle, E. J.,** Lymphocyte transformation, *Br. J. Dermatol.,* 85, 300, 1971.
58. **Macleod, T. M., Hutchinson, F., and Raffle, E. J.,** In vitro studies on blastogenic lymphokine activity in nickel allergy, *Acta Derm. Venereol.,* 62, 249, 1982.
59. **Powell, J. A., Whalen, J. J., and Levis, W. R.,** Studies on the contact sensitization of man with simple chemicals. IV. Timing of skin reactivity, lymphokine production and blastogenesis following rechallenge with dinitrochlorobenzene using an automated microassay, *J. Invest, Dermatol.,* 64, 357, 1975.
60. **Veien, N. K., Svejgaard, E., and Menne, T.,** In vitro lymphocyte transformation to nickel: a study of nickel-sensitive patients before and after epicutaneous and oral challenge with nickel, *Acta Derm. Venereol.,* 59, 447, 1979.
61. **Colvin, R. B., Johnson, R. A., Mihm, M. C., Jr., and Dvorak, H. F.,** Role of the clotting system in cell-mediated hypersensitivity. I. Fibrin deposition in delayed skin reactions in man, *J. Exp. Med.,* 138, 686, 1973.
62. **Nelson, D. S.,** The effects of anticoagulants and other drugs on cellular and cutaneous reactions to antigen in guinea-pigs with delayed-type hypersensitivity, *Immunology,* 9, 219, 1965.
63. **Feinman, L., Cohen, S., and Becker, E. L.,** The effect of fumaropimaric acid on delayed hypersensitivity and cutaneous Forssman reactions in the guinea pig, *J. Immunol.,* 104, 1401, 1970.
64. **Rickles, F. R., Hardin, J. A., Pitlick, F. A., Hoyer, L. W., and Conrad, M. E.,** Tissue factor activity in lymphocyte cultures from normal individuals and patients with hemophilia A, *J. Clin. Invest.,* 52, 1427, 1973.
65. **Edwards, R. L., Rickles, F. R., and Bobrove, A. M.,** Mononuclear cell tissue factor: cell of origin and requirements for activation, *Blood,* 54, 359, 1979.
66. **Geczy, C. L. and Hopper, K. E.,** A mechanism of migation inhibition in delayed-type hypersensitivity reactions. II. Lymphokines promote procoagulant activity of macrophages in vitro, *J. Immunol.,* 126, 1059, 1981.
67. **Geczy, C. L. and Meyer, P. A.,** Leukocyte procoagulant activity in man: an in vitro correlate of delayed-type hypersensitivity, *J. Immunol.,* 128, 331, 1982.
68. **Aldridge, R. D., Milton, J. I., and Thompson, A. W.,** Leukocyte procoagulant activity as an in vitro index of nickel contact hypersensitivity, *Int. Arch. Allergy Appl. Immunol.,* 76, 350, 1985.
69. **Isaacs, A. and Lindenmann, J.,** Virus interference. I. The interferon, *Proc. R. Soc. London Ser. B,* 147, 258, 1957.
70. **Wardle, E. N.,** Interferon-gamma: actions and importance, *Br. J. Hosp. Med.,* 37, 446, 1987.
71. **Green, J. A., Cooperband, S. A., and Kibrick, S.,** Immune specific induction of interferon production in cultures of human blood lymphocytes, *Science,* 164, 1415, 1969.
72. **McKimm-Breschkin, J. L., Mottram, P. L., Thomas, W. R., and Miller, J. F.,** Antigen-specific production of immune interferon by T cell lines, *J. Exp. Med.,* 155, 1204, 1982.
73. **McMillan, C., Adamson, H., Burrows, D., and McNeill, T.,** Lymphokine antiviral activity in contact dermatitis, presented at 8th Int. Symp. Contact Dermatitis, Cambridge, 1986, 5.

74. **Von Blomberg-van der Flier, B. M. E., Bruynzeel, D. P., and Scheper, R. J.,** In vitro diagnosis of allergic contact dermatitis, presented at 17th World Congr. Dermatology, Vol. 2, Berlin, 1987, 55.
75. **Al-Tawil, N. G., Berggren, G., Emtestam, L., Fransson, J., Jernselius, R., and Marcusson, J. A.,** Correlation between quantitative in vivo and in vitro responses in nickel-allergic patients, *Acta Derm. Venereol.,* 65, 385, 1985.
76. **Boerrigter, G. H., Vos, A., and Scheper, R. J.,** Direct measurement of in vitro antibody production using an enzyme-linked immunosorbent assay, *J. Immunol. Methods,* 61, 377, 1983.
77. **Milner, J. E.,** In vitro tests for delayed skin hypersensitivty: lymphokine production in allergic contact dermatitis, in *Dermatotoxicology,* 3rd ed., Marzulli, F. N. and Maibach, H. I., Eds., Hemisphere, New York, 1986, 217.

Chapter 9

CLINICAL CONCEPTS IN NICKEL TESTING

K. Lammintausta and H. I. Maibach

TABLE OF CONTENTS

I. Introduction ... 92

II. Test Indications ... 92
- A. Common Clinical Picture of Nickel-Allergic Contact Dermatitis ... 92
- B. Hand Dermatitis and Nickel Allergy ... 92
- C. Noneczematous Symptoms in Nickel Allergics ... 93

III. Patch Testing ... 93
- A. Different Nickel Salts ... 93
- B. Vehicles ... 93
- C. Concentration ... 94
- D. Occlusion Time ... 94
- E. Test Site ... 94
- F. False-Positive Reactions ... 94
- G. False- and Unexpected Negative Reactions ... 95

IV. Intradermal Testing ... 95

V. Open Testing ... 95

VI. Systemic Exposure ... 95

VII. *In Vitro* Testing ... 96

VIII. Dimethylglyoxime Spot Test ... 96

IX. Conclusions ... 96

References ... 96

I. INTRODUCTION

Adequate and reliable systems to reveal nickel sensitivity are highly important because exposure and sensitization to nickel is common and unusual diagnostic difficulties are not infrequent. It is no exaggeration to say that the pattern of contact dermatitis often raises the suspicion of nickel allergy, which is the most common source of allergic contact dermatitis in the U.S. and Europe.[1-5] We cannot avoid nickel in today's world and the incidence of sensitization has been increasing[6-8] in spite of extensive information and guidance.

Nickel allergy is mainly a problem of females and 15- to 30-year-olds account for the majority of the patch test-positive females[9-11] in clinical surveys. However, test positivity may not be found until in the seventh or eighth decade. Cases in 4- to 14-month-old children with a dermatitis at metal contact sites and proven nickel allergy have been documented.[12-14] Allergic nickel dermatitis in children and in elderly subjects often lacks the characteristic features, while it simultaneously is less expected than in the typical age groups. Elderly subjects may react more slowly and allergic reactions are often rather persistent.[15] Children may be susceptible to nonspecific patch test reactions[16,17] and lowered concentrations have been suggested. Using standard concentration for both adults and children has also been preferred.[18] However, controls done with corresponding age groups are needed.

II. TEST INDICATIONS

A. Common Clinical Picture of Nickel-Allergic Contact Dermatitis

A typical clinical picture in the site of exposure is not infrequent in nickel allergics. Most typically the dermatitis is located in the site of nickel contact, from cheap jewlery, zippers, or metal buttons most often.[7] In those cases, the diagnosis can be done on the basis of the clinical picture and by taking a short dermatologic history.

Nickel-allergic contact dermatitis is not infrequently manifested with less typical skin symptoms. The complicated clinical course of the dermatitis is also reflected in the most common locations of the symptoms in test-positive patient groups. Hands are the most common location followed by head, upper and lower extremities (often flexures), and other parts of the body.[9,11,19] The untypical forms of nickel allergic contact dermatitis apparently increase the importance of adequate and reliable test systems. However, the accuracy and completeness of the dermatologic history has to be suspected when no direct contact explains the dermatitis although secondary spread occurs.[20,21] Hidden, unrevealed nickel sources may explain at least some part of this nontypical, secondary spread nickel dermatitis.[9,22,23]

B. Hand Dermatitis and Nickel Allergy

Although hands are the most common location of dermatitis in nickel allergics,[9,11,19] direct allergen contact only infrequently explains this cincal picture. In a nickel-allergic hand dermatitis population in Sweden, four fifths of the patients had pompholyx-type hand dermatitis.[24,25] Episodic appearance of itching vesicles appearing in the palmar aspects of the hands were typical. These patients did not react with their typical dermatitis to a topical application of nickel-containing objects onto the palms, whereas a burst of vesicles was induced by oral ingestion of extensive amounts of nickel sulfate[25] (Chapter 14). Corresponding results and related increase of urinary nickel excretion have been reported by others.[26] However, these findings were not verified in later nickel feeding studies[27,28] and the amount of nickel needed to provoke the symptoms (2.5 mg) is too large to correspond to even maximal ingestion in normal life. The controversy of this topic became still more apparent when nickel-negative pompholyx patients were also reported to react in a similar way to oral nickel exposure.[30] It appears that the reaction occurs in certain circumstances in nickel-allergic pompholyx patients. The dose required and the frequency of occurrence are unknown.

C. Noneczematous Symptoms in Nickel Allergics

Nickel-induced systemic symptoms include chronic urticaria,[31] acute urticaria,[32] vasculitis,[33] anaphylactoid exanthema,[34] and other mucosal symptoms.[35,36] The mechanisms of these reactions are unrevealed. Precipitating antibodies[32,36] may contribute.

Nickel has induced local contact urticaria with a positive scratch chamber or prick test.[37,38] It is unknown whether or not the mechanism was immunologic.

III. PATCH TESTING

In spite of historical features, patch testing is still the most important diagnostic tool for clinical contact dermatitis. Since the 1890s when Jadasson introduced the method, the main principles have remained the same; only minor modifications have been done. Different devices for application are available,[7,39] but no apparent superiorities between them seems to exist.

The allergens used in standard test batteries have their own characteristic patterns leading to misinterpretation of results. Nickel is one of those allergens eliciting false-positive reactions, while false negatives also occur. When true positive reactions develop in 7 to 12% of the tested populations[9,19,40] and the history of metal allergy is even more common,[41,42] the evaluation of nickel patch test results requires knowledge, practice, and experience.

A. Different Nickel Salts

Nickel sulfate with six molecules of crystallized water in petrolatum is the most commonly used patch test material today. Although metallic nickel is not a sensitizer, sweat appears to corrode nickel effectively,[43] while sodium chloride may be a key assistant for the release of nickel into the skin.[39] Increasing temperature and lowered pH further enhance the penetration.[43] Nickel chloride, generally in aqueous solution, has been used for clinical patch tests. It seems to induce positive reactions more effectively than nickel sulfate.[44,45] Simultaneously, it is also more irritating, often producing follicular irritant reactions.[46] This irritation potential has limited its use as a standard allergen. However, nickel chloride in water is more irritating than nickel chloride in petrolatum and those irritant reactions are easily recognizable.[44] Also, considerable interindividual variation in cutaneous irritation to aqueous nickel chloride (2.5%) has been demonstrated.[45]

Occluded application of metallic nickel alloys produces positive patch test reactions, depending on the degree of individual sensitivity and on the amount of nickel released.[47,48] It may elicit positive reactions for some patients, even in cases where the practical spot test with dimethylglyoxime remains negative, since only about 10 μg or more released nickel can be detected with dimethylglyoxime.[49] Testing with metallic objects does not have any superiority over standard patch test methods. It may be useful in some exceptional cases where certain objects are suspected and their nickel content cannot be verified, as well as for experimental purposes.

B. Vehicles

Petrolatum is the general vehicle for nickel sulfate. However, its value has been questioned. Water may also be a more reliable vehicle for nickel sulfate,[44,50,51] but mainly technical and practical superiority of petrolatum has made petrolatum more popular.

Uneven spreading of nickel in petrolatum[52] and poor penetration from petrolatum may explain the less reliable results as compared with water-diluted allergen. When dimethylsulfoxide was compared with petrolatum and water, it appeared superior to both as a vehicle for nickel.[53] With the thin layer rapid use epicutaneous test, where the allergen is dispersed in a special gel, the results for nickel have appeared good or better than with petrolatum-dispersed test materials.[54] Softisan, a new wool alcohol substitute, was compared with petrolatum.[55] Unequivocally, it appeared superior to petrolatum, eliciting positive reactions significantly more often and with lower concentrations. The clinical value of these new vehicle systems needs to be proven in larger studies.

C. Concentration

The concentration of the test substance is an important factor in patch testing. The optimal concentration has to be found between too low and too high levels, which correspondingly lead to wrong-negative and wrong-positive reactions. If the test material is not homogeneous, locations of high allergen concentration occur and nonspecific reactions are elicited.[52,56] In addition to standardized methods in the preparation of the test substances, the homogenous concentration is also dependent on the storage temperature. Higher room temperature seemed to decrease the concentration of nickel sulfate in petrolatum, whereas that did not happen at 4°C.[57]

Nickel sulfate is generally tested at 5% in petrolatum. However, 80% of sensitized subjects react to 2.5% test material[7] and a considerable number react to even lower concentrations.[58,59] Occasional cases of nickel-allergic patients only reacting to 10% nickel sulfate in petrolatum were described.[60] The degree of sensitivity may be the main factor in the concentration level required for eliciting a positive reaction. Because nonreactivity to 5% test material is uncommon,[7,39] testing with different vehicles is preferred over increased concentration. Instead, follicular or other nonspecific appearing reactions need to be retested with lower concentrations.

D. Occlusion Time

The most generally used application time is 48 h. It is apparent that a shorter 24-h application would be much more practical and it has been used occasionally. Since, however, about 20% of positive reactions may be lost when using 24-h occlusion time,[45,61] no doubt exists that the less practical 48-h occlusion time must be preferred at present.

E. Test Site

Regional variations in patch test responses have been demonstrated, suggesting the upper back[62] and particularly the middle scapula[63] to be most preferable for testing. Optimal reactivity in the back may be due to good penetration induced by pressure,[64] but pressure has also been shown to have no influence on the reactivity.[65] It appears possible that true anatomical differences exist in cutaneous reactivity, favoring patch testing in the back.

F. False-Positive Reactions

Manifest dermatitis induces nonspecific reactions,[66-69] but it is uncertain how long a time the nonspecific cutaneous irritability will persist after the dermatitis has cleared. Bjornberg[68] found that 3 months after complete healing of hand dermatitis, cutaneous reactivity had normalized. Individual strong positive reactions may sometimes induce nonspecific reactivity in close-by test sites,[67,69,70] and thus retesting may be needed. Cases may also need retesting where suspicion of hyperactivity due to existing or previous dermatitis exists. While the variation in cutaneous irritability is wide.[68,70] and an objective distinction between allergic and irritant reactions is often impossible, the probability of a false-positive patch test reaction should considered in each case of a positive reaction.

Follicular test reactions represent a peculiar type of skin reactions, the origin of which is most often irritant.[7,39] Nonspecific, follicular nickel reactions occur frequently in atopics.[7,71] This may be most common in dermatitic or traumatized skin.[72] However, a follicular reaction may also allergic,[58,59] and thus retesting with a dilution series may be helpful. Although follicular, nonspecific reactions are typical for metals, individual propensity to this type of reaction may also be reflected in those reactions[59] sometimes elicited as an expression of contact sensitivity in patch test sites.

G. False- and Unexpected Negative Reactions

It is impossible to evaluate the true incidence of false-negative reactions. Usually, only the situation with a positive history of metal dermatitis without a positive patch test reaction evokes the suspicion. This may suggest atopy.[71] It may be due to latent sensitivity or to the degree of sensitivity being too low to elicit a positive reaction.[7] When 2.5% nickel sulfate was used instead of 5%, 20% of true nickel sensitivities were missed,[7] suggesting that the concentration needed is dependent on the degree of sensitivity.

UVB and PUVA have been shown to decrease cutaneous reactivity,[73,74] although it may not only be related to specific allergic reactions.[75] UVB or PUVA has induced significant decreases in allergic nickel reactions,[74] but contradictory results were also reported.[76] Seasonal variation in patch test reactivity[77] may be associated with different preceding sun exposure. False negative should be suspected in strongly tanned skin. The etiology of this alteration in patch test reactivity is not wholly understood. Local decrease of Langerhans cells at the site of UV exposure probably plays a role in addition to nonspecific factors. Technical, often occasional factors also need to be considered in the evaluation of possible false-negative reactions.

IV. INTRADERMAL TESTING

Intradermal nickel tests with nickel sulfate or nickel chloride can be used to reveal cases of unexpected negative reactions. This method may be useful in occasional cases.[78,79] The concentration of the nickel salt should not be too high — 1:10,000 appears sufficient — to avoid irritant reactions.[61]

A persistent, delayed-type nickel reaction in intradermal test associated with a negative or papular patch test reaction has been found typical for so-called dermal nickel sensitivity.[80] The existence of this particular dermal sensitivity is not generally accepted.[7,39]

V. OPEN TESTING

Repeated open application testing has been employed for cosmetics utilizing lower (non-irritating) concentrations[81] and a corresponding method might be useful in studies on nickel allergy. A single 2.5 to 20% nickel chloride (in 95% alcohol) application on the forearm was shown to elicit positive reactions in nickel allergics, who always reacted to 20% nickel chloride applications; 40% also reacted to open 2.5% nickel chloride application.[61] For nickel, this open application method may, however, be less informative than, for example, for cosmetics. Intimate contact with nickel-containing objects is more typical.

VI. SYSTEMIC EXPOSURE

Oral nickel exposure has been used in cases where suspicion of the contribution by nickel ingestion has been elicited. With 2.5 mg of nickel sulfate, the majority (75%) of nickel-allergic pompholyx patients reacted with vesicular eruptions[25,26] (Chapter 14), but negative exposure results were reported recently[28,29] and the specificity of this reaction type may be questionable.[30] In any case, daily nickel intake in food is always much below 2.5 mg. Pompholyx is characterized by periodic symptoms, and thus the evaluation of appearing symptoms is difficult. The importance of oral nickel has been analyzed by employing nickel avoidance diets and oral nickel chelating agents, which were[82,83] or were not[29,84] found to be beneficial. Since specific T-cells can persist for several months at a former site of an allergic contact dermatitis,[85] it appears possible that allergen exposure via a different route may reactivate those cells in some cases.

VII. *IN VITRO* TESTING

Lymphocyte transformation testing produces relatively reliable results today[86-89] (Chapter 8). Methodological improvements have minimized toxic effects of nickel on cells, thus developing nonspecific reactions.[90] Still, this method can only be used in association with patch testing. Development work has to be done before it is useful for practical purposes.

In leukocyte migration tests nickel-protein complexes have shown significant inhibition of leukocyte migration.[91,92] Another *in vitro* test which might be worth developing is a leukocyte aggregation test. Blood of nickel-sensitive donors increased the number of leukocyte aggregates significantly when combined with nickel sulfate as compared with the influence by blood of nickel-negative donors.[93] No helpful *in vitro* testing is available for clinical studies of nickel sensitivity today.

VIII. DIMETHYLGLYOXIME SPOT TEST

Dimethylglyoxime spot testing can be used for detecting nickel in metal objects by applying one drop of 1% aqueous solution of dimethylglyoxime and another drop of 10% ammonium hydroxide on the surface of the metal.[39] Pink color is an expression of nickel content in the metal. The sensitivity of this test, even in optimal conditions, may, however, be too low for detecting the lower-level nickel content clinically detectable by immunological mechanisms in the skin of some nickel allergics.[39,94] Dimethylglyoxime spot testing is still useful in most cases where sources of nickel need to be verified.

IX. CONCLUSIONS

If a typical nickel-allergic contact dermatitis occurs and the nickel source is apparent, no skin test may be needed. However, any contact dermatitis may be affected by nickel allergy. If an untypically spread nickel-allergic contact dermatitis is suspected, and in all long-lasting hand dermatitis patch testing is indicated, reliable nickel testing is critical.

Clinical practice patch testing is usually done with 5% nickel sulfate in petrolatum. Testing with nickel sulfate or nickel chloride in aqueous solution as well as with dilution series may be useful if false-negative results are suspected.

The dermatologic history, clinical picture, and clinical course of the dermatitis, test results, and reactivity to topical or systemic exposure are considered when making the synthesis for the clinical diagnosis.

REFERENCES

1. North American Contact Dermatitis Group, Epidemiology of contact dermatitis in North America, *Arch. Dermatol.*, 108, 537, 1972.
2. **Fregert, S., Hjorth, N., Magnusson, B., Bandmann, H.-J., Calnan, C. D., Malten, K. E., Meneghini, C. L., Pirila, V., and Wilkinson, D. S.,** Epidemiology of contact dermatitis, *Trans. St. John's Hosp. Dermatol. Soc.*, 55, 17, 1969.
3. **Hjorth, S.,** Occupational dermatitis in a 10-year material, *Contact Dermatitis*, 4, 17, 1975.
4. **Magnusson, B., Blohm, S. G., Fregert, S., Hjorth, N., Hovding, G., Pirila, V., and Skog, E.,** Routine patch testing. IV, *Acta Derm. Venereol.*, 48, 110, 1968.
5. **Van der Burg, C. K. H., Bruynzeel, K. J. J., Vreeburg, von Blumberg, M. E., and Scheper, J.,** Hand eczema in hairdressers and nurses; a prospective study. I. Evaluation of atopy and nickel hypersensitivity at the start of the apprenticeship, *Contact Dermatitis*, 14, 275, 1986.
6. **Edman, B. and Moller, H.,** Trends and forcasts for standard allergens in a 12-year patch test material, *Contact Dermatitis*, 12, 87, 1982.

7. **Cronin, E.**, Metals, in *Contact Dermatitis*, Cronin, E., Ed., Churchill Livingstone, Edinburgh, 1980, 279.
8. **Menne, T., Borgen, O., and Green, A.**, Nickel allergy and hand dermatitis in a stratified sample of Danish population: an epidemiological study including statistic appendix, *Acta Derm. Venereol.*, 62, 35, 1982.
9. **Schubert, Berova, N., Czernielewski, A., Hegyi, E., Jirasek, L., Kohanka, V., Korossy, S., Michailov, P., Nebenfuhrer, L., Prater, E., Rothe, A., Rudzki, E., Straski, L., Suss, E., Tarnick, M., Temesvari, E., Ziegler, V., and Zschunke, E.**, Epidemiology of nickel allergy, *Contact Dermatitis*, 16, 122, 1987.
10. **Menne, T.**, The prevalence of nickel allergy among women, *Dermatosen*, 26, 23, 1978.
11. **Gawkrodger, D. J., Vestey, J. P., Wong, W.-K., and Buxton, B. K.**, Contact clinic survey of nickel sensitive subjects, *Contact Dermatitis*, 14, 165, 1986.
12. **Reiffers, J., Hunziker, N., Brun, R., and Vidmar, B.**, Sensibilisationes cutanees allergiques per communes, *Dermatologica*, 148, 285, 1974.
13. **Finn, O. A.**, Nickel sensitivity in infancy, *Contact Dermatitis Newsl.*, 16, 512, 1974.
14. **Ho, V. C. and Johnson, H. M.**, Nickel dermatitis in infants, *Contact Dermatitis*, 15, 270, 1986.
15. **Lejman, E., Stoudemayer, T., Grove, G., and Kligman, A. M.**, Age differences in poison ivy dermatitis, *Contact Dermatitis*, 11, 163, 1984.
16. **Fisher, A. A.**, Childhood allergic contact dermatitis, *Cutis*, 15, 635, 1975.
17. **Hjorth, N.**, Contact dermatitis in children, *Acta Derm. Venereol.*, 95, 36, 1981.
18. **Veien, N. K., Hatil, T., Just, O., and Norrholm, A.**, Contact dermatitis in children, *Contact Dermatitis*, 8, 373, 1982.
19. **Peltonen, L.**, Nickel sensitivity in the general population, *Contact Dermatitis*, 5, 27, 1979.
20. **Calnan, C. D.**, Nickel dermatitis, *Br. J. Dermatol.*, 68, 229, 1956.
21. **Wilson, H. T. H.**, Nickel dermatitis, *Practitioner*, 177, 303, 1956.
22. **Fisher, A. A.**, Nickel dermatitis at home and abroad, *Cutis*, 10, 719, 1973.
23. **Marcussen, P. V.**, Spread of nickel dermatitis, *Dermatologica*, 115, 596, 1957.
24. **Christensen, O. B. and Moller, H.**, Nickel allergy and hand eczema, *Contact Dermatitis*, 1, 129, 1975.
25. **Christensen, O. B., and Moller, H.**, External and internal exposure to the antigen in hand eczema of nickel allergics, *Contact Dermatitis*, 1, 136, 1975.
26. **Menne, T., and Thorboe, A.**, Nickel dermatitis-nickel excretion, *Contact Dermatitis*, 2, 353, 1976.
27. **Jordan,W. P. and King, S. E.**, Nickel feeding in nickel sensitive patients with hand eczema, *J. Am. Acad. Dermatol.*, 1, 506, 1979.
28. **Burrows, D., Creswoll, S., and Merrett, J. D.**, Nickel, hands and hip prostheses, *Br. J. Dermatol.*, 105, 437, 1981.
29. **Gawkrodger, D. J., Ell, G. S., and Hunter, J. A. A.**, Nickel dermatitis: the reaction to oral nickel challenge, *Br. J. Dermatol.*, 51, 141, 1985.
30. **Veien, N, K. T., Hellel Justesen, Norrholm, A.**, Oral challenge with metal salts. I. Vesicular, patch test negative eczema, *Contact Dermatitis*, 9, 402, 1983.
31. **Warin, P. P.**, Chronic urticaria, *Contact Dermatitis*, 8, 117, 1982.
32. **Veien, N. K., Christiansen, A. H., Svejgaard, E., and Kaaber, K.**, Antibodies against nickel-albumin in rabbits and man, *Contact Dermatitis*, 5, 378, 1979.
33. **Hjorth, N.**, Nickel vasculitis, *Contact Dermatitis*, 2, 256, 1976.
34. **Stoddar, T. J. C.**, Nickel sensitivity as a cause of infusion reactions, *Lancet*, 2, 741, 1966.
35. **Block, G. T. and Yeung, M.**, Asthma induced by nickel, *J. Am. Med. Assoc.*, 247, 1600, 1982.
36. **McConnel, L. H., Fink, J. N., and Schlueter, D. P.**, Asthma caused by nickel sensitivity, *Ann. Intern. Med.*, 78, 888, 1973.
37. **Osmundsen, P. E.**, Contact urticaria from nickel and plastic additives (butyl hudroxy toluene oleyl amide), *Contact Dermatitis*, 6, 452, 1980.
38. **Tosti, A., Melino, M., Labanca, M., and Ragnazzi**, Immediate hypersensitivity to nickel, *Contact Dermatitis*, 15, 95, 1986.
39. **Fisher, A. A.**, *Contact Dermatitis*, Lea & Febiger, Philadelphia, 1986.
40. **Wilkinson, D. S., Bandmann, H.-J., Calnan, C. D., Cronin, E., Fregert, S., Hjorth, N., Magnusson, B., Maibach, H. I., Malten, K. E., Meneghini, C. L., and Pirila, V.**, The role of contact allergy in hand eczema, *Trans. St. John's Hosp. Dermatol. Soc.*, 56, 19, 1970.
41. **Kieffer, M.**, Nickel sensitivity: relationship between history and patch test reaction, *Contact Dermatitis*, 5, 398, 1979.
42. **Menne, T.**, Nickel allergy — reliability of patch test evaluated in female twins, *Dermatosen*, 29, 156, 1981.
43. **Hemingway, J. D. and Molokhia, M. M.**, The dissolution of metallic nickel in artificial sweat, *Contact Dermatitis*, 16, 99, 1987.
44. **Wall, L. M. and Calnan, C. D.**, Occupational nickel dermatitis in the electroplating industry, *Contact Dermatitis*, 6, 414, 1980.

45. **Kalimo, K. and Lammintausta, K.,** 24 and 48 h allergen exposure in patch testing, *Contact Dermatitis,* 10, 25, 1984.
46. **Vandenburg, J. J. and Epstein, W. L.,** Experimental nickel-contact sensitization in man, *J. Invest. Dermatol.,* 41, 413, 1963.
47. **Cavalier, C., Foussereau, J., and Massin, M.,** Nickel allergy: analysis of metal clothing objects and patch testing to metal samples, *Contact Dermatitis,* 12, 65, 1985.
48. **Menne, T., Brandrup, E., Thestrup-Pedersen, K., Veien, N. K., Andersen, J. K., Yding, F., and Valeur, G.,** Patch test reactivity to nickel alloys, *Contact Dermatitis,* 16, 255, 1987.
49. **Fiscer, T., Fregert, S., Gruvberger, B., and Rystedt, I.,** Nickel release from ear piercing kits and earrings, *Contact Dermatitis,* 10, 39, 1984.
50. **Wahlberg, J. E.,** Petrolatum — a reliable vehicle for metal allergens?, *Contact Dermatis,* 6, 134, 1980.
51. **Van Ketel, W. G.,** Petrolatum again: an adequate vehicle in cases of metal allergy?, *Contact Dermatitis,* 5, 192, 1979.
52. **Fischer, T. and Maibach, H. I.,** Amount of nickel applied with a standard patch test, *Contact Dermatitis,* 11, 285, 1984.
53. **van Ketel, W. G.,** Patch testing with nickel sulphate in DMSO, *Contact Dermatitis,* 4, 167, 1978.
54. **Fischer, T. I. and Maibach, H. I.,** The thin layer rapid use epicutaneous test (TRUE-test). A new patch test method with high accuracy, *Br. J. Dermatol.,* 112, 63, 1985.
55. **Vaananen, A. and Hannuksela, M.,** Softisan — a new vehicle for patch testing, *Contact Dermatitis,* 14, 215, 1987.
56. **Marzulli, F. N. and Maibach, H. I.,** Further studies of effects of vehicles and elicitation concentration in experimental contact sensitization testing on humans, *Contact Dermatitis,* 6, 131, 1980.
57. **Goh, C. L. and Ckwok, S. F.,** The influence of temperature on the concentration homogenicity of patch test materials, *Contact Dermatitis,* 15, 231, 1986.
58. **Kalimo, K., Lammintausta, K., Teuho, J., Jansen, C., and Maki, J.,** Nickel penetration in allergic individuals bioavailability versus X-ray microanalysis detection, *Contact Dermatitis,* 12, 255, 1985.
59. **Rystedt, I. and Fischer, T.,** False positive follicula and irritant patch test reactions to metal salts, *Contact Dermatitis,* 12, 93, 1985.
60. **Dooms Goossens, A., Naert, C., and Chrispeels, M. I.,** Is a 5 percent nickel sulphate patch test concentration adequate?, *Contact Dermatitis,* 6, 232, 1980.
61. **Christensen, O. B. and Wall, L. M.,** Open, closed and intradermal testing in nickel allergy, *Contact Dermatitis,* 16, 21, 1987.
62. **Magnusson, B. and Hersle, K.,** Patch test methods. II. Regional variation of patch test responses, *Acta Derm. Venereol.,* 45, 257, 1965.
63. **Flannigan, F. A., Smith, R. E., and McCovern, J. P.,** Intraregional variation between contact irritant patch test sites, *Contact Dermatitis,* 10, 123, 1984.
64. **Hornstein, O. P. and Kienlein-Kletscka, B. M.,** Improvement of patch test allergen exposure by short term local pressure, *Dermatologica,* 165, 607, 1982.
65. **Gollhausen, R. and Kligman, A. M.,** Effects of pressure on contact dermatitis, *Am. J. Ind. Med.,* 8, 323, 1985.
66. **Mitchell, J. C.,** The angry back syndrome: eczema creates eczema, *Contact Dermatitis,* 1, 193, 1975.
67. **Bruynzeel, D. P. and Maibach, H. I.,** Excited skin syndrome (angry back), *Arch. Dermatol.,* 122, 323, 1986.
68. **Bjornberg, A.,** Skin reactions to primary irritants in patients with hand eczema, *Oscar Isaacsson's,* Goteborg, Sweden, 1968.
69. **Mitchell, J. C.,** Multiple positive patch test reactions, *Contact Dermatitis,* 3, 315, 1977.
70. **Lammintausta, K. and Maibach, H. I.,** Exogenous and endogenous factors in skin irritation, *Int. J. Dermatol.,* in press.
71. **Moller, H. and Svensson, A.,** Metal sensitivity: positive history but negative test indicates atopy, *Contact Dermatitis,* 14, 57, 1986.
72. **Uehara, M., Takahashi, C., and Ofuji, S.,** Pustular patch test reactions in atopic dermatitis, *Arch. Dermatol.,* 111, 154, 1975.
73. **Thorvaldsen, J. and Volden, G.,** PUVA-induced diminution of contact allergy and irritant skin reactions, *Clin. Exp. Dermatol.,* 5, 43, 1980.
74. **Kalimo, K., Koulu, L., and Jansen, C. T.,** Effect of a single UVB or PUVA exposure on immediate and delayed skin hypersensitivity reactions in humans, *Arch. Dermatol. Res.,* 85, 374, 1983.
75. **Fjellner, B. and Hagemark, B.,** Influence of ultraviolet light on itch and flare reactions in human skin induced by histamine and the histamine liberator comp 48/80, *Acta Derm. Venereol.,* 62, 137, 1982.
76. **Sjovall, P., Christensen, O. B., and Moller, H.,** Single exposure to UV irradiation and elicitation of human allergic contact dermatitis, *Acta Derm. Venereol.,* 65, 93, 1985.
77. **Bruze, M.,** Seasonal influence in routine patch test results, *Contact Dermatitis,* 14, 184, 1986.

78. **Marcussen, P. V.,** Comparison of intradermal test and patch test using nickel sulphate and formaldehyde, *J. Invest. Dermatol.,* 40, 263, 1963.
79. **Meneghini, C. and Angelini, G.,** Intradermal test in contact allergy to metals, *Acta Derm. Venereol.,* 59 (Suppl. 85), 123, 1979.
80. **Epstein, S.,** Epidermal and dermal reactions in a case of sensitivity to nickel, *J. Invest. Dermatol.,* 38, 37, 1962.
81. **Hannuksela, M. and Salto, H.,** The repeated open application test (ROAT), *Contact Dermatitis,* 14, 221, 1986.
82. **Kaaber, K., Sjolin, K. E., and Menne, T.,** Elbow eruptions in nickel and chromate dermatitis, *Contact Dermatitis,* 9, 213, 1983.
83. **Christensen, O. B. and Kristensen, M.,** Treatment with disulfiram in chronic nickel hand dermatitis, *Contact Dermatitis,* 8, 59, 1982.
84. **Burrows, D., Rogers, S., Beck, M., Kellet, J., McMaster, D., Merrett, D., and Ecdy, D. J.,** Treatment of nickel dermatitis with trientine, *Contact Dermatitis,* 15, 55, 1986.
85. **Scheper, R. J., von Blomberg, M., Oerrigter, Bruynzeel, D., van Dichter, A., and Vos, A.,** Induction of immunologic memory in the skin. Role of local T cell retention, *Clin. Exp. Immunol.,* 51, 141, 1983.
86. **Hutchinson, F., Raffle, E. J., and McLeod, T. M.,** The specificity of lymphocyte transformation in vitro by nickel salts in nickel sensitive subjects, *J. Invest. Dermatol.,* 58, 362, 1972.
87. **Millikan, L. E., Conway, F., and Foste, J. E.,** In vitro studies of contact hypersensitivity: lymphocyte transformation in nickel sensitivity, *J. Invest. Dermatol.,* 60, 88, 1973.
88. **Svejgaard, E., Morling, N., Vejgaard, A., and Veien, N. K.,** Lymphocyte transformation induced by nickel sulphate; an vitro study of subjects with and without a positive nickel patch test, *Acta Derm. Venereol.,* 58, 245, 1978.
89. **Silvennoinen-Kassinen, S.,** Lymphocyte transformation in nickel allergy; amplification of T-lymphocyte responses to nickel sulphate by macrophages in vitro, *Scand. J. Immunol.,* 12, 61, 1980.
90. **Pappas, A., Orfanos, C. E., and Bertram, R.,** Non-specific lymphocyte transformation in vitro by nickel sulphate, *J. Invest. Dermatol.,* 55, 198, 1970.
91. **Thulin, H.,** The leucocyte migration test in nickel contact dermatitis, *Acta Derm. Venereol.,* 56, 377, 1976.
92. **Mirza, A. M., Perera, M. G., Maecia, C. A., Dzjubynsky, O. G., and Bernstein, I. L.,** Leucocyte migration inhibition in nickel dermatitis, *Int. Arch. Allergy Appl. Immunol.,* 49, 782, 1975.
93. **MacLeod, T. M., Hutchinson, F., and Raffle, E. J.,** Leucocyte aggregation in subjects with nickel dermatitis, *Clin. Exp. Immunol.,* 26,, 528, 1976.
94. **Fischer, T., Fregert, S., Gruvberger, B., and Rystedt, I.,** Contact sensitivity to nickel in white gold, *Contact Dermatitis,* 10, 23, 1984.

Chapter 10

GENETIC ASPECTS OF NICKEL SENSITIZATION*

T. Menné and N. V. Holm

TABLE OF CONTENTS

I. Introduction 102

II. Individual Susceptibility to Sensitization 102

III. Genetic Markers Including Human Leukocyte Antigen (HLA), Blood, Serum, and Enzyme Group Determination 102

IV. Family Studies 103

V. Twin Studies 104

VI. Environmental Factors Evaluated in Twin Studies 105

VII. Conclusion 106

References 106

* Adapted from Menné, T. and Holm, N. V., Genetic susceptibility to human allergy contact sensitization, *Semin. Dermatol.*, 5, 301, 1986.

I. INTRODUCTION

The development of contact allergy requires both an allergen and presupposes a genetically susceptible individual. Research has mostly focused on exposure and evaluation of the sensitizing capacity of chemicals in animal and human experimental models. The importance of genetic factors for allergic contact sensitization has only been investigated to a limited extent.

Breeding experiments combined with sensitization tests have shown that contact sensitivity in inbred laboratory animals is genetically controlled.[1] In humans, the importance of genetic influence in the development of contact allergy may be evaluated by epidemiological studies including migration and association studies, family studies including segregation and linkage analysis, twin studies, or a combination of all these approaches. Because susceptibility to contact allergy does not seem to follow simple Mendelian inheritance with complete penetrance, all of the methods may provide useful information on the importance of genetic factors. Furthermore, studies including both experimental and accidental sensitization may give additional information. Because nickel sensitivity affects approximately 10% of the female population, most genetic studies in the field include a large number of nickel-sensitive persons.

II. INDIVIDUAL SUSCEPTIBILITY TO SENSITIZATION

If a group of individuals is exposed to a potent sensitizing chemical (for example poison ivy or epoxy resin), only a proportion is sensitized by the first exposure. Repeated exposures will sensitize still more, but probably not all individuals in the group. Sulzberger and Rostenberg[2] and Landsteiner et al.[3] exposed the same 82 individuals to both *p*-nitrosodimethylaniline (NDMA) and 2,4-dinitrochlorobenzene (DNCB). Of the sample, 37% became sensitized to NDMA and 50% to DNCB by the first exposure. Further, some persons who were highly susceptible to sensitization with one chemical showed little or no susceptibility to sensitization with the other chemical. This specificity of the ability to develop contact sensitization to different chemicals has also been demonstrated experimentally in animals.[4,5] Thus, in both human and animal studies, great individual variations in susceptibility to sensitization are found, and in certain individuals there seems to be a complete refractoriness to develop contact allergy.

III. GENETIC MARKERS INCLUDING HUMAN LEUKOCYTE ANTIGEN (HLA), BLOOD, SERUM, AND ENZYME GROUP DETERMINATION

When a disease and a genetic marker occur more frequently or rarely than expected by chance together in a population, the disease and genetic marker are said to be associated positively or negatively, respectively. This might indicate the importance of genetic factors in the etiology of the disease. In laboratory animals, genes controlling the delayed hypersensitivity reaction have been limited to those controlling the major histocompatibility antigens.[6] It must, however, be stressed that if an association between a genetic marker and a disease cannot be demonstrated in association studies, this does not exclude genetic factors to be of etiologic importance. Beckman et al.[7] studied the association between a total of 13 genetic marker systems (blood and serum groups and red cell enzyme types) in 195 patients with allergic contact dermatitis. A number of statistically significant associations were found. However, due to the many comparisons made, the final evaluation must await confirmative studies as mentioned by the authors. The HLA genes are located on the short arm of chromosome number 6. The HLA genes code for cell membrane components of importance for the immune response.[8,9]

Table 1
STUDIES OF THE HLA GENES IN ALLERGIC CONTACT SENSITIZATION

Studied HLA genes	Contact sensitizers	Study population	Number of patients	Associations: Statistical (significant)	Associations: Trends	Ref.
A, B	European standard	Eczema patients	149	A2, A3, A9, B7, B8, B28	—	10
A, B, C	Chromate	Bricklayers	50	—	—	11
A, B, C, D	Nickel	Eczema patients	36	—	Bw22	12
A, B	Nickel	Eczema patients	60	—	—	13
A, B	Nickel, chromate, formaldehyde, balsam of Peru,	Eczema patients	129	—	B7	14
A, B	Nickel, chromate	Eczema patients, families	Ni 39, Cr 16	B21 —	— —	15
A, B, C	"Para group"	Eczema patients	40	—	—	16
A, B	Nickel	Twins, eczema patients	112	—	—	17
A, B, L, DR	Nickel	Eczema patients	53	—	—	18
A, B, C, DR	Cobalt	Eczema patients	34	—	—	19
A, B, C, DR, MT	Nickel	Eczema patients	38	—	—	20
A, B, C, DR	Nickel	Eczema patients	56	—	—	21
A, B, DR	Nickel	Eczema patients	26	B35	—	22

Presently, four loci termed A, C, B, and D/DR are known. The HLA genes at each locus are alleles, and each therefore carries two allelic genes. The high number of allelic genes at each locus opens for extensive association studies and linkage analyses. Strong HLA associations have mostly been found in disease groups, where immunological abnormalities are involved, e.g., Reiter's disease and anchylosing spondylitis. The HLA-association studies in allergic contact dermatitis have generally been disappointing as no consistent statistically significant association or even any consistent trend has been established. These studies are summarized in Table 1.[10-22] In some of the studies, certain associations have been found, but none of them has been reproducible. The lack of association between the HLA genes and allergic contact sensitization does not exclude the importance of genetic factors. Hitherto unknown HLA genes might be associated with allergic contact dermatitis, there might be genetic heterogeneity in allergic contact dermatitis, and/or allergic contact dermatitis might not be associated or linked to the HLA region.

IV. FAMILY STUDIES

If genetic factors are of importance for the development of allergic contact sensitization, close relatives of patients manifesting this trait are expected to have a higher risk of developing contact allergy as compared to unrelated individuals.

However, close relatives also share a set of common environmental factors that might also explain an accumulation of affected family members. Therefore, casuistic reports of one or a few pedigrees demonstrating accumulation of contact allergy as, for example, the study of Kapoor-Pillaisette et al.[15] including three families with an accumulation of nickel allergy, cannot without caution be taken as an indication of genetic factors being of causal importance. The sampling frame of such families is unknown and they are reported only because of an ''interesting'' high accumulation of cases and, therefore, do not reflect the situation in the population in general. Similarly, warnings must be issued concerning the collection and interpretation of series of casuistic reports.

To obtain unbiased estimates of segregation parameters it is of utmost importance to have well-defined proband criteria.[23] Forsbeck et al.[29] implemented patch testing, including 23 contact allergens, to all first-degree relatives (parents, siblings, and children) of 94 probands, representing a consecutive series of patients with contact allergy and 2 groups of unrelated individuals. The frequency of persons reacting positively to the patch tests was higher among the relatives of the probands than among the control groups, supporting the hypothesis that the genetic constitution may be of importance for the development of delayed hypersensitivity reaction. Walker et al.[25] studied contact sensitization in parents and children in 99 families, with a total of 301 individuals. Children whose parents were sensitized to NDMA became sensitized at a significantly higher rate than children of nonsensitized parents. In families where both parents were DNCB sensitized, 90% of the children were sensitized as compared to only 53% where only one of the parents was sensitized. Further analyses of 295 individuals with nonequivocal test results showed a statistically significant interaction between DNCB and NDMA sensitivity, supporting the findings of Sulzberger and Rostenberg[2] that individuals sensitive to one allergen more easily become sensitized to others.

Mendelian inheritance with complete penetrance could not be demonstrated in either of these studies. One of the shortcomings of family studies of traits or diseases without a complete genetic determination is the difference in exposure time to different substances for the parents and their offspring, as environmental exposure is then partly responsible for the manifestation. The possible genetic influence may then be underestimated because the offspring have had a shorter period of exposure. Furthermore, the exposure pattern to common sensitizers in the daily environment may change from one generation to the next. These reservations make it difficult to achieve conclusive evidence by studying accidentally occurring contact sensitivities in families.

A different approach to the problem has been made by Silvennoinen-Kassinen.[26] In this study, 10 families, including 14 nickel sensitives and 46 nonnickel sensitives, were tested with standard epicutaneous test and *in vitro* test (lymphocyte-transformation test) for nickel sensitivity. None of the family members demonstrated a subclinical nickel sensitivity which could only be recognized by an *in vitro* test.

V. TWIN STUDIES

If neither a Mendelian inheritance nor linkage to genetic markers can be demonstrated from family studies, it is unknown to what extent genetic factors contribute to the liability of developing contact allergy. Twin studies, which focus on the risk in cotwins of monozygotic (MZ) and dizygotic (DZ) probands (the classical twin method), are then the only study design existing to give a solid answer to this problem, as MZ twin partners are genetically identical and DZ twin partners correspond to oridinary siblings and have, therefore, on the average 50% of their genes in common. Any difference between MZ twins must then be caused by environmental factors, whereas the differences between two DZ partners may be caused by environmental and/or genetic factors. The classical twin method for evaluating the degree of genetic determination for a certain condition or disease is based on a comparison of concordance rates for a representative sample of MZ and DZ twin pairs.[27] If the intrapair difference of environment is considered to be the same for MZ and DZ twins, a statistically significant difference in the concordance rates between the two groups of twins can be taken as evidence that genetic factors are of importance for the trait or disease in question. However, to obtain unbiased estimates of the relative importance of genetic factors, the twin study requires well-defined and described proband criteria and detailed analyses of the validity of the study.[28] Included in the evaluation of the validity is that the zygosity classification should be made accurately and without consideration of disease status and concordance/discordance.

Forsbeck et al.[29] patch tested 101 male and female twin pairs from the Swedish Twin Register with 23 common haptens;[32] (16%) of the 202 individuals tested had a positive reaction to at least one of the substances. The tendency in the results was a higher but insignificant concordance rate among the MZ pairs as compared to the DZ pairs. However, the number of twin pairs with positive reaction in one or both twin partners was too small for any definite conclusion to be made. All of the 202 twins were evaluated with DNCB sensitization; 96 (48%) of the 202 were sensitized based on the spontaneous flare-up reaction. Retesting was not made, and a number of positive reactions might then have been lost. No difference in the sensitization pattern between the MZ and DZ pairs was observed, in contrast to the study of Walker et al.[25]

The genetic predisposition of nickel allergy has been evaluated on a population based on twin material obtained from the Danish Twin Register.[31,31] Basically, 1546 female twin pairs, born between 1906 and 1930, were included. Based on a mailed questionnaire, twins from 115 pairs where one or both partners had possible history of nickel allergy were further investigated with clinical examination and patch testing. A statistically significant difference between the concordance rates for nickel allergy among the MZ and DZ pairs was found. Based on the prevalence rates of nickel allergy in the Danish female population and the proband concordance rates, the heritability was calculated according to the method of Smith.[32] Heritability is defined as the proportion of the total phenotypic variance (genetic and non-genetic) which is due to additive genetic variance. In other terms, heritability expresses the relative importance of genetic factors compared to environmental factors.

Based on the study of nickel-sensitive female twins, the heritability for nickel allergy was calculated to be about 60%. Thus, for a medium strong sensitizer such as nickel, genetic factors seem not to be a negligible factor for the development of allergic contact sensitization.

VI. ENVIRONMENTAL FACTORS EVALUATED IN TWIN STUDIES

The principle of the classical twin method can also be used to evaluate the causal relation between different diseases,[28] e.g., nickel allergy and hand eczema. In 30 MZ twin pairs where one or both had nickel allergy and/or hand eczema, the hand eczema predominantly occurred among those with nickel allergy.[33] If hand eczema in this group of nickel-sensitive twins was due to genetic factors, the occurrence of hand eczema would have been independent of the nickel allergy. Furthermore, in case genetic factors seem to be of more limited importance, twin studies of discordant MZ twin pairs (the twin control method) can elucidate the weight of specific environmental risk factors for the development of a given disease.[28] This study design is unique in the sense that no other design exists in humans which can eliminate genetic factors as confounders. In the Danish twin study of nickel allergy[30] this design has been applied to evaluate the possibility of prophylaxis. Only the twin partner with a history of dermatitis in relation to cheap jewelry, metal clasps, etc. had a positive patch test to nickel. The twin partner without such a history did not react to patch testing with $NiSO_4$ 5% in petrolatum. Therefore, even if the individual has a genetic predisposition, nickel sensitization only takes place if there is a relative massive cutaneous exposure.

VII. CONCLUSION

The available evidence on genetic factors in allergic contact sensitization is sparse and partly contradictive, but, in general, most studies find genetic factors of some importance. It is likely that this genetic predisposition is directed against specific haptens since the highest risk to close relatives of contact-sensitized patients was obtained from family and twin studies dealing with a single allergen only. None of the genetic markers investigated to date has revealed any consistent association to allergic contact sensitization. The studies of contact allergy in groups of individuals indicate that some persons are more easily sensitized than others, but the total number sensitized in a group or population depends on the degree of cutaneous exposure, as demonstrated from the study of discordant MZ twins. The knowledge and study of genetic factors in allergic contact sensitization is therefore of major importance with respect to general prophylactic measures of allergic contact dermatitis. Further twin and family studies concentrating on one or a few allergens may provide important new information leading to a better understanding of human allergic contact sensitization, with the application of DNA recombinant techniques perhaps at present being the most promising.

REFERENCES

1. **Parker, D., Sommer, G., and Turk, J. L.,** Variation in guinea pig responsiveness, *Cell. Immunol.*, 18, 233, 1975.
2. **Sulzberger, M. B. and Rostenberg, A., Jr.,** Acquired specific supersensitivity (allergy) to simple chemicals, *J. Immunol.*, 36, 17, 1939.
3. **Landsteiner, K., Rostenberg, A., Jr., and Sulzberger, M. B.,** Individual differences in susceptibility to eczematous sensitization with simple chemical substances, *J. Invest. Dermatol.*, 2, 25, 1939.
4. **Chase, M.,** Inheritance in guinea pigs of the susceptibility to skin sensitization with simple chemical compounds, *J. Exp. Med.*, 73, 711, 1941.
5. **Polak, L., Barnes, J. M., and Turk, J. L.,** The genetic control of contact sensitization to inorganic metal compounds in guinea pigs, *Immunology*, 14, 707, 1968.
6. **McDevitt, H. O. and Benacerrat, B.,** Genetic control of specific immune response, *Adv. Immunol.*, 11, 31, 1969.
7. **Beckman, G., Bekman, L., Cedergren, B., Göransson, K., and Lidén, S.,** Blood groups, serum groups and red cell enzyme types in allergic contact dermatitis, *Hum. Hered.*, 31, 54, 1981.
8. **Svejgaard, A., Hauge, M., Jersild, C., Platz, P., Ryder, L. P., Nielsen, L.S., and Thomsen, M.,** *The HLA System. An Introductory Survey,* S. Karger, Basel, 1979.
9. **Vogel, F. and Motulsky, A. G.,** *Human Genetics. Problems and Approaches,* Springer-Verlag, New York, 1982.
10. **Lidén, S., Beckman, L., Cedergren, B., Göransson, K., and Nyquist, H.,** HLA antigens in allergic contact dermatitis, *Acta Derm. Venereol. Supply.* 79, 53, 1978.
11. **Roupe, G., Rydberg, L., and Swanbeck, G.,** HLA-antigens and contact hypersensitivity, *J. Invest. Dermatol.*, 72, 131, 1979.
12. **Silvennoinen-Kassinen, S., Ilonen, J., Tiilikainen, A., and Karvonen, J.,** No significant association between HLA and nickel contact sensitivity, *Tissue Antigens,* 14, 459, 1979.
13. **Dumont-Fruytier, M., Van Neste, D., De Bruyere, M., Tennstedt, D., and Lachapelle, J. M.,** Nickel contact sensitivity in women and HLA antigens, *Arch. Dermatol. Res.*, 269, 205, 1980.
14. **Lidén, S., Beckman, L., Cedergren, B., Groth, O., Görensson, K., and Wählby, L.,** Lack of association between allergic contact dermatitis and HLA antigens of the A and B series, *Acta Derm. Venereol.*, 61, 155, 1981.
15. **Kapoo-Pillarisetti, A., Mowbray, J. F., Brostoff, J., and Conin, E. A.,** HLA dependence of sensitivity to nickel and chromium, *Tissue Antigens,* 17, 261, 1981.
16. **Valsecchi, R., Bontempelli, M., Vicari, O., Scudeller, G., and Cainelli, T.,** HLA antigens and contact sensitivity, *Arch. Dermatol.*, 118, 533, 1982.
17. **Hansen, H., Menné, T., and Larsen, S. O.,** HLA antigens in nickel sensitive female twins, *Tissue Antigens,* 19, 305, 1982.

18. **Braathen, L. R., Haavelsrud, O., and Thorsby, E.,** HLA antigens in patients with allergic contact sensitivity to nickel, *Arch. Dermatol. Res.,* 275, 355, 1983.
19. **Fischer, T., Rystedt, I., Säfwenberg, J., and Egle, I.,** HLA A, B, C and DR antigens in individuals with sensitivity to cobalt, *Acta Derm. Venereol.,* 64, 121, 1984.
20. **Karvonen, J., Silvennoinen-Kassinen, S., Ilonen, J., Jakkula, H., and Tiilikainen, A.,** HLA antigens in nickel allergy, *Ann. Clin. Res.,* 16, 211, 1984.
21. **White, S. I., Friedmann, P. S., and Stratton, A.,** HLA antigens and Langerhans cell density in contact dermatitis, *Br. J. Dermatol.,* 115, 447, 1986.
22. **Walton, S., Keczhes, K., Learoyd, P. A., and Rajah, S. M.,** HLA-A,-B and DR antigens in nickel sensitive females, *Clin. Exp. Dermatol.,* 11, 636, 1986.
23. **Morton, N. E.,** Genetic tests under incomplete ascertainment, *Am. J. Hum. Genet,* 11, 1, 1959.
24. **Forsbeck, M., Skog, E., and Ytterborn, K. H.,** Allergic diseases among relatives of patients with allergic contact dermatitis, *Acta Derm. Venereol.,* 51, 123, 1971.
25. **Walker, F. B., Smith, P. D.,and Maibach, H. I.,** Genetic factors in human allergic contact dermatitis, *Int. Arch. Allergy,* 32, 453, 1967.
26. **Silvennoinen-Kassinen, S.,** The specificity of nickel sulphate reaction in vitro: a family study and a study of chromium-allergic subjects, *Scand. J. Immunol.,* 13, 231, 1981.
27. **Harvald, B. and Hauge, M.,** Hereditary factors elucidated by twin studies, in *Genetics and the Epidemiology of Chronic Diseases,* Neel, J. V., Shaw, M. G., and Schull, W. J., Eds., No. 1163, U.S. Public Health Service, Washington, D.C., 1965, 61.
28. **Holm. N. V.,** Population-based twin studies: an irreplaceable tool in etiological cancer research, in *Familial Cancer,* 1st Int. Res. Conf., Müller, H. J. and Weber, W., Eds., S. Karger, Basel, 1985, 208.
29. **Forsbeck, M., Skog, E., and Ytterborn, K. H.,** Delayed type of allergy and atopic disease among twins, *Acta Derm. Venereol.,* 48, 192, 1968.
30. **Menné, T. and Holm N. V.,** Nickel Allergy in a female twin population, *Int. J. Dermatol.,* 22, 22, 1983.
31. **Hauge, M.,** The Danish Twin Register, in *Prospective Longitudinal Research: An Empirical Basis for the Primary Prevention of Psychosocial Disorders,* Mednick, S. A., Baert, A. E., and Bachmann, B. P., Eds., Oxford University Press, Oxford, 1981, 217.
32. **Smith, C.,** Concordance in twins: methods and interpretations, *Am. J. Hum. Genet.,* 26, 454, 1974.
33. **Menné, T. and Holm, N. V.,** Hand eczema in nickel sensitive female twins. Genetic predisposition and environmental factors, *Contact Dermatitis,* 9, 289, 1983.

Chapter 11

EPIDEMIOLOGY OF NICKEL DERMATITIS

T. Menné, J. Christophersen, and A. Green

TABLE OF CONTENTS

I. Introduction 110

II. History 110

III. Patch Test Studies 110

IV. Population Studies 111

V. Conclusion 114

References 114

I. INTRODUCTION

Epidemiology is the scientific discipline used when describing disease occurrence in a population and assessing the nature, as well as the distribution, of risk factors. Thus, epidemiologic research provides the basis for prevention of disease and planning of health care. Next to poison ivy, nickel is the most common cause of allergic contact sensitization, mainly due to nickel alloys worn in direct skin contact. Accordingly, research in the epidemiology of nickel sensitization and dermatitis is essential for understanding, treating, and preventing this condition at the individual patient level as well as in the population at large.

II. HISTORY

Nickel dermatitis was clinically recognized as "Das Galvanizierekzem" in 1889.[1] Not until 1925 was nickel proven by patch testing to be the etiologic factor for nickel itch in the electroplating industry.[2]

Occupational nickel dermatitis was common in the 1920s and 1930s,[3-6] while consumer nickel dermatitis was first reported in the early 1930s. Bonnevie recognized nickel sensitization as a large-scale consumer problem in 1936.[10] Among 800 consecutive male and female patients, 45 were patch test positive to nickel. It was found that 33 of the 40 nickel-sensitive women had been sensitized from nickel-plated suspenders.

III. PATCH TEST STUDIES

Most knowledge on the epidemiology of allergic contact sensitization derives from clinical patch test studies. Collection of data from consecutive patch-tested patients is an easy and inexpensive method of monitoring the sensitization pattern in a population. Longitudinal data of clinical patch test results from Copenhagen and Malmö, Sweden indicate that sensitization is a steadily increasing problem, particularly in the female population.[11,12] At St. John's Hospital in London, the percentage of nickel-positive results rose from 12 to 21% among the females tested in the decade 1967 to 1976. The percentage of positive males fluctuated between 1 and 4% without a significant rise.[13] Results from the patch test clinic at Gentofte Hospital, Copenhagen from 1973 to 1986 are given in Figure 1. The same pattern as observed in other centers is illustrated: an increasing number of positives in females and a constant number in males.[14]

The epidemiology of nickel allergy has been extensively investigated and patch test results are available from a number of countries (Table 1).[15-25] Most studies confirm the known pattern, with a predominance of female patients due to exposure to costume jewelry. However, exceptions do occur. In Kuwait,[16] for instance, nickel allergy was most common in males, and in Nigeria[23] and Japan[21] it was equally frequent between the two sexes. Local dressing habits and differences in occupational nickel exposure might influence the results.

In Nigeria,[23] cheap nickel-plated wristwatches and watchbands from Hong Kong are the main source of nickel sensitization in men. The hot and humid climate increases nickel corrosion from nickel-plated articles and nickel alloys. If the industry continues to use inappropriate surface treatment for metal items designed to be used for prolonged skin contact, an increasing number of nickel-sensitized persons can be expected, particularly in the tropical areas of the world.

In the 1950s,[26,27] most patients were sensitized from suspenders, with the mean age for sensitization being 30 to 35 years. The mean age for sensitization has gradually declined in the last decades, and today most females develop their contact sensitivity to nickel as teenagers. This early sensitization probably raises the risk of subsequent occupational nickel hand eczema (Chapter 12).

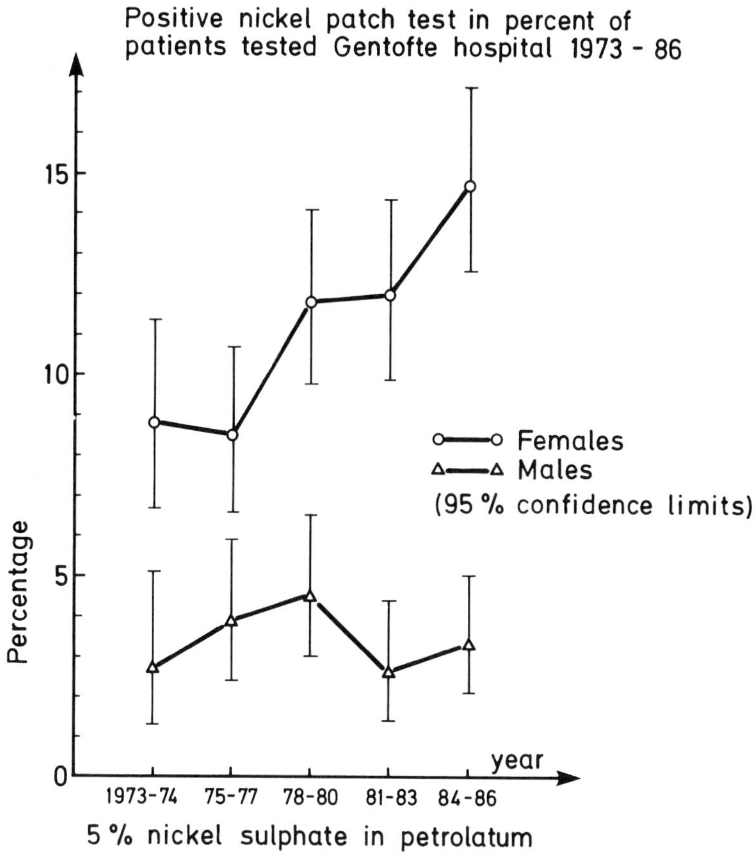

FIGURE 1. Positive nickel patch test in percent of patients tested, Gentofte Hospital, 1973 to 1986.

As data are often based on badly defined populations and imprecise diagnostic criteria, comparisons of results between centers must be done with caution. A number of variables such as sex, age, occupation, hand eczema, leg ulcer, and atopy are known to influence the sensitization pattern in a population. Patch test results from 2166 consecutive patients in the Danish Contact Dermatitis Group[26] were evaluated with respect to these variables using a computerized logistic regression model. This analysis showed the risk of being nickel sensitized to be associated with young age, female sex, and hand eczema and other types of eczema, whereas atopy decreased the risk of being nickel sensitized. No systemic occupational relation was recognized in this study. As the composition of patient materials varies with respect to the mentioned variables, comparisons can be misleading if these points are not taken into account. Another bias associated with clinical patch test results is selection with respect to the indication for patch testing. Some centers patch test all cases of eczema, while others only patch test those with a high degree of clinical suspicion. Standardized criteria for patch testing are only applicable to a certain degree. A number of centers have computerized their patch test results. Introduction of standardized statistical programs for evaluation of the results will make clinical patch test results more reliable parameters of preventive measures.

IV. POPULATION STUDIES

Population studies have disclosed an unexpected high prevalence of nickel allergy in the

Table 1
CLINICAL PATCH TEST STUDIES ON NICKEL ALLERGY

Study	Number of patients tested			Percentage positive to nickel		
	M	F	Total	M	F	Total
Fregert et al. Europe[15] 1969	2,039	2,786	4,825	1.8	10.2	6.7
Kanan Kuwait[16] 1969	—	—	389	24[a]	10[a]	34[a]
Oleffe et al. Belgium[17] 1972	184	116	300	3.8	18.1	9.3
NACDG U.S.[18] 1973	509	691	1,200	5.6	15.0	11.0
Husain Scotland[19] 1977	603	709	1,312	5.0	26.0	16.0
Moriearty et al. Brazil[20] 1978	271	265	536	2.6	11.7	7.1
Sugai et al. Japan[21] 1979	291	447	738	4.8	4.3	4.5
Hammershøy Denmark[22] 1980	1,451	1,774	3,225	—	—	6.4
Olumide Nigeria[23] 1985	230	223	453	11.0	12.4	11.7
Angelini et al. Italy[24] 1986	3,587	3,483	7.070	4.6	20.0	12.0
Schubert et al. Eastern Europe[25] 1987	913	1,487	2,400	2.1	10.5	7.3

Note: M = male; F = female.

[a] Total number.

general population. Before these studies were undertaken it was assumed that clinical patch test studies gave a reliable estimate of the problem in general.

The prevalence of sensitization is given by the proportion of sensitized persons at a given point of time in the total population at risk. Prevalence data are obtained by cross-sectional studies of populations or population groups and provide an estimate of the numerical size of the current disease problem. Prevalence data for nickel allergy are summarized in Table 2. The outcome of the studies are rather uniform, with a prevalence among females between 7 to 10%. The different studies must be compared with caution because the investigated groups are more or less selected and the age structure differs among the studies. School-children are included in three of the studies and exhibit the same high prevalence rate of nickel sensitization as adults, most likely indicating an increasing disease problem over time. In the young age groups ear piercing is the main reason for nickel sensitization.

Table 2
PREVALENCE OF NICKEL ALLERGY IN DIFFERENT POPULATION GROUPS

Author	Study population	Number of persons investigated			Prevalence (%)	
		M	F	Total	M	F
Menné[29] 1978	Female nondermatological inpatients	—	213	213	—	9.4[a]
Prystowsky et al.[30] 1979	Paid adult volunteers	460	698	1,158	0.9	9.0
Peltonen[31] 1979	General population	478	502	980	0.8	8.0
Kieffer[32] 1979	Veterinary students	247	168	415	2.8	9.8
Magnusson and Möller[33] 1979	Orthopedic patients	106	168	274	1.0	10.0
Menné et al.[34] 1982	General population	—	1,976	1,976	—	14.5[a]
Boss and Menné[35] 1982	Hairdressers (school)	—	53	53	—	20.0
Menné et al.[36] 1983	Female twins	—	1,546	1,546	—	9.6
Larsson-Stymme and Widström[37] 1985	Schoolgirls	—	960	960	—	9.0
Schmiel[38] 1985	Surgical inpatients and hospital staff	244	259	503	1.6	7.7
von Hums[39] 1986	Schoolchildren	135	131	266	—	7.0
van der Burg et al.[40] 1986	Nursing (school)	29	188	217	—	13.0
	Hairdressers (school)	12	74	86	17.0	27.0

Note: M = male; F = female.

[a] Nickel allergy diagnosed by history only.

Young females starting training as hairdressers have a particularly high prevalence of nickel sensitization as compared to other groups and the general population.[35,40] This might be explained by social factors (Chapter 18), habits of costume jewelry use, and age at first ear piercing. This high preoccupational nickel sensitization in hairdressers is of relevance to the problem of hand eczema in hairdressers (Chapter 12).

As previously stated, the prevalence proportion of nickel-sensitized individuals provides a cross-sectional picture of the size of this particular health problem at a given point in time. An alternative measure, the incidence rate, provides a measure of the dynamics of the health problem, i.e., how often the disease in question develops as a population is followed over time. The incidence rate is defined as the number of new cases developed within a defined period divided by the number of persons (or more precisely, the number of persons' years at risk) observed during this period.

Incidence data are obtained from following populations (fixed cohorts or, alternatively, so-called dynamic populations) over time with regard to the development of new cases. The cohort studies represent the classical epidemiological designs applied, for example, when assessing the effect of cigarette smoking on the development of lung cancer and cardiovascular disease. Incidence studies based on observations in dynamic populations represent the more traditional design by which incidence rates of diseases, e.g., cancers, are obtained.

So far, the incidence of nickel sensitization has not been studied systematically by means of these designs. However, since it is assumed that the condition is not associated with excess mortality, incidence data may to a certain extent be obtained from prevalence (cross-sectional) studies, in which the individuals investigated comprise a series of retrospectively defined cohorts.

This design option has been applied to a Danish female population sample including 2552 individuals stratified with respect to age, geographical distribution, and occupational structure.[34] The study confirms that nickel allergy is an increasing problem, particularly in the young age groups. In the analysis, stratification with respect to known important variables has been performed, which makes reliable comparisons possible with future studies.

V. CONCLUSION

Nickel is the most common man-made contact sensitizer, and nickel sensitization and dermatitis appear to be an increasing public health problem. Undoubtedly, industry will develop nickel alloys which are safe with respect to sensitizations. Furthermore, nickel sensitization is to a large extent preventable by reducing occupational exposures and changing social habits like the wearing of certain clothing and jewelry.

Dermatologists are currently obliged to evaluate the efforts made to reduce the problem of nickel sensitization by using appropriate epidemiological methods. However, in order to adequately handle all the methodological problems associated with epidemiological research, collaboration with statisticians and epidemiologists is necessary.[41]

REFERENCES

1. **Blasko, A.,** Die Berufsdermatosen der Arbeiter. Das Galvanizierekzem, *Dtsch. Med. Wochenschr.,* 15, 925, 1889.
2. **Schittenhelm, A. and Stockinger, W.,** Über die Idiosynkrasie gegen Nickel (Nickelkrätze) und ihre Beziehung zur Anaphylaxie, *Z. Gesamte Exp. Med.,* 45, 58, 1925.
3. **Bulmer, F. M. R.,** Studies in the control and treatment of "nickel rash", *J. Ind. Hyg. Toxicol.,* 8, 517, 1926.
4. **Jadahssohn, W. and Schaaf, F.,** Über die Häufigkeit des Vorkommens von Nickelekzem, *Arch. Dermatol. Syphilol.,* 157, 572, 1929.
5. **Du Bois, C.,** La dermatite du nickel, *Schweiz. Med. Wochenschr.,* 12, 278, 1931.
6. **Wedroff, N.,** Über Ekzeme bei Vernicklern, *Arch. Gewerbepathol. Gewerbehyg.,* 6, 179, 1935.
7. **McAlester, A. W., Jr. and McAlester, A. W., III,** Nickel sensitization from white gold spectacle frames, *Am. J. Ophthalmol.,* 14, 925, 1931.
8. **Goldman, L.,** Nickel eczema, *Arch. Dermatol. Syphilol.,* 28, 688, 1933.
9. **Foster, P. D. and Ball, F. I.,** Sources of nickel eczema, *Arch. Dermatol. Syphilol.,* 31, 461, 1935.
10. **Bonnevie, P.,** Der klinische Wert der Ekzemproben, und der Nickelidiosynkrasie erläutert, *Acta Derm. Venereol.,* 17, 376, 1936.
11. **Marcussen, P. V.,** Eczematous allergy to metals, *Acta Allergol.,* 17, 311, 1962.
12. **Edman, B. and Möller, H.,** Trends and forecasts for standard allergens in a 12-year patch test material, *Contact Dermatitis,* 8, 95, 1982.
13. **Cronin, E.,** *Contact Dermatitis,* Churchill Livingston, London, 1980, 345.
14. **Menné, T.,** unpublished, 1987.
15. **Fregert, S., Hjorth, N., Magnusson, B., Bandman, H.-J., Calnan, C. D., Cronin E., Malten, K., Meneghini, C. L., Pirilä, V., and Wilkinson, D. S.,** Epidemiology of contact dermatitis, *Trans. St. Johns Hosp. Dermatol. Soc.,* 55, 17, 1969.
16. **Kanan, M. W.,** Contact dermatitis in Kuwait, *J. Kuwait Med. Assoc.,* 3, 129, 1968.
17. **Oleffe, J., Nopp-Oger, M. J., and Achten, G.,** Batterie eruopéenne de tests epicutariés — bilan de 300 observations, *Berufs Dermatosen,* 20, 209, 1972.
18. NACDG, Epidemiology of contact dermatitis in North America 1972, *Arch. Dermatol.,* 108, 537, 1973.
19. **Husain, S. L.,** Contact dermatitis in the West of Scotland, *Contact Dermatitis,* 3, 327, 1977.
20. **Moriearty, P. L., Pereira, C., and Guimaraes, N. A.,** Contact dermatitis in Salvador, Brazil, *Contact Dermatitis,* 4, 185, 1978.
21. **Sugai, T., Takagi, T., Yamamoto, S., and Takahashi, Y.,** Age distribution of the incidence of contact sensitivity to standard allergens, *Contact Dermatitis,* 5, 383, 1979.

22. **Hammershøy, O.**, Standard patch test results in 3,225 consecutive Danish patients from 1973 to 1977, *Contact Dermatitis,* 6, 263, 1980.
23. **Olumide, Y. M.**, Contact dermatitis in Nigeria, *Contact Dermatitis,* 12, 241, 1985.
24. **Angelini, G., Vena, G. A., Fiordalisi, F., Giglio, G., and Meneghini, C. L.**, Allergia da contatto al nickel. Rilievi epidemiologici e clinici, *G. Ital. Dermatol. Venerol.,* 121, 121, 1986.
25. **Schubert, H., Berova, N., Czernielewski, A., Hegyl, E., Jirásek, L., Kohánka, V., Korossy, S., Michailov, P., Nebenführer, L., Prater, E., Rothe, A., Rudzki, E., Stranski, L., Süss, E., Tarnick, M., Temesvári, E., Ziegler, V., and Zschunke, E.**, Epidemiology of nickel allergy, *Contact Dermatitis,* 16, 122, 1987.
26. **Calnan, C. D.**, Nickel dermatitis, *Br. J. Dermatol.,* 68, 229, 1956.
27. **Wilson, H. T. H.**, Nickel dermatitis, *Practitioner,* 177, 303, 1956.
28. **Christophersen, J., Menné, T., and Tanghøj, P.**, Evaluation of patch test results in a logistic regression model, unpublished.
29. **Menné, T.**, The prevalence of nickel allergy among women, *Berufs Dermatosen,* 26, 123, 1978.
30. **Prystowsky, S. D., Allen, A. M., Smith, R. W., Nonomura, J H., Odom, R. B., and Akers, W. A.**, Allergic contact hypersensitivity to nickel, neomycin, ethylenediamine and benzocaine, *Arch. Dermatol.,* 115, 959, 1979.
31. **Peltonen, L.**, Nickel sensitivity in the general population, *Contact Dermatitis,* 5, 27, 1979.
32. **Kieffer, M.**, Nickel sensitivity: relationship between history and patch test reaction, *Contact Dermatitis,* 5, 398, 1979.
33. **Magnusson, B. and Möller, H.**, Contact allergy without skin disease, *Acta Derm. Venereol.,* 59 (Suppl. 89), 113, 1979.
34. **Menné, T., Borgan, Ø., and Green, A.**, Nickel allergy and hand dermatitis in a stratified sample of the Danish female population: an epidemiological study including a statistic appendix, *Acta Derm Venereol.,* 62, 35, 1982.
35. **Boss, A. and Menné, T.**, Nickel sensitization from ear piercing, *Contact Dermatitis,* 8, 27, 1982.
36. **Menné, T. and Holm, N. V.**, Nickel allergy in a female twin population, *Int. J. Dermatol.,* 22, 22, 1983.
37. **Larsson-Stymme, B. and Widström, L.**, Ear piercing — a cause of nickel allergy in schoolgirls?, *Contact Dermatitis,* 13, 289, 1985.
38. **von Schmiel, G.**, Häufigkeit von Nickel-Kontaktallergien am unausgewählten Patientengut im Raum München, *Berufs Dermatosen,* 33, 92, 1985.
39. **Hums, R.**, Nickelexposition — Nickelsensibilisierung, *Dermatol. Monatsschr.,* 172, 697, 1986.
40. **van der Burg, C. K. H., Brunzeel, D. P., Vreeburg, K. J. J., von Blomberg, B. M. E., and Scheper, R. J.**, Hand eczema in hairdressers and nurses: a prospective study. I. Evaluation of atopy and nickel hypersensitivity at the start of apprenticeship, *Contact Dermatitis,* 14, 275, 1986.
41. **Miettinen, O. S. and Cook, E. F.**, Confounding. Essence and detection, *Am. J. Epidemiol.,* 114, 593, 1981.

Chapter 12

OCCUPATIONAL NICKEL DERMATITIS

Torkel Fischer

TABLE OF CONTENTS

I. Introduction 118

II. Nickel Exposure 118
- A. Nickel Manufacturing 118
- B. Occupational Nickel Exposure 118
  - 1. Metal Manufacturing 118
  - 2. Cement 118
  - 3. Detergents 119
  - 4. Hairdressing 119
  - 5. Food and Water 119
  - 6. Coins 119
  - 7. Other 119
- C. Nickel in Intimate Contact with the Skin 119

III. Background of Allergic Nickel Dermatitis 120
- A. Sensitizing and Eliciting Dose of Nickel Dermatitis 120
- B. The Sensitized Patient 121
- C. Clinical Expression of Nickel Hand Dermatitis 121
- D. Pompholyx 122
- E. Contact Urticaria 122

IV. Association between Nickel-Sensitization Hand Eczema and Work 122
- A. Nickel Manufacturing 123
- B. Nickel Plating 123
- C. Metal manufacturing — Cutting Fluids 124
- D. Cement and the Building and Construction Industry 124
- E. Hairdressing 124
- F. Home Work, Cleaning, and Hospital Wet Work 125
- G. Business and Coin Contacts 125
- H. Rare and Curious Reports of Occupational Nickel Dermatitis 126

V. Prevention 126

References 127

I. INTRODUCTION

Nickel is one of the most frequent contact sensitizers worldwide. With rare exceptions, nickel allergy is more common in women than men; the prevalence in women is 5 to 15% and in men 0.5 to 1%. Most young women are already sensitized by their early teens, and maximal sensitization is found in ages below 30 years.[1-19]

Although the majority of nickel sensitivity derives from nonoccupational sources, work-related nickel dermatitis is a predominant diagnosis in reports on permanent disability due to skin diseases.[7,20-31]

II. NICKEL EXPOSURE

We are continuously exposed to small amounts of nickel found in our home and working environment.[32,33]

A. Nickel Manufacturing

Nickel is processed from sulfide or oxide ores via complex concentration, roasting, smelting, converting, leaching processes, and final steps of refining, either electrolytic or via the nickel carbonyl process. The different processes expose the workers to nickel sulfide, nickel sulfate, nickel oxides, and metallic nickel. Most of these are water insoluble. Nickel carbonyl is an extremely toxic gas which is kept in closed compartments separate from human contact.[33-37]

B. Occupational Nickel Exposure

Nickel plating and electroforming of nickel imply high and continuous exposure to both nickel and nickel salts in percent concentrations.[35,38-40] Other branches of the industry which use high concentrations of nickel compounds are storage battery production, some branches of enamel and glass production, and chemical plants where nickel catalysts are used for many organic reactions.[34,41]

Moderate exposure occurs in mechanical engineering and many other industrial branches where nickel-alloyed metal is worked up and where the handles of the machines and tools are made from nickel alloy or are nickel plated. Cement contains nickel but the soluble fraction is low.

Minor amounts of nickel are found in many technical products such as detergents, and even in our food and drinking water.[4,33,42-51] Individuals in wet work, e.g., hairdressers, hospital workers, food handlers, and cleaning personnel, are continuously exposed to small amounts of nickel, but this low exposure may have significance in causing skin disease.[4,7,24,26,28,48,52-66]

1. Metal Manufacturing

Wahlberg et al. found the content of nickel in 24 of 27 unused cutting fluids to be below 0.10 μg/g. One extreme exception was a fluid with 19.4 μg/g Ni. In cutting fluids used for steel alloys or hard metal for 14 d, the nickel concentration in centrifuged samples was 0.10 to 0.15 μg/g. The nickel concentration of unused mold oils was in general lower than in cutting fluids. In only 6 of 28 such oils the nickel concentration exceeded 0.10 μg/g and none exceeded 0.92 μg/g.[46,67]

2. Cement

Nickel in cement exists mainly as insoluble salts. The nickel content of Swedish cement was found to be 5 to 59 μg/g. In another investigation the total nickel content of 52 samples of Portland cement from 9 different countries ranged between 32 and 116 μg nickel per gram. Calcium aluminate cement contained 672 μg nickel per gram. The nickel was non-soluble. Asian cement contains nearly identical concentrations of nickel: 15 to 29 μg/g and a free fraction of 0.4 to 0.7 μg/g.[46,68,69]

3. Detergents

Nickel in detergents and soaps has long been suspected of causing hand dermatitis in nickel-sensitive women.[50,60] Nickel content of soap is low and nearly undetectable in the concentrations applied for cleaning.[21] The nickel content of concentrated detergent solutions seldom exceeds 10 μg/g. The normal user concentration is 0.1 to 1% detergent, and the skin is thus only exceptionally exposed to more than 0.1 μg/g.[21,48,49,70] Almost all detergents for household use contain complex binding agents which inactivate the nickel to compounds of low water solubility.[70] A significant increase in nickel concentration in cleaning water was observed after 30 min of hospital cleaning.[21]

4. Hairdressing

Hairdressers and barbers use tools made from stainless steel; the clips are generally nickel plated. Hair dressers are also in contact with permanent wave liquids which contain ammonium thioglycolate. Thioglycol acid can form soluble complexes with nickel ions.[54,71] Dahlquist and Fregert found nickel-plated objects to release about 21 to 129 μg nickel in commercial permanent wave liquids and 76 to 170 μg nickel in ammonium thioglycolate, as compared to 4 to 14 μg nickel in synthetic sweat. Nickel in stainless steel should in fact be safe as it is not corroded by the thioglycolate solution.[56]

5. Food and Water

The nickel concentration of vegetables on a dry weight basis is in general 0.05 to 5 μg/g. Beans, peas, cabbage, beets, tea, and cocoa are especially rich in nickel. The nickel content of plants is highly dependent on the nickel concentration of the soil where they are grown. Most nickel in vegetables is chelated to compounds of low solubility. Seafood, meat, and eggs hold 0.02 to 1 μg/g nickel.[33,46]

Cooking food, especially acid food, in stainless-steel pots will result in a significant nickel enrichment.[72,73]

Nickel in rivers ranges from 0.005 to 0.05 μg/g and in municipal water supplies from 0.003 to 0.065 μg/g. The first tap water may be contaminated from pipes and fittings, and concentrations as high as 1.3 μg/g have been reported.[42-44,73]

The normal daily oral nickel intake is 0.1 to 0.3 mg.[4]

6. Coins

Shop assistants and cashiers handle nickel continuously, and this may result in high nickel contamination of the skin. Paper money is often kept in contact with nickel coins; nickel release from used bank notes is in the same range as from coins.[74]

7. Other

In addition to the skin, mucous membranes of the mouth, gut, and respiratory system are exposed to nickel from food and air pollution. In industries with high nickel dust exposure, blood and urinary nickel levels of workers can be increased 10 to 20 times. Such exposure is reported to cause asthma and rhinitis.[41,75-81]

C. Nickel in Intimate Contact with the Skin

Metallic nickel and many nickel alloys release soluble nickel under humid conditions. The salts and amino acids in human sweat will increase the corrosive action of water. Nickel-plated objects such as buttons, zippers, buckles, and clasps release identical amounts of nickel as the metal. Gold-plated objects are often first plated with nickel and then gold.[2,9,51,82,83] Nickel penetrates the thin gold layer, and the surface concentration of nickel is almost as high as metallic nickel.[84,85]

Stainless steel contains about 8% nickel and is resistant to corrosion because the included chromium will form resistant surface oxides.[86,87]

Details about nickel release from alloys and objects are presented in Chapter 4.

III. BACKGROUND OF ALLERGIC NICKEL DERMATITIS

A. Sensitizing and Eliciting Dose of Nickel Dermatitis

No one is sensitive to nickel at birth. The sensitization starts through skin contact with metallic nickel and nickel salts. Nickel must be in a soluble state to pass the horny layer and enter the epidermis where it can cause skin disease. The surface of metallic nickel is corroded by water and sweat. Nickel ions thus formed enter the skin. The corrosion resistance of different nickel alloys is variable, and this will affect their sensitizing properties. The stratum corneum barrier restricts most of the metal ions and only minimal amounts of nickel will pass an undamaged horny layer.[59,88-91]

Nickel is a moderately strong allergen. The risk of becoming sensitized is related to the concentration of allergen, the area exposed, and the time of exposure. Other factors of significance are the mode of exposure of the skin to the allergen, humidity, pH, bacterial flora of the skin, and the presence of other chemical substances.

An intact skin surface is important in preventing nickel penetration. The risk of becoming sensitized is significantly increased if the skin is injured. Skin damaged by irritation will prime nickel sensitization. Wet work also impairs the barrier function of the skin and increases the penetration of nickel.

Detergents increase both the irritant dermatitis and the penetration of nickel.[21,92] This is illustrated by the fact that if a detergent is added to a patch test solution, the eliciting nickel concentration is decreased by a factor of 5 to 100.[93]

Sensitization to allergens other than nickel may be the background which increases te risk of becoming sensitized. Nickel and cobalt are present together in most alloys and patients frequently become sensitized to both. They then react more readily to either allergen and this results in a more severe dermatitis,[94.102] also verified in guinea pig experiments.[103]

Presently we have insufficient knowledge about the minimum dose of nickel which can trigger eczema in a sensitized individual. There are no human reports concerning the minimum exposure necessary for sensitization.

Malten and Spruit considered an eliciting safety limit concentration of nickel ions of 0.05 μg nickel per gram from the fact that few nickel-sensitive individuals react with positive patch tests below 0.5 μg nickel per gram. They also suggested, as a working hypothesis, the use of the same sensitizing safety limit concentration of nickel ions.[60]

The amount of soluble nickel released from metallic nickel and nickel-plated objects exceeds this limit. The amount released from white gold is barely in the region between sensitizing and eliciting concentration.[88,104-107] Opinions differ concerning the sensitizing risk of stainless steel. The quality of the steel determines its corrosive properties. Surgical stainless steel is generally regarded as safe and not producing dermatitis insensitive patients,[72,86,87] but some qualities release enough nickel to provoke dermatitis in nickel-sensitive patients.[85,108]

Most cases of nickel sensitization in women are nonoccupational and a result of ear piercing or intimate skin contact with metal jewelry, clips, clasps, buckles, zippers, etc.[2,3,9,51,82,83,101] Occupational sensitization is considered uncommon in women, but regarded as more common in men.[21,22,26,30,31,57,58,62-64,71,109,110] The increase of ear piercing in men may have changed this pattern.

It is often easy to determine from the patient history the time of sensitization following nickel exposure from earrings or from a jeans button.[2] Dermatitis will start at the site of the nickel contact. Sensitization from occupational contacts is more difficult to prove or dismiss because such eczemas usually start at the hands and are, as a rule, of multifactorial origin.

Significant occupational sensitization, however, occurs in work with high nickel exposure. The sensitization frequency in low-exposure occupations has been subject to much debate. The occupational background of the dermatitis is unmistakable when the dermatitis starts at the site of nickel-plated rivets in gloves or clothing. It is more difficult to evaluate the role of nickel-plated tools, workbenches, or fittings in direct and long-lasting contact with the skin where a dermatitis suddenly starts at the site of contact. The question remains, however, as to how frequently occupations with wet work and low nickel exposure will cause sensitization. It is seldom verified that patients become nickel sensitized from low-exposure jobs.

In 166 Polish nickel-sensitive patients, Rudzki et al.[57] found the following objects with occupational significance to cause primary sensitization to nickel or recurrences of nickel dermatitis: electrical assembly, locksmith tools, cufflinks, dental equipment, chemical reagent, galvanic solution, typewriters, scissors, knitting machines, and knitting needles.

B. The Sensitized Patient

The relationship between nickel allergy and contact dermatitis of the hands has been a source of confusion to dermatologists ever since Calnan reported a high incidence of nickel sensitivity in patients with hand eczema.[3] The incidence of nickel sensitization among patients with hand eczema is 12 to 21%. This increased frequency cannot be explained by the single factor of handling nickel objects.[111-115]

Hand dermatitis has a complex background of constitutional, environmental, and sometimes allergic factors. The proportions between these factors vary individually.[116-119]

Nickel sensitization does not invariably lead to hand eczema. Despite the high prevalence of nickel allergy in the general female population (10%), hand eczema is much less common. If, however, hand dermatitis develops, it is periodically disabling and has pronounced chronicity.[3,4,17,120-124] A reinvestigation of women with occupational hand eczema and nickel allergy showed that one half had periodic and one half continuous dermatitis; no one was healed.[22] Christensen found only 30% healing in women with hand nickel dermatitis.[125]

Most nickel-sensitive individuals become sensitized before they have left their teens. They rarely have occupational skin problems during the first years of sensitization. The nickel sensitivity then appears as an eczema at the site of skin contact with nickel objects — earlobes, wrists, or mid-stomach (jeans button).

After leaving school, women in particular get jobs which are wet and traumatizing to their hands. In such occupations almost everyone will experience a mild irritant hand dermatitis. In addition, there may be wear and tear of the skin from household duties or hobby work. This is why many individuals function at the limit of their skin resistance. Previously the nickel sensitivity was no problem, but the combination of irritant dermatitis and wet work decreases the barrier function of the skin. Nickel from the environment will enter the epidermis and the nickel sensitivity may turn a mild irritant dermatitis into a severe hand eczema. A fulminant dermatitis will develop, especially in those sensitized against both nickel and cobalt.[94-101] Others may have only moderately increased problems even if they have the same type of work. There is presently no good explanation for this variation in the expression of the dermatitis.

C. Clinical Expression of Nickel Hand Dermatitis

Dorsal hand eczemas are, as a rule, of exogenous, environmental origin,[126] while endogenous factors such as atopy, psoriasis, pustulosis palmoplantaris, or tylotic eczema are involved in the majority of volar hand dermatitis. A more detailed clinical grouping was used by Squire et al.,[127] which has since been modified:[4,120,128,129]

- Type I: Red papulo-vesicles on forearms and dorsa of the hands (contact dermatitis type)
- Type II: Papulo-vesicles on palms and interdigitally (pompholyx type)
- Type III: Chronic lichenified and fissured palmar eczema
- Type IV: Discoid eczema on the dorsa of hands and forearms (nummular type/atopic)

The clinical picture of nickel dermatitis is discussed in detail in Chapter 13.

D. Pompholyx

Pompholyx, or dyshidrotic eczema (Type II), is the predominant type of nickel allergy hand eczema.[125] It is, however, disputed whether nickel allergy is an important background factor in that type of hand eczema.[112,113,130,131]

Dyshidrotic eczema is regarded to be essentially endogenous, and the 5-year healing is low.[125] Oral nickel provocation sometimes increases the severity of the dyshidrotic hand eczema.[99,113,131-134] There are reports of improvement after low nickel diet or attempts to reduce body nickel by medication[135,136] (Chapter 14).

This could imply a risk of dyshidrotic eczema in occupations with high ingestion, inhalation, and skin exposure to nickel. However, no such reports have been found. On the contrary, there is at least one report that the pompholyx type of hand eczema was not overrepresented in an industry with high nickel exposure and several cases of nickel dermatitis.[39]

The background of dyshidrotic eczema is thus probably essentially endogenous. Wet work, occlusion of the hands by gloves and pore-occlusive ointments, heat, and greasy work seem to worsen this type of hand eczema. Such environmental factors also reduce the barrier function of the skin, which results in an increased nickel penetration from the environment, and this may explain the severity and chronicity of the dermatitis.

E. Contact Urticaria

Occupational contact urticaria has been increasingly recognized as relative to chemicals and food products. Reported cases of contact urticaria from metal compounds concern mainly cobalt and platinum salts.[137] A few pulmonary type I reactions to nickel have been reported and verified by inhalation challenge tests.[138-140] On oral challenge some nickel-sensitive patients respond with urticaria, and there is also an example of urticaria due possibly to nickel from an infusion needle.[141] Patch tests read after 1 h in 56 patients with chronic urticaria gave urticarial response in 4, but only 1 had a positive 48-h patch test reaction to nickel.[140] Nickel urticaria is rarely described, and further research is necessary to determine the clinical and occupational significance of contact urticarial reactions to nickel.

IV. ASSOCIATION BETWEEN NICKEL-SENSITIZATION HAND ECZEMA AND WORK

With few exceptons, nickel sensitivity is equally common in all types of occupations. There are at present few occupational problems from nickel allergy in occupations with high exposure to metallic nickel and nickel salts. The significance of low exposure to nickel as the basis of dermatitis has long been debated and remains an unsolved question.

Occupations reported to be common among nickel-sensitized women are home work, cleaning, catering and kitchen work, cashiering and shop assistance, hairdressing, nursing and hospital work, dressmaking, and office work, while nickel sensitized men are employed

Table 1
OCCUPATIONS WITH SIGNIFICANT EXPOSURE TO NICKEL

Auto mechanic	Ink maker
Butcher	Jeweler
Carpenter	Mechanic
Cashier	Metal worker and welder
Ceramic worker	Office worker
Cleaner and hospital worker	Painter
Dentist and veterinarian	Plumber
Dryer	Printer
Electrician	Radio and television electronic repairmen
Electroplater	Rubber worker
Enamel worker	Shop assistant
Food manufacturer and restaurant personnel	Spark plug maker
Hairdresser and barber	Textile worker
Household worker	

in engineering and building industries (Table 1). Only a few men have a definite occupational exposure.[7,21,24,26-28,58,61,63-66,142,143] In most office workers the basis of the hand dermatitis is probably nonoccupational. It is not uncommon for individuals to choose office work because of hand eczema, but even office workers may be exposed to nickel and a typewriter has been described as a cause of nickel dermatitis.[58]

Moisture and/or hand sweating are factors which may contribute to hand dermatitis in most occupations. Dry and clean occupations with moderate or even intense contact to nickel objects will seldom alone provoke dermatitis. Only extreme exposure to metallic nickel, e.g., that of cashiers, will give problems. Even this extreme nickel exposure has been debated as background factor of hand dermatitis.[74,114,144-146]

A. Nickel Manufacturing

Manufacturing of nickel will contaminate workers' skin heavily with the metal and metal salts. Most individuals become hardened after several months of exposure in this demanding work. Those who cannot tolerate the environment are rapidly eliminated, whereas those hardened can go on working for a long time. Among old workers in this area, there are few problems of irritant and allergic dermatitis. The background is probably the early selection of tolerant workers.[35,147]

B. Nickel Plating

Electroplating and electroforming of nickel implies heavy and continuous exposure to nickel and nickel salts and a significant risk of becoming sensitized.[35,38,40] Workers with such occupations handle hot nickel salt solutions with risk of nickel absorption through mucous membranes, lungs, and skin and have regularly increased levels of serum and urinary nickel.[17,38,41,78,148]

Until 1930, nickel dermatitis was common among nickel platers. During the next 10 years there was a successive decrease in nickel sensitization, largely due to improved industrial hygiene. In a series of 621 nickel-sensitive patients from Denmark, only 25 were nickel platers.[28] Since 1960, reports of nickel sensitivity in the electroplating industry have been sparse.[38] Schubert et al. found only 2 nickel-sensitive electroplaters among 176 nickel-sensitive individuals.[30]

Wall and Calnan recently reported an outbreak of nickel dermatitis in an electroforming industry.[39] Electroforming is a process which involves controlled electrodeposition of nickel onto a form of predetermined shape. A high percentage of the workers were continuously exposed to nickel solutions. Improvements in industrial hygiene led to an immediate decrease in the incidence of dermatitis. Interestingly, there was a high percentage of nickel sensitivity among those with hand eczema, but no clinical evidence of an associated palmar eczema.

C. Metal Manufacturing — Cutting Fluids

The most common dermatological problem in metal manufacturing is irritant dermatitis from cutting fluids and cutting oils. Mechanical injury from metal fragments and sharp edges is also common. The role of nickel contamination in cutting fluids as a background cause of nickel contact dermatitis in the metalworking industry is under debate.[150] Contact sensitization from metallic nickel or nickel dissolved in cutting fluids is probably rare, and nickel allergy itself is a minor problem in the metallurgic environment. When sensitivity is reported, corrosion inhibitors, bacteriostatic agents, and breakdown products are the common causes.[110,129,151-154] Most investigators agree that nickel dermatitis is as frequent among metalworkers as in the general population.[102,110,152] Rycroft, in a modern summary of dermatologic effects of cutting oils, does not even discuss the problem of nickel sensitivity or nickel contamination of cutting fluids.[154]

In an investigation of skin diseases in a hard metal factory,[101,155] 0.5% of the men and 10% of the women were found to be sensitive to nickel. Hard metal manufacturing implies a high cobalt and some nickel exposure as well as significant irritation, contamination, and wet exposure of the hands. More than half of the nickel-sensitized women gave a history of time relationship between ear piercing and nickel sensitivity, and many of them had the start of hand dermatitis within 6 months of their nickel sensitization. Less than half had hand dermatitis. In these cases it was obvious that sensitization occurred from the ear piercing, while the hand eczema had a complex background of irritant exposure from work and home environments and nickel allergy, frequently in combination with cobalt allergy. All workers with combined nickel-cobalt sensitivity had moderate or severe hand eczema and a more severe dermatitis than those sensitized to either metal.

Welding fumes and direct contact with nickel-alloyed electrodes may exacerbate eczema in nickel-sensitized individuals.[156]

Nickel may be incorporated in plastic material to mimic the metal, and this material can provoke nickel dermatitis.[157,158]

In conclusion, nickel sensitization in metalworkers increases their risk of hand eczema and possibly also sensitization to cobalt and other allergens. This may in turn result in severe long-lasting hand dermatitis.

D. Cement and the Building and Construction Industry

Few investigators consider nickel an important cause of dermatitis in the building and construction industry. Most investigators do not even discuss nickel sensitivity in reports on occupational dermatoses among construction workers.[96,159-161] In a recent report where 272 construction workers were nickel tested, there was no relation between isolated nickel sensitivity and hand eczema; 1 patient with hand eczema and nickel allergy was also sensitive to chromium and cobalt.[69]

Pirilä concluded, from the literature and his own experience, that nickel allergy was relatively rare in cement workers suffering from contact dermatitis. Even in this group he found a female dominance among the nickel-allergic patients. He assumes that strong nickel sensitization is required for the development of nickel-caused cement eczema partly due to the poor solubility of nickel in cement.[96]

E. Hairdressing

Several investigators report a higher proportion of hand eczema among hairdressers and barbers than in the general population. The background of this hand eczema is interesting and a good example of the complexity of occupational dermatitis.[55,71,111,142,162-171]

At the start of their apprenticeship, junior hairdressers have a high frequency of nickel sensitivity (27%) which cannot be attributed to occupational exposure,[55] but is probably the result of frequent ear piercing and exposure to metallic nickel at an early age.

Barbers and hairdressers are intermittently exposed to a wet environment, permanent wave liquids, and other chemicals which destroy the keratin layer. Irritant exposure to detergents also reduces the barrier function of the skin and increases the penetration of nickel. There is also exposure to several potent allergens, such as fragrances, rubber chemicals, and thioglycolates, in addition to nickel.

Water will dissolve nickel from tools, clips, and other metallic objects, especially in the presence of thioglycolate, oxidizers in permanent liquids, amino acids, and proteins from hair, skin, sweat, and bacteria.[54,56,142,172]

Nickel exposure from stainless steel seems to be of minor importance, since the contact surfaces against metal tools, scissors, etc. are seldom the location of eczema.[162]

Probably all the factors mentioned contribute to the high frequency and severity of hand eczema in nickel-sensitive hairdressers.

F. Home Work, Cleaning, and Hospital Wet Work

Hand eczema is frequent in cleaning and hospital wet workers, especially among individuals sensitive to nickel.[14,21,55,58,92,110,173-175] Fregert found that among 595 women considered to have occupational dermatoses, 82 had nickel reactions. Of these, one fifth were caused by cleaning work, while the other reactions occurred in individuals in 11 other occupations, e.g., hairdressers, office workers, and shop workers. A change in occupation did not effect healing more than continuing the same work.[22]

Among hospital workers, individuals with any type of sensitization to an allergen, not only nickel, have a higher frequency of hand eczema than nonsensitized individuals, i.e., 60% as compared to 46% of the entire population of hospital wet workers examined. Nickel gave the same increase in risk of developing hand eczema as most other allergens. Combined cobalt-nickel allergy increased the risk to 70%. Nickel was the most common individual sensitizer.[58]

The frequent occurrence of hand dermatitis in cleaning personnel is probably due to the irritant effect of detergents and water. This will start the hand dermatitis, which often becomes chronic in nickel-sensitized individuals. Nickel in detergents and cleaning water may be of relevance in causing the dermatitis.[21,58,92,174-176]

An unusual cause of nickel dermatitis in cleaning personnel was exposure to nickel-releasing steel fiber pads. The pads contained 0.12% nickel.[177]

G. Business and Coin Contacts

Nickel-alloyed coins will give a positive patch test reaction when taped to the skin.[144,145] Exposure to coins, both nickel and copper, will result in a significant nickel contamination of the skin.[74]

The relation between nickel sensitivity, coin exposure, and hand eczema remains unsolved. Nickel coins probably do not elicit dermatitis in undamaged skin unless the patient is strongly allergic.[114] Christensen and Möller did not succeed in inducing flares of eczema among nickel-sensitive patients with pompholyx by making them handle coins for 20 min.[113] Most investigators agree with Fisher that nickel dermatitis of the hands from normal handling of coins rarely occurs because of the protection by the thick horny layer of the palms.[178] My experience from a nickel-sensitive daughter and a bingo hostess regularly attending our clinic, who both flare on intense contact with coins, has convinced me, however, that coin contacts can both start and maintain hand dermatitis in nickel-sensitive individuals.

It is doubtful whether a patient with a moderately strong nickel allergy should always be advised against working as a cashier. In my opinion, it is more important when counseling nickel-sensitive young people to stress the risks with wet work.

H. Rare and Curious Reports of Occupational Nickel Dermatitis

Detailed reports on occupational exposure and nickel dermatitis outside metal manufacturing and wet work are rare. Most such cases are only mentioned among occupations in tables from epidemiological investigations as individuals sensitive to nickel. There is then a lack of detail in the description of the relation between nickel sensitivity and work.

Nickel-sensitive tailors and other sewers develop dermatitis on the first and second fingers of their sewing hand, and this heals when they change to stainless-steel knitting needles.[178] Recently a woman garment worker developed a papular eruption on the third and fourth fingers at the contact site of her scissors, but also a general multiform eruption, probably of type III, which disappeared on avoidance of nickel but reappeared with a patch test.[179]

Pirilä found that a combined sensitivity to nickel and cobalt occurred in 7 of 44 cobalt-sensitive pottery workers — suspender dermatitis excluded. Cobalt is the main sensitizer in this industry and nickel sensitivity may be secondary to cobalt sensitization.[180]

In margarine production, nickel is used as a catalyst. This may sensitize the workers, and even the consumers may flare.[16,181]

Among butchers, nickel eczema may appear under clasps and buttons of protective gloves. Dermatitis appears to be most common in female workers, indicating that the primary sensitization started outside work.[182]

Some musicians are at risk for nickel dermatitis. Fiddler's neck (an area of erythema, pigmentation, lichenification, and acne-like changes below the jaw) has been described to be at least partly caused by metal clamps rubbing the chin.[183] A cellist presented recurrent dyshidrosiform vesicles and red scaly eczema on the second and third fingers of his right hand. The location matched his contact with metallic parts of the bow.[184]

Nickel-sensitive doctors and veterinarians must be careful to not wear the stethoscope for too long, because metallic contact may cause dermatitis.[185]

White blackboard chalk may cause dermatitis, not only of the irritant type. A nickel-sensitive teacher developed a vesicular eruption on both hands. She had atopic diathesis, but was of the opinion that chalk caused her hand dermatitis. She improved on avoidance of chalk. Patch testing with chalk was negative. The dimethylglyoxime test was positive and chemical analysis indicated 12.5 μg nickel per gram of chalk.[186]

V. PREVENTION

The prevention of nickel sensitization is essentially a nonoccupational task. Information should be given about the risks of ear piercing and of buttons and cheap jewelry in direct contact with the skin.

Find out and inform where nickel is found in the occupational environment to reduce skin contact with nickel. Plastic and wooden handles should be preferred over nickel plated and stainless. Gloves and clothing should be free from nickel-plated rivets and hooks in direct contact with the skin.

When working with nickel solutions and nickel salts, protection with vinyl gloves is preferred because nickel penetrates rubber gloves.[187]

General hygienic measures should be instituted to prevent irritant dermatitis, which is the precursor to most sensitization, by improvement of skin protection, shielding, and elimination and reduction of wet work, dirt exposure, and solvent contacts.

Cutting fluids in general do not dissolve high concentrations of nickel, but keep control of the nickel content; the use of cutting fluids with low nickel-dissolving properties is recommended. The addition of chelating agents may be instituted.

A proper, nonirritant skin care program should be developed.

Attention should be paid at the preemployment examination or during occupational counseling to nickel sensitivity. It should be pointed out that there is an increased risk of hand eczema in irritant and wet work.

Most investigators have found no relation between atopy and nickel allergy. In individuals with atopy and nickel allergy, however, there is a high risk of hand eczema independent of choice of work, and these individuals should choose dry and clean work outside nickel manufacturing and metal industry.[3,11,18,51,55,188-191]

Distinctly unsuitable occupations are haircutting, barbering, food manufacturing, butchering, and cooking. In most other work areas it is possible to find suitable niches. In hospital work such sectors as operative, laboratory, and cleaning work are unsuitable, but there are many dry and clean sectors which are compatible with nickel allergy.

REFERENCES

1. **Black, H.,** Analysis of the application of routine battery May 1970—December 1971, *Contact Dermatitis Newsl.*, p. 323, 1972.
2. **Brandrup, F. and Schultz Larsen, F.,** Nickel dermatitis provoked by buttons in blue jeans, *Contact Dermatitis*, 5, 148, 1979.
3. **Calnan, C. D.,** Nickel dermatitis, *Br. J. Dermol.*, 68, 229, 1956.
4. **Christensen, O. B.,** Nickel Allergy and Hand Eczema in Females, Disputation, Department of Dermatology, University of Lund, Sweden, Infotryck Malmö, 1981.
5. **Dooms-Goosens, A., Ceuterick, A., Vanmaele, N., and Degreef, H.,** Follow-up study of patients with contact dermatitis caused by chromates, nickel and cobalt, *Dermatologica*, 160, 249, 1980.
6. **Gaul, L. E.,** Incidence of sensitivity to chromium, nickel, gold, silver and copper compared to reactions to their aqueous salts including cobalt sulphate, *Ann. Allergy*, 12, 429, 1954.
7. **Gawkrodger, D. J., Vestey, J. P., Wong, W.-K., and Buxton, P. K.,** Contact clinic survey of nickel-sensitive subjects, *Contact Dermatitis*, 14, 165, 1986.
8. **Lachapelle, J. M. and Tennstedt, D.,** Epidemiological survey of occupational contact dermatitis of the hands in Belgium, *Contact Dermatitis*, 5, 244, 1979.
9. **Larsson-Stymne, B. and Widström, L.,** Ear piercing — a cause of nickel allergy in schoolgirls?, *Contact Dermatitis*, 13, 289, 1985.
10. **Marcussen, P. V.,** Spread of nickel dermatitis, *Dermatologica*, 115, 596, 1957.
11. **Menné, T., Borgan, Ö., and Green, A.,** Nickel allergy and hand dermatitis in a stratified sample of the Danish female population: an epidemiological study including a statistic appendix, *Acta Derm. Venereol.*, 62, 35, 1982.
12. **Menné, T. and Holm, N. V.,** Hand eczema in nickel sensitive female twins. Genetic predisposition and environmental factors, *Contact Dermatitis*, 9, 289, 1983.
13. **Menné, T.,** The prevalence of nickel allergy among women. An epidemiological study in hospitalized female patients, *Dermatosen*, 26, 123, 1978.
14. **Peltonen, L.,** Nickel sensitivity in the general population, *Contact Dermatitis*, 5, 27, 1979.
15. **Rudner, E. J. et al.,** Epidemiology of contact dermatitis in North America: 1972, *Arch. Dermatol.*, 108, 537, 1973.
16. **Rudzki, E.,** Some peculiarities of contact dermatitis in Poland, *Dermatosen*, 29, 98, 1981.
17. **Wilson, H. T. H.,** Nickel dermatitis, *Practitioner*, 177, 303, 1956.
18. **Kanan, M. W.,** Contact dermatitis in Kuwait, *J. Kuwait Med. Assoc.*, 3, 129, 1969.
19. **Romaguera, C. and Grimalt, F.,** Statistical and comparative study of 4600 patients tested in Barcelona (1973-1977), *Contact Dermatitis*, 6, 309, 1980.
20. **Burrows, D. and Corbett, J. R.,** Industrial dermatitis in Northern Ireland, *Contact Dermatitis*, 3, 148, 1977.
21. **Clemmensen, O. J., Menné, T., Kaaber, K., and Solgaard, P.,** Exposure of nickel and the relevance of nickel sensitivity among hospital cleaners, *Contact Dermatitis*, 7, 14, 1981.
22. **Fregert, S.,** Occupational dermatology in a 10-year material, *Contact Dermatitis*, 1, 96, 1975.
23. **Malten, K. E., Fregert, S., Bandmann, H.-J., Calnan, C. D., Cronin, E., Hjorth, N., Magnusson, B., Maibach, H. I., Meneghini, C. L., Pirilä, V., and Wilkinson, D. S.,** Occupational dermatitis in five European dermatological departments, *Berufs Dermatosen*, 19, 1, 1971.
24. **Menné, T. and Bachmann, E.,** Permanent disability from hand dermatitis in females sensitive to nickel, chromium and cobalt, *Dermatosen*, 27, 129, 1979.
25. **Menné, T. and Bachmann, E.,** Permanent disability from skin diseases. A study of 564 patients registered over a six year period, *Dermatosen*, 27, 37, 1979.

26. **Schmiel, G.**, Häufigkeit von Nickel-Kontaktallergien am anausgewählten Patientgut im Raum München, *Dermatosen*, 33, 92, 1985.
27. **Camarasa, J. M. G.**, First epidemiological study of contact dermatitis in Spain — 1977, *Acta Derm. Venereol.*, 59 (Suppl. 85), 33, 1977.
28. **Marcussen, P. V.**, Ecological considerations on nickel dermatitis, *Br. J. Ind. Med.*, 17, 65, 1960.
29. **Rostenberg, A. and Perkins, A. J.**, Nickel and cobalt dermatitis, *J. Allergy*, 22, 466, 1951.
30. **Schubert, H., Bernova, N., Czernielewski, A., Hegyi, E., Jirasek, L., Kohanka, V., Korossy, S., Michailov, P., Nebenfuhrer, L., Prater, E., Rothe, A., Stranski, L., Suss, E., Tarnick, M., Temesvari, E., Ziegler, V., and Zschunke, E.**, Epidemiology of nickel allergy, *Contact Dermatitis*, 16, 122, 1987.
31. **Schubert, H., Bernova, N., Hegyi, E., Jirasek, L., Kalensky, J., Korossy, S., Michailov, P., Nebenfuhrer, L., Prater, E., Rothe, A., Rudzki, E., Temesvari, E., Ziegler, V., and Zschunke, E.** Das allergische Kontaktekzem. Analyse einer Stickprobe in fünf sozialistischen Ländern Europas, *Dermatol. Monatsschr.*. 168, 613, 1982.
32. **Lagerwerff, J. V. and Specht, A. W.**, Contamination of roadside soil and vegetation with cadmium, nickel, lead and zinc, *Environ. Sci. Technol.*, 4, 583, 1970.
33. **Sunderman, F. W., Coulston, F., Eichhorn, G. L., Fellows, J. A., Mastromatteo, E., and Samitz, M. H.**, *Nickel*, National Academy of Sciences, Washington D.C., 1975.
34. **Adams, R. M.**, *Occupational Skin Disease*, Grune & Stratton, New York, 1983, 225.
35. **Schwartz, L., Tulipan, L., and Birmingham, D. J.**, *Occupational Skin Diseases of the Skin*, 3rd ed., Lea & Febiger, Philadelphia, 1957, 274.
36. **Morgan, L. G.**, Manufacturing processes. Refining of nickel, *J. Soc. Occup. Med.*, 29, 33, 1979.
37. **Cronin, E.**, *Contact Dermatitis*, Churchill Livingstone, Edinburgh, 1980.
38. **Mathur, A. K.**, Occupational dermatitis and absorption in a metal plater, *Contact Dermatitis*, 10, 530, 1984.
39. **Wall, L. M. and Calnan, C. D.**, Occupational nickel dermatitis in the electroforming industry, *Contact Dermatitis*, 6, 414, 1980.
40. **Blaschko, A.**, Die Berufsdermatosen der Arbeiterer. Ein Beitrag zur Gewerbehygiene. I. Das Galvaniseur — Ekzem., *Dtsch. Med. Wochenschr.*, 15, 925, 1889.
41. **Bernacki, E. J., Zygowicz, E., and Sunderman, F. W.**, Fluctuations of nickel concentrations in urine of electroplating workers, *Ann. Clin. Lab. Sci.*, 10, 33, 1980.
42. **Andersen, K., Nielsen, G. D., Flyvholm, A.-M., Fregert, S., and Gruvberger, B.**, Nickel in tap water, *Contact Dermatitis*, 9, 140, 1983.
43. **Gammelgaard, B. and Andersen, J. R.**, Nickel in tap water, *Contact Dermatitis*, 12, 123, 1985.
44. **Fregert, S.**, Nickel in tap water, *Contact Dermatitis Newsl.*, p. 202, 1971.
45. **Rudzki, E. and Grzywa, Z.**, Exacerbation of nickel dermatitis by margarine, *Contact Dermatitis*, 3, 344, 1977.
46. **Wahlberg, J. E., Lindstedt, G., and Einarsson, Ö**, Chromium, cobalt and nickel in Swedish cement, detergents and cutting oils, *Berufs Dermatosen*, 25, 220, 1977.
47. **Veien, N. K. and Anderson, M. R.**, Nickel in Danish food, *Acta Derm. Venereol.*, 66, 502, 1986.
48. **Allenby, C. F. and Goodwin, B. F. J.**, Influence of detergent washing powders on minimal elicting patch test concentrations of nickel and chromium, *Contact Dermatitis*, 9, 491, 1983.
49. **Vilaplana, J., Grimault, F., Romaguera, C., and Mascaro, J. M.**, Cobalt content of household cleaning products, *Contact Dermatitis*, 16, 139, 1987.
50. **Kroepfli, P. and Schuppli, R.**, Beiträge zum Waschmittelekzem, *Dermatologica*, 110, 1, 1955.
51. **Watt, T. L. and Baumann, R. R.**, Nickel earlobe dermatitis, *Arch. Dermatol.*, 98, 155, 1968.
52. **Bandmann, H.-J. and Fuchs, G.**, Über die Kobaltkontaktallergie, ihre Beiziehung zur Bichromat und Nickelkontaktallergie, sowie ihre gewerbedermatologische Bedeutung, *Hautarzt*, 14, 207, 1986.
53. **Borelli, S. and Dungemann, H.**, Aktuelle Kontaktekzem-Ursachen in der Metallindustrie, *Berufs Dermatosen*, 12, 1, 1964.
54. **Brandao, F. M.**, Release of nickel by permanent wave liquids, shown by the dimethylglyoxime test, *Contact Dermatitis*, 5, 406, 1979.
55. **van der Burg, C. K. H., Bruynzeel, D. P., Vreeburg, K. J. J., von Blomberg, B. M. E., and Scheper, R. J.**, Hand eczema in hairdressers and nurses: a prospective study. I. Evaluation of atopy and nickel hypersensitivity at the start of apprenticeship, *Contact Dermatitis*, 14, 275, 1986.
56. **Dahlquist, I., Fregert, S., and Gruvberger, B.**, Release of nickel from plated utensils in permanent wave liquids, *Contact Dermatitis*, 5, 52, 1979.
57. **Rudzki, E., Rebandel, P., Napiorkowska, T., Dynowska, D., and Legiéc, C.**, Nickel objects causing sensitization and relapses, *Contact Dermatitis*, 13, 335, 1985.
58. **Lammintausta, K., Kalimo, K., and Havu, V. K.**, Occurrence of contact allergy and hand eczema in hospital wet work, *Contact Dermatitis*, 8, 84, 1982.

59. **Lloyd, G. K.,** Dermal absorption and conjugation of nickel in relation to the induction of allergic contact dermatitis, in *Nickel Toxicology,* Brown, S. S. and Sunderman, F. W., Jr., Eds., Academic Press, London, 1980, 145.
60. **Malten, K. E. and Spruit, D.,** The relative importance of various environmental exposures to nickel in causing contact hypersensitivity, *Acta derm. Venereol.,* 49, 14, 1969.
61. **Müller, R. and Breucker, G.,** Kobalt als arbeitsbedingtes Ekzematogen und als Kopplungsallergen mit Chrom und Nickel, *Dermatol. Wochenschr.,* 34, 276, 1968.
62. **Nebenführer, L., Kohanka, V., Vincze, E., Korossy, S., Szakaly, I., and Gozony, M.,** Sources of nickel allergy in Budapest, *Contact Dermatitis,* 10, 116, 1984.
63. **Reichenberger, M., Ebke, M., and Patiri, C.,** Zur Nickelsensibilisierung bei Frauen und ihre Relevanz zur beruflichen Tätigkeit, *Berufs Dermatosen,* 24, 91, 1976.
64. **Skog, E. and Thyresson, N.,** The occupational significance of some common occupational contact allergens, *Acta derm. Veneraol.,* 33, 65, 1953.
65. **Cronin, E.,** The significance of nickel sensitivity in women, *Br. J. Dermatol.,* 84, 96, 1971.
66. **Malten, K., Fregert, S., Bandmann, H.-J., Calnan, C. D., Cronin, E., Hjorth, N., Magnusson, B., Maibach, H. I., Meneghini, C. L., Pirilä, V., and Wilkinson, D. S.,** Occupational dermatitis in five European dermatological departments, *Berufs Dermatosen,* 19, 1, 1971.
67. **Einarsson, Ö., Kylin, B., Lindstedt, G., and Wahlberg, J. E.,** Chromium, cobalt and nickel in used cutting fluids, *Contact Dermatitis,* 1, 182, 1975.
68. **Fregert, S. and Gruvberger, B.,** Chemical properties of cement, *Berufs Dermatosen,* 20, 238, 1972.
69. **Goh, C. L., Kwok, S. F., and Gan, S. L.,** Cobalt and nickel content of Asian cements, *Contact Dermatitis,* 15, 169, 1986.
70. **Ebner, H., Luger, T., Binder, R., and Machata, G.,** Zur allergologischen Bedeutung von Metallspuren in Haushaltsprodukten, *Wien. Klin. Wochenschr,* 90, 311, 1978.
71. **Lindemayr, H.,** Das Friseurekzem, *Dermatosen,* 32, 5, 1984.
72. **Katz, S. A. and Samitz, M. H.,** Leaching of nickel from stainless steel consumer commodities, *Acta Derm. Venereol.,* 55, 113, 1975.
73. **Christensen, O. B. and Möller, H.,** Release of nickel from cooking utensils, *Contact Dermatitis,* 4, 343, 1978.
74. **Bang-Pedersen, N., Fregert, S., Brodelius, P., and Gruvberger, B.,** Release of nickel from silver coins, *Acta Derm. Venereol.,* 54, 231, 1974.
75. **Malo, J.-L., Cartier, A., Doepner, M., Nieboer, E., Evans, S., and Dolovich, J.,** Occupational asthma caused by nickel sulphate, *J. Allergy Clin. Immunol.,* 69, 55, 1982.
76. **Niordson, A.-M.,** Nickel sensitivity as a cause of rhinitis, *Contact Dermatitis,* 7, 273, 1981.
77. **Novey, H. S., Habib, M., and Wells, I. D.,** Asthma and IgE antibodies induced by chromium and nickel salts, *J. Allergy Clin. Immunol.,* 72, 407, 1983.
78. **Tola, S., Kilpiö, J., and Virtamo, M.,** Urinary and plasma concentrations of nickel as indicators of exposure to nickel in an electroplating shop, *J. Occup. Med.,* 21, 184, 1979.
79. **Block, G. T. and Yeung, M.,** Asthma induced by nickel, *JAMA,* 247, 1600, 1982.
80. **Bernacki, E. J., Parsons, G. E., Roy, B. J., Mikac-Devic, M., Kennedy, C. D., and Sunderman, F. W.,** Urine nickel concentrations in nickel exposed workers, *Ann. Clin. Lab. Sci.,* 8, 184, 1978.
81. **Högetveit, A. C., Barton, R. T., and Kostöl, C. O.,** Plasma nickel as a primary index of exposure in nickel refining, *Ann. Occup. Hyg.,* 21, 113, 1978.
82. **Fisher, A. A.,** Metal dermatitis — some questions and answers, *Cutis,* 19, 156, 1977.
83. **Gaul, L. E.,** Development of allergic nickel dermatitis from earrings, *JAMA,* 200, 186, 1967.
84. **Cavelier, C., Fossereau, J., and Massin, M.,** Nickel allergy: analysis of metal clothing objects and patch testing to metal samples, *Contact Dermatitis,* 12, 65, 1985.
85. **Fischer, T., Fregert, S., Gruvberger, B., and Rystedt, I.,** Nickel release from ear piercing kits and earrings, *Contact Dermatitis,* 10, 39, 1984.
86. **Fisher, A. A.,** Safety of stainless steel in nickel sensitivity, *JAMA,* 221, 1279, 1972.
87. **Menné, T., Brandrup, F., Thesdrup-Pedersen, K., Veien, N. K., Andersen, K., Yding, F., and Valeur, G.,** Patch test reactivity to nickel alloys, *Contact Dermatitis,* 16, 255, 1987.
88. **Fischer, T. and Maibach, H. I.,** Recovery of nickel from a standard patch test, *Contact Dermatitis,* 11, 134, 1984.
89. **Kalimo, K., Lammintausta, K., Mäki, J., Teuho, J., and Jansen, C.,** Nickel penetration in allergic individuals: bioavailability versus X-ray microanalysis detection, *Contact Dermatitis,* 12, 255, 1985.
90. **Nörgaard, O.,** Investigations with radioactive Ni 57 into the resorption of nickel through the skin in normal and nickel sensitive persons, *Acta Derm. Venereol.,* 34, 111, 1954.
91. **Cotton, D. W. K.,** Studies on the binding of protein by nickel with special reference to its role in nickel sensitivity, *Br. J. Dermatol.,* 76, 99, 1964.
92. **Kienlein-Kletschka, B.,** Feuchtarbeit als konditionierender Faktor bei der Genese berufsbedingter Dermatosen, *Dermatosen,* 32, 14, 1984.

93. **Kvorning, S. A. and Borup Svendsen, I.**, A synthetic detergent as a provocative agent in patch tests, *J. Invest. Dermatol.*, 26, 421, 1956.
94. **Fregert, S. and Rorsman, H.**, Allergy to chromium, nickel and cobalt, *Acta Derm. Venereol.*, 46, 144, 1966.
95. **van Joost, T. and van Everdingen, J. J. E.**, Sensitization to cobalt associated with nickel allergy: clinical and statistical studies, *Acta Derm. Venereol.*, 62, 525, 1982.
96. **Pirilä, V. and Kajenne, H.**, Sensitization to cobalt and nickel in cement eczema, *Acta Derm. Venereol.*, 45, 9, 1965.
97. **Rystedt, I.**, Evaluation and relevance of isolated test reactions to cobalt, *Contact Dermatitis*, 5, 233, 1979.
98. **Menné, T.**, Relation between cobalt and nickel sensitization in females, *Contact Dermatitis*, 6, 337, 1980.
99. **Veien, N. K. and Kaaber, K.**, Nickel, cobalt and chromium sensitivity in patients with pompholyx (dyshidrotic eczema), *Contact Dermatitis*, 5, 371, 1979.
100. **Fischer, T. and Rystedt, I.**, Cobalt allergy in hard metal workers, *Contact Dermatitis*, 9, 115, 1983.
101. **Rystedt, I. and Fischer, T.**, Relationship between nickel and cobalt sensitization in hard metal workers, *Contact Dermatitis*, 9, 195, 1983.
102. **Menné, T. and Kaaber, K.**, Treatment of pompholyx due to nickel allergy with chelating agents, *Contact Dermatitis*, 4, 289, 1978.
103. **Lammintausta, K., Pitkänen, O.-P., Kalimo, K., and Jansen, C. T.**, Interrelationship of nickel and cobalt contact sensitization, *Contact Dermatitis*, 13, 148, 1985.
104. **Menné, T. and Solgaard, P.**, Temperature-dependent nickel release from nickel alloys, *Contact Dermatitis*, 5, 82, 1979.
105. **Fischer, T., Fregert, S., Gruvberger, B., and Rystedt, I.**, Contact sensitivity to nickel in white gold, *Contact Dermatitis*, 10, 23, 1984.
106. **Cormia, F. E.**, Nickel dermatitis: a report of eleven cases, *Can. Med. Assoc. J.*, 32, 270, 1935.
107. **Foster, P. D. and Ball, F. I.**, Sources of nickel eczema, *Arch. Dermatol. Syphilol.*, 31, 461, 1935.
108. **Eun, H. C.**, Nickel in acupuncture needles, *Contact Dermatitis*, 7, 334, 1981.
109. **Ippen, H.**, Allergische Hautschäden bei der Metallbearbeitung, *Dermatosen*, 27, 71, 1979.
110. **Agathos, M. and Bernecker, H. A.**, Handdermatitis bei medizinischem Personal, *Dermatosen*, 30, 43, 1982.
111. **Cronin, E. and Kullavanijaya, P.**, Hand dermatitis in hair dressers, *Acta derm. Venereol.*, 59 (Suppl. 85), 47, 1977.
112. **Burrows, D., Creswell, S., and Merrett, J. D.**, Nickel, hands and hip prostheses, *Br. J. Dermatol.*, 105, 437, 1981.
113. **Christensen, O. B. and Möller, H.**, External and internal exposure to the antigen in the hand eczema of nickel allergy, *Contact Dermatitis*, 1, 136, 1975.
114. **van Ketel, W. G.**, Occupational contact with coins in nickel-allergic patients, *Contact Dermatitis*, 12, 108, 1985.
115. **Prytstowsky, S. D., Allen, A. M., Smith, R. W., Nonomura, J. H., Odom, R. B., and Akkers, A.**, Allergic contact hypersensitivity to nickel, neomycin ethylenediamine and benzocain, *Arch. Dermatol.*, 115, 959, 1979.
116. **Bettley, T. R.**, The irritant effect of detergents, *Trans. St. Johns Hosp. Dermatol. Soc.*, 58, 65, 1972.
117. **Calnan, C. D.**, Hand dermatitis in housewives, *Br. J. Dermatol.*, 82, 543, 1970.
118. **Feuerman, E. J.**, Housewives eczema and the role of chromates, *Acta Derm. Venereol.*, 49, 288, 1969.
119. **Wilkinson, D. S.**, Contact dermatitis of the hands, *Trans. St. Johns Hosp. Dermatol. Soc.*, 58, 261, 1972.
120. **Christensen, O. B. and Möller, H.**, Nickel allergy and hand eczema, *Contact Dermatitis*, 1, 129, 1975.
121. **Harrison, P. V.**, A postal survey of patients with nickel and chromate dermatitis, *Contact Dermatitis*, 5, 229, 1979.
122. **Wubs, P. L. and Spruit, D.**, Course of nickel dermatitis, *Contact Dermatitis*, 5, 57, 1979.
123. **Morgan, J. K.**, Observations on the persistence of skin sensitivity with reference to nickel eczema, *Br. J. Dermatol.*, 65, 84, 1953.
124. **Wagmann, B.**, Beitrag zur Klinik des Nickelekzems, *Dermatologica*, 119, 197, 1959.
125. **Christensen, O. B.**, Prognosis in nickel allergy and hand eczema, *Contact Dermatitis*, 8, 7, 1982.
126. **Fregert, S.**, *Manual of Contact Dermatitis*, 2nd ed., Year Book Medical Publishers, Chicago, 1981.
127. **Squire, J., Cruickshank, C. N. D., and Topley, E.**, Occupational skin disease, *Br. Med. Bull.*, 7, 28, 1950.
128. **Alomar, A., Conde-Salazar, L., and Romaguera, C.**, Occupational dermatoses from cutting oils, *Contact Dermatitis*, 12, 129, 1985.
129. **Calnan, C. D.**, Dermatology and industry, *Clin. Exp. Dermatol.*, 3, 1, 1978.
130. **Christensen, O. B. and Kristensen, M.**, Treatment with disulfiram in chronic nickel hand dermatitis, *Contact Dermatitis*, 8, 59, 1982.
131. **Gawkrodger, D. J., Cook, S. W., Fell, G. S., and Hunter, J. A. A.**, Nickel dermatitis: the reaction to oral nickel challenge, *Br. J. Dermatol.*, 115, 33, 1986.

132. **Jordan, W. P. and King, S. E.,** Nickel feeding in nickel-sensitive patients with hand eczema, *J. Am. Acad. Dermatol.,* 1, 506, 1979.
133. **Tilsley, D. A. and Rotstein, H.,** Sensitivity caused by internal exposure to nickel, chrome and cobalt, *Contact Dermatitis,* 6, 175, 1980.
134. **Cronin, E. di Michiel, A. D., and Brown, S. S.,** Oral challenge in nickel-sensitive women with hand eczema, in *Nickel Toxicology,* Brown, S. S. and Sunderman, F. W., Jr., Eds., Academic Press, London, 1980, 149.
135. **Kaaber, K., Menné, T., Tjell, J. C., and Veien, N.,** Antabuse treatment of nickel dermatitis, Chelation — a new principle in the treatment of nickel dermatitis, *Contact Dermatitis,* 5, 221, 1979.
136. **Kaaber, K., Veien, N. K., and Tjell, J. C.,** Low nickel diet in the treatment of patients with chronic nickel dermatitis, *Br. J. Dermatol.,* 98, 197, 1978.
137. **Lahti, A. and Maibach, H.,** Immediate contact reactions (contact urticaria syndrome), in *Occupational and Industrial Dermatology,* 2nd ed., Maibach, H. I. and Gellin, G. A., Eds., Year Book Medical Publishers, Chicago, 1987, 32.
138. **Mitchell, J. C.,** Type I without type III and type IV hypersensitivity reaction to nickel, *Contact Dermatitis,* 7, 270, 1981.
139. **Menné, T. and Thorboe, A.,** Nickel dermatitis — nickel excretion, *Contact Dermatitis,* 2, 353, 1976.
140. **Warin, R. P. and Smith, R. J.,** Chronic urticaria. Investigation with patch and challenge tests, *Contact Dermatitis,* 8, 117, 1982.
141. **Stoddart, J. C.,** Nickel sensitivity as a cause of infusion reaction, *Lancet,* 2, 741, 1960.
142. **Wahlberg, J. E.,** Nickel allergy and atopy in hair dressers, *Contact Dermatitis,* 1, 161, 1975.
143. **Hansen, K. S.,** Occupational dermatoses among hospital cleaning women, *Contact Dermatitis,* 9, 343, 1983.
144. **Preininger, T.,** Überempfindlichkeit gegen Nickelgeld, *Dermatol. Wochenschr.,* 99, 1082, 1934.
145. **Rothman, S.,** Überempfindlichkeit gegen Hartgeld, *Dermatol. Wochenschr.,* 90, 98, 1930.
146. **Bettly, F. R.,** Nickel coin dermatitis, *Contact Dermatitis Newsl.,* 9, 198, 1971.
147. **Bulmer, F. M. R. and Mackenzie, E. A.,** Studies in the control and treatment of "nickel rash", *J. Ind. Hyg.,* 8, 517, 1926.
148. **Sunderman, F. W.,** A review of the metabolism and toxicology of nickel, *Ann. Clin. Lab. Sci.,* 7, 377, 1977.
149. **Hodgson, G.,** Cutaneous hazards of lubricants, *Trans. Soc. Occup. Med.,* 19, 9, 1969.
150. **Fisher, A. A.,** Allergic contact dermatitis of the hands due to industrial oils and fluids, *Cutis,* 23, 131, 1979.
151. **Samitz, M. H. and Katz, S. A.,** Skin hazards from nickel and chromium salts in association with cutting oils, *Contact Dermatitis,* 1, 158, 1975.
152. **Arndt, K. A.,** Cutting fluids and the skin, *Cutis,* 5, 143, 1969.
153. **Gellin, G. A.,** Cutting fluids and skin disorders, *Ind. Med.,* 39, 65, 1970.
154. **Rycroft, R. J. G.,** Cutting fluids oils and lubricants, in *Occupational and Industrial Dermatology,* 2nd ed., Maibach, H. I., Ed., Year Book Medical Publishers, Chicago, 1987, 286.
155. **Fischer, T. and Rystedt, I.,** Hand eczema among hard-metal workers, *Am. J. Ind. Med.,* 8, 381, 1985.
156. **Weiler, K. J.,** Nickelkontaktekzem durch Elektroschweissarbeiten, *Dermatosen,* 27, 142, 1979.
157. **Handfield-Jones, S., Boyle, J., and Harman, R. R. M.,** Contact allergy caused by metal sprays, *Contact Dermatitis,* 16, 44, 1987.
158. **Fregert, S.,** Plastic and metal — a combination that stimulates the sale, *Contact Dermatitis Newsl.,* 12, 329, 1972.
159. **Coenraads, P. J., Nater, J. P., Jansen, H. A., and Latinga, H.,** Prevalence of eczema and other dermatoses of the hands and forearms in construction workers in the Netherlands, *Clin. Exp. Dermatol.,* 9, 149, 1984.
160. **Varigos, G. A. and Dunt, D. R.,** Occupational dermatitis. An epidemiological study in the rubber and cement industries, *Contact Dermatitis,* 7, 105, 1981.
161. **Weiler, K. J. and Russel, H. A.,** Branchenspezifike Einwirkungsprobleme von Zement Kalk und Gips auf die Haut, *Dermatosen,* 28, 182, 1980.
162. **Kuhl, M.,** Zur Nickelallergie im Friseurgerwerbe, *Zentralbl. Arbeitsmed. Arbeitsschutz,* 29, 159, 1979.
163. **Landthaler, M.,** Farbstoff und Nickelallergie im Friseurgewerbe, *Hautarzt,* 32, 281, 1981.
164. **Lindemayr, H.,** Eczema and nickel allergy in hairdressers, *Hautarzt,* 35, 292, 1984.
165. **Ludwig, E.,** Für allergische Berufsekzeme der Friseure verantwortlige Allergene, *Dermatosen,* 30, 159, 1982.
166. **Lynde, C. W. and Mitchell, J. C.,** Patch test results in 66 hairdressers 1973-81, *Contact Dermatitis,* 8, 302, 1982.
167. **Nethercott, J. R., MacPherson, M., Choi, B. C. K., and Nixon, P.,** Contact dermatitis in hairdressers, *Contact Dermatitis,* 14, 73, 1986.
168. **Wilkinson, D. S. and Hambly, E. M.,** Prognosis of hand eczema in hairdessing apprentices, *Contact Dermatitis,* 4, 63, 1978.

169. **Mattheus, A. and Heise, H.,** Berufstauchlichketisuntersuchungen bei Friseurbewerben, *Dermatol. Monatsschr.,* 172, 23, 1986.
170. **Reichenberger, M.,** Katamnestische Untersuchungen bei 69 Friseuren, *Hautarzt,* 23, 252, 1972.
171. **Reichenberger, M.,** Befunde bei Erstuntersuchungen von Hautkranken im Friseurgewerbe unter besonderer Berucksichtigung der Dyshidrosis, *Berufs Dermatosen,* 20, 124, 1972.
172. **Hemingway, J. D. and Molokhia, M. M.,** The dissolution of metallic nickel in artificial sweat, *Contact Dermatitis,* 16, 99, 1987.
173. **Lammintausta, K., Kalimo, K., and Havu, V. K.,** Contact allergy in atopics, who perform wet work in hospitals, *Dermatosen,* 30, 184, 1982.
174. **Nilsson, E., Mikaelsson, B., and Andersson, S.,** Atopy, occupation and domestic work as risk factors for hand eczema in hospital workers, *Contact Dermatitis,* 13, 216, 1985.
175. **Singgih, S. I. R., Lantinga, H., Nater, J. P., Woest, T. E., and Kruyt-Gaspersz, J. A.,** Occupational hand dermatoses in hospital cleaning personnel, *Contact Dermatitis,* 14, 14, 1986.
176. **Rystedt, I.,** Factors influencing the occurrence of hand eczema in adults with a history of atopic dermatitis in the childhood, *Contact Dermatitis,* 12, 185, 1985.
177. **Dawber, R. and Sonnex, T.,** Nickel dermatitis due to steel fibre and soap cleaning pads, *Contact Dermatitis,* 8, 342, 1982.
178. **Fisher, A. A., Ed.,** *Contact Dermatitis,* 3rd ed., Lea & Febiger, Philadelphia, 1986.
179. **Friedman, S. J. and Perry, H. O.,** Erythema multiforme associated with contact dermatitis, *Contact Dermatitis,* 12, 21, 1985.
180. **Pirilä, V. and Förström, L,** Pseudo-cross-sensitivity between cobalt and nickel, *Acta Derm. Venereol.,* 46, 40, 1966.
181. **Lada-Jagas, R.,** Allergie auf Chrom und Nickel in Margarinefabrik, *Przegl. Dermatol.,* Suppl. 63, 97, 1976.
182. **Heber, W.,** Nickel allergy in butchers, *Berufsgenossenschaft,* 8, 452, 1986.
183. **Pincelli, C., Monolese, A., and Pincelli, L.,** Fiddler's neck and nickel dermatitis, *Contact Dermatitis,* 13, 37, 1985.
184. **Machackova, J. and Pock, L.,** Occupational nickel dermatitis in a cellist, *Contact Dermatitis,* 15, 41, 1986.
185. **Kieffer, M.,** Nickel sensitivity: relationship between history and patch test reaction, *Contact Dermatitis,* 5, 398, 1979.
186. **Raith, L. and Jaeger, K.,** The nickel content of chalk — cause of contact dermatitis?, *Contact Dermatitis,* 14, 61, 1986.
187. **Wall, L. M.,** Nickel penetration through rubber gloves, *Contact Dermatitis,* 6, 461, 1980.
188. **Caron, G. A.,** Nickel sensitivity and atopy, *Br. J. Dermatol.,* 76, 384, 1964.
189. **Wahlberg, J. E. and Skog, E.,** Nickel allergy and atopy; threshold of nickel sensitivity and immunoglobulin E determinations, *Br. J. Dermatol.,* 85, 97, 1971.
190. **Rystedt, I.,** Contact sensitivity in adults with atopic dermatitis in the childhood, *Contact Dermatitis,* 13, 1, 1985.
191. **Blondeel, A., Achten, G., Dooms Goossens, A., et al.,** Atopy and allergic contact dermatitis, *Ann. Dermatol. Venereol.,* 114(2), 203, 1987.

Chapter 13

NICKEL ALLERGY AND HAND ECZEMA

D. S. Wilkinson and J. D. Wilkinson

TABLE OF CONTENTS

I. Introduction 134

II. Nickel Allergy in the General Population 134

III. Hand Eczema in the General Population 135

IV. The Relationship Between Hand Eczema and Nickel Allergy 136
- A. Nickel Sensitivity in Patients with Hand Eczema 136
- B. Hand Eczema in Nickel-Sensitive Patients 138
  - 1. Sex Distribution 139
  - 2. Time Relationship between Nickel Sensitivity and Hand Eczema 139
- C. Conclusions 140

V. Factors Involved in Hand Eczema in the Nickel-Sensitive Patient 141
- A. Factors Relating to Hand Dermatitis 141
- B. Factors Related to Nickel Sensitivity 142

VI. Factors Common to Hand Eczema and Nickel Sensitivity 143
- A. Wet Work 143
- B. Atopy 144

VII. Sources of Nickel-Induced Hand Eczema 146
- A. Sources of Primary Sensitization 147
  - 1. Occupational 147
  - 2. Environmental 147
- B. Sources of Elicitation 147
  - 1. Acceptable 147
  - 2. Questionable 148
    - a. Coins 148
    - b. Stainless-Steel Sinks and Pans 149
    - c. Detergents 149

VIII. Clinical Patterns 149
- A. Pompholyx, Dyshidrotic Eczema, Dyshidrosis 151
  - 1. Nickel Allergy and Pompholyx 152
  - 2. The Role of Systemic Nickel in Patients with Pompholyx 152
- B. Secondary Spread 153
- C. Contact Urticaria 153

IX. Diagnosis 153

X. Treatment 154

XI. Course and Prognosis 155

References 158

I. INTRODUCTION

An aspect of nickel dermatitis which is both difficult and controversial will be considered in this chapter. It is difficult because eczema of the hands is a multifactorial disease[1-3] and controversial because of differences in the material used to justify conclusions on the importance of nickel sensitization in this context.

The terms "eczema" and "dermatitis" have often caused confusion, especially to those not brought up in the European tradition of dermatology, where the term "eczema" is well entrenched. We agree with Calnan[3] that both terms can, for most practical purposes, be considered synonymous and interchangeable. However, a slight nuance is evident in published reports, particularly when the hands are involved. Authors who are dealing particularly with exogenous or occupational causes tend to use the term "dermatitis", while those who are concerned more with endogenous or constitutional factors prefer the term "eczema". The difficulty of separating these two groups of factors is a further argument in favor of synonymity of the terms. We shall also touch briefly in this chapter on some conditions that cannot rightly be considered eczematous, e.g., contact urticaria and the "irritant reaction".

Before we consider the relationship between nickel sensitization and hand eczema, we must determine, as far as is possible, the prevalence of both in the community at large. Most, though not all, of the reports indicating a close relationship between the two are based on material that is doubly selected: first, by being seen at hospital and, second, by being referred for patch testing. Few such authors are able to match their data with age- and sex-matched patch test-negative patients or with those seen with hand eczema in general dermatological departments.[4] Nor is the incidence of nickel sensitivity in noneczematous age-matched control material always taken into account when assessing the significance of the levels of nickel allergy found in these patients with hand eczema.[5] However, there have been sufficient epidemiological studies on the prevalence of both nickel allergy and hand eczema to be able to arrive at a reasonable estimate of the prevalence of both.

II. NICKEL ALLERGY IN THE GENERAL POPULATION

Epidemiological studies based on patch test material obviously do not reflect the prevalence of nickel allergy in the general population because the material will be biased heavily towards patients with suspected contact dermatitis or with hand eczema. Such figures vary widely between countries and even between different centers in the same country[6-8] (Chapter 11, Table 1). In the case of nickel, they also fluctuate as a result of changes in occupational patterns, fashion, culture, and habits[8] (Chapter 12). Finally, the figures differ considerably between the sexes and in different age groups. Thus, in a recent analysis of 1245 females patch tested at Wycombe from 1983—87, the total figures of 16.8% sensitive to nickel hid age-related differences from 9.2 to 24.3%.[9] This is important in relating the prevalence of nickel sensitivity in the age groups in which hand eczema is most common. Although the frequency of nickel sensitivity in women in the general population can be taken to be approximately 10%[4,10-13] (Chapter 11, Table 2), it is higher than this at the period of life in which hand eczema commonly develops. It is this figure — perhaps as much as 15 to 18% — that must be taken as the "background" level of nickel sensitivity which has to be used in assessing the relevance of positive nickel patch tests in patients with hand eczema. However, it is not clear from published reports what proportion of these patients with a relatively high level of nickel sensitivity are patients with hand eczema at the time of testing.

Other epidemiological studies are based on questionnaires,[4,10-13] either alone or followed by examination and perhaps by patch testing, which correct inaccuracies in patients' beliefs, but which may introduce their own errors. Even "general population" studies may be based on poorly selected material, as Menné et al. point out.[4]

III. HAND ECZEMA IN THE GENERAL POPULATION

This is even more difficult to assess than the prevalence of nickel sensitivity, which can be estimated by a questionnaire and then confirmed by patch testing. There have been few population studies of hand eczema, and in those that have been carried out, a clear definition of what is included as "hand eczema" has not always been specified. Patients' views about this are notably inexact and even dermatologists differ in their interpretation. Some will include transient cases of the "irritant reaction" of recurrent chapping and fissuring; some exclude hyperkeratotic dermatitis of the palms. Even the definition of "pompholyx" is open to different constructions. Most authors confine their studies to those whose hands alone are affected, but others may include the wrists, forearms, or other sites.

The troublesome and often persistent nature of hand eczema is reflected in the frequency of hospital attendance for this condition, even though these account for only a minority of all cases occurring in the general population. Hand eczema accounted for over 30% of cases of eczema seen in eight European centers[6] and 21% of all cases of eczema seen in another series.[1]

Some studies are concerned with point prevalence and others with past incidence, or "period prevalence",[14] over a defined number of years. The most frequently cited estimate of prevalence is that of Agrup,[15] whose enormous and painstaking field study still remains an important source of information. She estimated the prevalence in a population of 141,000 as between 1.2 and 2.4%, with a female-to-male sex ratio of 2:1. Menné[4,10,16] in more recent studies calculated a prevalence in women of 2.3 to 6.2% with a cumulated prevalence of 22%. The extensive North American HANES study[17] was not specifically attuned to hand eczema as such; for example, "dyshidrosis" (more common in males) was distinguished from other forms of "hand eczema" and atopic and contact dermatitis were treated separately. The prevalence was therefore understandably lower.

In a study of the adult population of a part of the Netherlands,[18] subjects were re-examined after an interval of 3 years. In 1979, 5.8% had a 3-year prevalence of eczema of the hands and forearms; on re-examination in 1982, it was 7.1%, but the method of examination was not strictly comparable. Manifest hand eczema at the time of examination was found in 3.4%. They calculated the incidence of new cases of hand eczema to be 0.8% annually (0.5% for males and 1.1% for females). In a further analysis of 3140 adults in the Netherlands, the 3-year period prevalence of eczema of the hands and/or forearms was 4.6% for males and 8.0% for females.[19]

During a survey of coronary risk factors in northern Norway,[20] 14,667 men and women 18 to 55 years old were asked about "allergic hand eczema" during the previous 12 months; 4.9% of the men and 13.2% of the women replied affirmatively. These high figures might, the authors felt, be related to the particular climatic conditions in that region. Housewives showed a slightly higher incidence than women employed outside the home.

During a pre-employment examination of 2452 applicants for hospital work, Nillson et al.[21] found a history of (well-defined) hand eczema in 21% of 2342 predominantly female applicants and in 8.2% of 100 predominantly male applicants.

Finally, in a response rate of 83% of 20,000 inhabitants of an industrial city, 11% considered themselves to have had hand eczema in the previous 12 months; $^{2}/_{3}$ were women.[14] Of these, 71% were examined. A point prevalence rate of 5.4% was found.

Summarizing these rather diverse figures drawn predominantly from Scandinavian countries and which may, therefore, not be applicable to different races, countries, climates, and cultures, we can come to the following conclusions:

1. Hand eczema is more common in females than in males in an unselected population by a factor of two or three to one and particularly affects housewives. Analyses of occupational groups[22,23] may, of course, show a different ratio.
2. The prevalence of hand eczema in adult women is of the order of 2 to 4% at any one time in adult life.
3. The incidence of new cases annually is of the order of 0.5% in males and 1% in females.
4. In a 3-year period 5 to 8% of men and 13 to 15% of women will have had at least one episode of hand eczema.
5. About 20% of women will have had hand eczema at some time in their lives.
6. The figures for males are less well studied, but are certainly lower.

However, we must emphasize that these figures are derived by extrapolation from particular samples and may not be universally applicable. The surveys, although meticulously carried out, are concerned, in the main, with Scandinavian or northern European stock. They may not be applicable to other races, countries, climates, or cultures.

IV. THE RELATIONSHP BETWEEN HAND ECZEMA AND NICKEL ALLERGY

We must now consider the extent to which hand eczema and nickel allergy co-exist and whether there is a relationship that is closer and more significant that can be accounted for by the co-existence of two common conditions.

The problem chiefly concerns women. In men, the frequency of nickel sensitivity is much lower,[5] even in countries where the occupational risks are high.[8] Occupational hazards in platers and metal workers will obviously tend especially to affect the hands and forearms;[20,21] in such cases the connection is obvious. The problem is discussed fully in Chapter 12. We shall, however, consider "pompholyx" in males later in this chapter.

There are two ways of looking at this relationship: the incidence of nickel sensitivity in patients with hand eczema and the incidence of hand eczema in patients who are nickel sensitive.

A. Nickel Sensitivity in Patients with Hand Eczema (Table 1)

Here we have to rely for the most part on patch test material, which is inevitably selected to a greater or lesser degree. In some patch test clinics, most of those tested will have been referred only because of a strong suspicion of allergic contact dermatitis. In others, most cases of hand eczema — whatever the type — may have been referred as a matter of policy. The question to be asked is whether these patients show a higher rate of positive reactions than can be accounted for by the incidence of nickel sensitivity in the adult female population. There are, fortunately, some good field studies in which all patients in a population sample who claim to have or to have had hand eczema are subjected to testing. Agrup[15] found 12% of 462 women with hand eczema to give positive reactions to 5.4% nickel sulfate. She

Table 1
NICKEL SENSITIVITY IN PATIENTS WITH HAND ECZEMA (HE)

Location	Subjects	Positive (%)	Ref.
Sweden	462 F	12	15
	250 M	0	
Scandinavia	346 F	9	30
	288 M	3	
Europe	2383 F	11 (with HE) 9 (without HE)	25
Europe	283 F	10 (6 — 29)	26
Wycombe	177 F	17	27
	108 M	1	
Wycombe	129 F	17	13
	119 M		
Wycombe	233 F	15	28
	119 M	2	
London	238 F	18 (with HE) 12 (without HE)	24
London	263 F	28	29

believed the finding to be relevant in all but seven patients. However, Cronin[24] found the same percentage in 285 women *without* hand eczema, whereas 21% of 238 *with* hand eczema alone reacted positively to nickel. Such control groups are invaluable in considering the relevance of figures that may not be greatly in excess of the incidence of nickel sensitivity found in the general population.

Patch test studies of hospital material are distorted by many factors: local occupational patterns, cultural differences in adornment, selection of material, and, perhaps, climate.[20] In the first large European multicenter survey,[25] only 11% of women with hand eczema gave positive reactions to nickel, compared to 9% in those without hand eczema. The figures for men were 2% and 1%. In 283 housewives from the same material,[26] 10% were nickel sensitive, but there was a wide range between the centers (6 to 29%). In a continuing personal study of 177 females and 108 males with eczema of the hands,[27] only 17% of the women and 1% of the men gave positive reactions — figures very similar to those of Cronin.[24] Finally, a very recent computerized study from the same hospital[28] gave a figure of 15% of 233 women and 2% of 159 males with hands only affected.

When hands were affected with other sites, the figures were 16 and 3%, respectively. These figures, over a period of 15 years from the same center, are remarkably consistent, although there were considerable differences in different forms of hand eczema, a point we shall return to later. In contrast, 36% of women with clinically diagnosed contact dermatitis reacted to nickel.

Meanwhile, Cronin[29] had also carried out a further study of 263 women with hand eczema. She found that 28% reacted to nickel, about equally for four different site patterns of eczema. This rather high figure may be due to the nature of the material in this clinic or perhaps to differences between capital city and semirural environments. Such differences are often apparent in multicenter trials, even when the same protocol is used.[8,25,30]

Any conclusions to be drawn from these figures must be tentative. Further studies are needed on age- and sex-matched nickel-sensitive patients with and without hand eczema of specified type and severity. The difference in the frequency of nickel sensitivity between such comparisons as exist is not striking. Given that a figure of about 10% positive reactors is to be expected from any female adult population in the geographical regions studied, one can only state that this figure is somewhat higher in patients with hand eczema. Since the

Table 2
HAND ECZEMA IN NICKEL-SENSITIVE PATIENTS

Location	Subjects	Positive (%)	Ref.
London	400 F	20	31
London	85 F	17	32
Finland	Gen. pop.	20	11
U.S.	36 F	40	34
Sweden	621 (552 F)	43	35
Sweden	165	52	33
London	1321 F	56	5
	185 M	64	
Sweden	53 (49 F)	47	
Sweden	149 mostly pompholyx	47	36
Wycombe	83	43 (24 hands only; 19 hands, other)	13
Wycombe	209 F	29 (17 hands only; 12 hands, other)	28
Denmark	279	43	4,38
Denmark	79 twins (15/ 32 pompholyx)	41	40
Scotland	119 F	46	39
	15 M	73	

peak age for the acquisition of nickel sensitivity is earlier than that for hand eczema (except perhaps in young atopics), there appears to be a slightly greater risk for sensitized females to develop hand eczema. This is entirely understandable on the basis of elicitation by contact or, perhaps, by diet (Chapter 14).

B. Hand Eczema in Nickel-Sensitive Patients

This aspect of the interrelationship between nickel allergy and hand eczema has received more attention and has aroused much interest and speculation. Unfortunately, it is difficult to interpret the figures quoted (Table 2) since the time span involved is not always stated. Some refer to prevalence at the time of examination and some to a past history of hand eczema. As with the preceding figures discussed above, the nature and severity of the eczema is seldom well defined. In an open-ended question on the lines of "have you ever suffered from eczema (dermatitis) of the hands" subjects may well be more likely to answer "yes" than "no" if they have any doubt, even if such attacks have been transient episodes of chapping and soreness. Some of those who have studied the problem have specified a minimum period of involvement, e.g., 3 weeks. Others have indeed attempted to grade the severity. However, few studies have included a matched "control" group of women who are not nickel sensitive. Finally, it is not always clear whether figures refer to eczema of the hands alone or eczema that includes involvement of the hands.

Despite these criticisms, the figures quoted in Table 2 are not as disparate as they seem at first glance. They fall into two groups, and within these groups the figures are not inconsistent with one another. The first group[11,13,28,31,32] refers to the presence of hand eczema in females at the time of enquiry or patch testing. In these, the incidence of nickel sensitivity appears to lie between 17 and 24%. The second group[3,5,33-40] is concerned with past or present hand eczema. Here, the incidence is between 40 and 56%. All but two lie between 40 and 47%. The exceptions are one series from Sweden[33] and the large series from London,[5] which may be influenced by a larger proportion of female office workers, cleaners, etc.

Figures for males are seldom available in these surveys in sufficient numbers to justify comparison, but where they are specifically quoted[5,39] they are higher. This is compatible with a higher occupational incidence in men of both nickel sensitivity and irritant or allergic hand dermatitis and is considered in more detail in Chapter 12.

In comparing these findings with the figures given above for the prevalence or period prevalence of hand eczema in the general population, one must conclude that the nickel-sensitive female is more prone to have or to have developed hand eczema than the woman who is not sensitive to this metal.

1. Sex Distribution

Hand eczema is more common in women, if specific occupational risks are excluded. Agrup[15] found the female-to-male ratio in 827 persons examined to be 2.3:1 for allergic contact dermatitis and 3.2:1 for "traumiterative" (chronic irritant) dermatitis. Peltonen found a similar ratio in 980 subjects examined. Although the reason for this is unknown, we can probably assume that the woman's greater exposure to repeated wet work and household "wear and tear" is a significant factor. The sex difference becomes blurred in those of either sex who are engaged professionally in wet work.[41]

2. Time Relationship Between Nickel Sensitivity and Hand Eczema

It is accepted that the peak age for the development of nickel sensitivity in the female is in the second or into the third decade. Menné[38] found the highest incidence density before the age of 20, with a twofold increase over a period of 40 years. In a personal questionnaire survey,[13] over 50% of women who claimed to have become sensitive had done so by the age of 20 and over 70% by the age of 30.

Contact dermatitis of the hands has been less well studied, but it is generally accepted that it is chiefly an affliction of adults, with its main incidence in women in the childbearing age, but extending into the sixth or seventh decade. The peak period for the onset of housewives' dermatitis was found by Agrup to be between the ages of 20 and 40.[15] Few other studies have been concerned with the age of onset of hand eczema; even these have shown such a wide range on each side of the mean as to be of limited value. Agrup[15] found the mean age of onset in both sexes together to be 37 years in allergic contact dermatitis of the hands and 33 years in those cases due to irritants. In a study of nickel-sensitive twins with hand eczema, Menné and Holm[40] found the mean age of onset to be 45 years, but with a range of 20 to 70 years.

Hospital figures based on selected material are of even more limited value, since most deal only with age at presentation to the clinic. In an analysis of 124 females and 66 males presenting with hand eczema at Wycombe Hospital, 49% of the women and 45% of the men had first developed this before the age of 30 years and 19% of women and 23% of men after the age of 50 years.

"Pompholyx", however, may have a younger mean of age of presentation. In a personal series of 106 cases,[13] 60% of women and 26% of men had their first attack before the age of 30 years. This was most marked among the 16 of 41 females and the 15 of 65 males who gave a personal history of atopy. In recent survey from the same hospital,[28] 42% of females and 33% of males with "pompholyx" presented to the patch test clinic before this age.

With so wide a span in the age of onset of hand eczema added to the other usually unspecified variables of type, pattern, and severity, it is unlikely to be worthwhile to consider this general relationship further. Of more interest are the observations regarding the sequence of the development of nickel allergy and hand eczema, more especially in cases of "pompholyx", which shares with nickel sensitization a relatively early onset. Menné and Holm,[40] in their study of nickel-sensitive twins, found 32 with hand eczema; 15 of these had "pom-

Table 3
TIME-RELATIONSHIP BETWEEN ONSET OF NICKEL ALLERGY AND HAND ECZEMA

Material	Hand eczema (%)	Ref.
622 patients	(18 ''primary'', 25 ''secondary'')	35
53 patients	30 ''primary''	36
85 patients	6 ''primary''	32
198 patients	19 ''primary''	42

Material	Ni allergy first (%)	Hand eczema first (%)	Simultaneously (%)	Ref.
50 patients	46	14		24
66 females	73	1.5	25	37
32 twins	66	19	16	40

pholyx''. In 12 the ''pompholyx'' had appeared after the onset of nickel eczema and in only 1 before it. In contrast, in 11 subjects with a dorsal/web pattern of hand eczema, this had occurred in 4 subjects before the onset of nickel allergy.

Christensen's[37] series consisted of 66 patients, 51 of whom had a ''pompholyx''-type eczema. Their metal sensitivity had been noticed at a mean age of 25 years and the hand eczema at 32 years. In three fourths of the patients, the metal sensitivity preceded the hand eczema and in one fourth both appeared simultaneously. Cronin,[24] however, found that only 46% of her 50 patients developed their hand eczema after nickel sensitization and that it appeared first in 14%. The proportion of patients in whom nickel dermatitis appeared primarily on the hands differs widely in other accounts[32,36,42] (Table 3).

C. Conclusions

It is evident from these studies and from general experience that the following relationships may exist between metal sensitivity and hand eczema:

1. A primary allergic dermatitis of the hand occurs as a result of contact with soluble or solubilized nickel. This is more likely to occur from industrial contacts in males, when it may also involve the forearms or other sites. The degree to which it figures in reports depends greatly on the nature and selection of material. It may occur, although probably rarely, in housewives, cashiers, and office workers, but it is always difficult to be certain that they had not been previously sensitized by jewelry.
2. A primary sensitization occurs on hands which have developed an irritant dermatitis. The nickel sensitivity may become established rapidly, as may happen in hairdressers[43,44] (although it is often uncertain which comes first), and is more frequent in those who are engaged in wet work. This subject is discussed further below and, in more detail, in Chapter 12.
3. An allergic hand dermatitis is elicited in a patient already sensitized to nickel by contact elsewhere. If we can assume that about 10% of the adult female population is nickel sensitive, then these women should be at particular risk of developing such a reaction. In practice, we agree that it appears to be uncommon,[37] perhaps because patients may perceive the connection and desist from the contact.
4. An irritant hand eczema occurs as a result of the above. What was initially an allergic dermatitis may, by interfering with the normal barrier, allow penetration by irritants to have a greater effect. This can only be determined — if at all — by the history.

5. Atopic hand dermatitis occurs in nickel-sensitive females. These are usually young and may have had atopic eczema involving the hands in childhood.[45] Many of these cases may be hybrids or have a coincidental nickel allergy. The subject of atopy is discussed below.
6. A completely disconnected and fortuitous association of irritant dermatitis or an allergic contact dermatitis to other allergens than nickel occurs in a nickel-sensitive woman. This situation is particularly common in longstanding cases in which secondary allergens have been garnered in the course of attempted prevention or treatment, for example, to medicaments, perfumes, or rubber chemicals. We believe that this must account, on the basis of the prevalence of nickel allergy, for at least 10% of all cases of persistent hand eczema where positive reactions to nickel are found on patch testing. We believe that the relevance of this nickel sensitivity in these cases is often overrated and that, unless an extremely careful and exact history implicates this metal, the nickel allergy should be regarded as of ''unknown relevance''.
7. ''Pompholyx'' is a special form of hand eczema that is in a separate category although an acute, vesicular dermatitis of the volar surfaces of the hands may at times be due to contact sensitivity of Type IV[46] or even Type I.[47] However, the main interest in regard to nickel allergy lies in the assumed — but somewhat disputed — association with ingested nickel. This subject is discussed below and, more fully, in Chapter 14.

V. FACTORS INVOLVED IN HAND ECZEMA IN THE NICKEL-SENSITIVE PATIENT

Discussion of all the factors involved in the etiology of hand eczema is outside the remit of this chapter. These are discussed in standard texts and articles.[1-3,25,27,29,48] ''Hand dermatitis . . . is not a single entity but an affliction with multiple causes.''[48] The nickel-sensitive person incurs the same risks as those who are not sensitive to this metal as well as the extra risks associated with this sensitivity. We must accept, therefore without further discussion, the role of occupation, wear and tear, friction and abrasion[49,50] — often underestimated — and allergens other than nickel in initiating an exogenous eczema of the hands. We shall discuss here some less well-recognized factors that have been implicated and others that have been regarded as relevant in the development or elicitation of nickel sensitivity in patients with normal or with damaged hands.

A. Factors Relating to Hand Dermatitis

Pre-existing dermatitis — Rystedt[45,51] stresses the risk of hand eczema developing in patients who have had atopic dermatitis, especially on the hands in earlier years, or who have a persistent atopic dermatitis or dry, itchy skin elsewhere. Nilsson et al.[21] also found that hospital workers with pre-existing atopic dermatitis were more prone to develop hand eczema and that this tended to be more severe.

Adverse domestic background: young children — Nilsson et al. also found, in a multivariate regression analysis of risk factors, that the odds for developing hand eczema in an occupation were ''twice as high for subjects with the most unfavourable combination of domestic factors than for subjects with the most favourable combination.''[21] Among ''unfavourable'' factors were the pressures of young children under 4 years and lack of a dishwashing machine. It is common experience, of course, to obtain a history of hand dermatitis starting when the care of young children constitutes an extra load on the hands of a mother and housewife.

Hard water: change of domicile — Although far less important, all dermatologists have encountered patients whose hand dermatitis has begun soon after a move from a soft- to a hard-water area (e.g., northwest to southeast England). This change of abode may also involve extra decorating, painting, and wear and tear. We accept that this is a hypothesis with no evidence to support it other than that of clinical experience. However, it would fit well with Malten's concept of a repeated and gradually increasing cumulative irritant load.[52]

Low humidity and climatic conditions — The high incidence of hand eczema in a survey in northern Norway was thought to be due to the prevalent climatic conditions there.[20] Certainly, winter conditions favor the onset of chapping, dryness, and fissuring of the skin of the hands, a form of irritant reaction which may proceed to a true eczematous state.

The "eczematous diathesis" — The concept of a genetic or "constitutional" factor, apart from atopy, is at present entirely speculative. Careful twin studies[4] have not so far revealed any genetic predisposition to hand eczema. Nevertheless, those who have studied occupational dermatitis have pointed to the apparent marked differences in susceptibility between one worker and another when both are in contact with the same potential irritants, e.g., soluble oils.[53] This cannot always be explained on the basis of an atopic diathesis.

B. Factors Related to Nickel Sensitivity

We shall now deal with some factors that have been suggested, with varying degrees of confidence, as being important in the development of hand eczema in nickel-sensitive subjects.

Genetic — Here again, the evidence is oblique and speculative. It has been suggested[54] that sensitization to a particular chemical is only an "episode in an eczematous career" and that women who are genetically disposed to nickel allergy show symptoms of the disease early, whereas those who are not can tolerate nickel alloys directly on the skin throughout a lifetime.[10] Genetic predisposition to contact allergy is known in guinea pigs;[55] twin studies in humans[38,40] are not conclusive, but do suggest a possible susceptibility to the development of nickel allergy.

Hormonal — In an attempt to explain the great preponderance of females with nickel sensitivity, the effects of estradiol on lymphocytes was examined.[56] This was found to enhance nickel-induced blast transformation, perhaps because of a decrease in suppressor cell activity.

Friction, attrition, and pressure — Moderate sweating increases friction. By repeatedly exposing new surfaces, friction may increase the quantity of metal dissolved.[57] This is of obvious relevance to nickel sensitization at nickel contact sites. Trauma and friction were regarded as important in the development of cobalt allergy in metal workers.[58] The authors point out that cobalt is extremely soluble in plasma that could be produced during scaling and fissuring of the fingertips as part of a primary contact dermatitis. The same may be true of nickel.

Sweating — Although nickel is resistant to attack by air and water and is poorly released by alloys in saline solution or deionized water, it readily dissolves in whole blood and in synthetic sweat (Chapter 4). Menné et al.[59] found that the degree of release of nickel from alloys varied significantly and that those releasing more than $1 \mu g/cm^2$/week gave strong patch test reactions in nickel-sensitive patients, while those releasing less than $0.5 \mu g/cm^2$/week usually gave weak reactions. In a study of the dissolution of pure metallic nickel in artificial sweat, this was found to increase greatly with a lowering of the pH, with time, and by a factor of 2.5 for every 10°C rise in temperature.[57] The rates of dissolution may be more with metal that is impure or imperfectly plated.[60]

Humidity, temperature — It is a commonplace observation that excessive sweating or very humid conditions enhance the likelihood of developing signs of metal dermatitis at sites of contact. Among 354 women who claimed to be nickel sensitive, 4 stated that their dermatitis was only present (beneath watches or a ring) when they were in Hong Kong or Malaysia and that they had no further trouble after returning to England. The higher ambient temperature may also have accelerated nickel dissolution in their sweat. Other nickel-sensitive patients note a worsening of symptoms during hot summers. The increased dissolution of nickel with a rise in temperature, noted above, may also be relevant to housewives and cleaners.

Hemingway and Molokhia[57] raise an interesting point in speculating whether differences in the pH of sweat in nickel-sensitive patients may put them at risk, citing the analogy of the "rusters".[61] in whom corrosion rates of metal increased with increased sweating and an increased concentration of salt from evaporation.

Occlusion — Similarly, occlusion, i.e., by rubber or plastic gloves increases sweating and hydration of the horny layer, facilitating the penetration of allergens[62] such as nickel, the ions of which have been found to penetrate rubber[63] (although not PVC)[64] gloves. The effect of wearing rubber gloves for any length of time may therefore be to accelerate the passage of nickel ions through sweat ducts and follicles and thus to elicit or even initiate nickel dermatitis on the hands. It has been shown experimentally[65] that the rate of permeability of aqueous nickel chloride through skin membranes was very greatly enhanced under occlusion, although it occurs slowly.

VI. FACTORS COMMON TO HAND ECZEMA AND NICKEL SENSITIVITY

We must now consider two important aspects of the interrelationship between nickel sensitivity and hand eczema: wet work and atopy.

A. Wet Work

Although cleaners, and particularly hospital cleaners, who are easily accessible for hospital studies, form the material for most of the evidence presented, housewives are obviously also involved insofar as much of their daily work falls into this category. What is true of "wet" service workers is therefore also applicable — although to a less extent — to the housewife. This becomes more relevant when young children in the household are the cause of a greater load of washing, bathing, and hairwashing.

The role of wet work as a risk factor in the development of irritant hand dermatitis needs no emphasis, although it is only one of many factors involved.[52] Cleaners, hospital workers, and others engaged in wet work form a high proportion of female patients in occupational studies.[23,25,66] The risk is enhanced by previous atopic dermatitis, hand eczema, and metal dermatitis.[21,67,68] Although this is dealt with more thoroughly in Chapter 12, it is worth noting here that wet work "doubles the odds" of developing hand eczema compared to dry office workers[21] or shop assistants,[69] but that an unfavorable combination of domestic parameters would expose dry office workers to as great a degree of risk as wet workers with a favorable combination of domestic factors.[21] This may explain the divergence between the high prevalence rates found in some studies[21,70] compared to a lower prevalence in others.[71,72] One study exceptionally involved more males than females (268 to 88).[73] The 3-year period prevalence was 10% for the men and 19% for the women.

If, as most of these studies show, there is an increased risk for wet workers to develop hand eczema, is there also an increased risk of their developing nickel sensitivity? In most of the studies in which tests have been carried out on affected subjects, a positive test for nickel is taken to be relevant to their work. We may question whether this is necessarily always so: a rate of 10 to 12% of positive reactors, especially in younger age groups, may only represent the normal female population level of nickel allergy from earrings, jewelry, etc. The range of potential contact with nickel alloys in a hospital setting is not inconsiderable. One convincing study[74] demonstrated how the level of nickel in buckets and, especially, wash cloths increased during a period of hospital cleaning, exceeding the theoretical eliciting safety limits (1 μM Ni/l).[75] It must be accepted that a hospital cleaning environment does involve contact with many substances — detergents, oxidizing agents such as sodium perborate,[69] and alkalis — which can, theoretically at least, enhance nickel dissolution or facilitate penetration of nickel ions.

Figures for the incidence of nickel allergy in studies of wet workers vary. Fregert[66] found that 19% of 82 positive nickel reactions occurred in cleaners. Nilsson[67] found an incidence of 15.2% among 120 subjects, most of whom were women. Other authors[41,69,73] have found lower figures, sometimes only comparable to the level in the noneczematous population. Indeed, the relevance of the findings have been questioned in another study[76] in which 13.7% of 102 subjects (mostly women) were nickel positive. However, a higher figure was found in hospital kitchen workers.[77]

In the figures presently available, we do not find very convincing evidence that wet work is a major factor in inducing nickel sensitivity. However, we believe it is reasonable to accept that the conditions of such work are such as to facilitate the elicitation of nickel sensitivity in sensitized subjects. What starts, usually, as an irritant dermatitis of the hands may then become more persistent and more severe, as a combined irritant and allergic eczema.[41,67,74,76] However, social and environmental factors may be even more important than the work. ''Dyshidrosis'' is not a prominent pattern in these workers.[41]

B. Atopy

In discussing the role of atopy as a risk factor in the development of either hand eczema or nickel sensitization, one is faced with the same difficulties of selection and interpretation that have been mentioned above, as well as some further ones relevant to atopy itself. Although the major features of the atopic state (or diathesis) are well enough known,[78-80] there is no clear agreement on minor features such as childhood urticaria and migraine. Most recent authors divide their subjects into those with atopic dermatitis and those with mucosal (or respiratory) allergy. Some would only accept prick-test evidence for the atopic state. Some studies include those who have a family history as well as a personal one — and this may extend to nephews;[81] others separate the two. This would not matter if ''control'' groups were included. Such is not always the case. Finally, the existence of ''latent atopy'' means that all figures are likely to be incorrect for any point prevalence at any rate among younger subjects.

The true prevalence of atopy in the general population is unknown and varies from race to race.[79] A figure of 20%,[82] although 25 years old, is probably not excessive. A more recent study of 980 persons in Finland[83] gave an incidence of 16.1% for males and 18.1% for females. In a questionnaire survey of all patients attending a skin clinic,[13] 15 to 23% claimed to have or to have had atopic manifestations, but these were not checked by prick testing. Lower estimates are quoted by Rajka.[79]

One accepted feature of patients with a personal history of atopic dermatitis is an increased susceptibility to irritants[84] and one would not be surprised, therefore, to find a high prevalence of irritant hand dermatitis among atopics. In her extensive and careful follow-up studies of atopic children, Rystedt[45,51,78,84-86] found that 41% of those with a history of severe atopic dermatitis developed hand eczema and 25% of those had been moderately affected. These figures were far higher than in control groups (Table 4). In atopics, the hands are affected in one third to one half of the cases. Other figures confirm a high incidence of hand eczema, alone or with other sites, in patients with past or present atopic dermatitis.[87,88] A high incidence of hand eczema in patients with a personal or family history of atopy has also been found,[89] although Rystedt[45] regards this as of less importance than other endogenous and exogenous factors.

Figures for the incidence of atopy in patients with hand eczema vary, but most show an increase over that expected in the general population, particularly when a familial as well as a personal history is included[29,90] (Table 4).

If it is accepted that about 30 to 35% of the population will give a personal or a close

Table 4
HAND ECZEMA AND ATOPY (%)

Material	Personal atopy	Personal atopic dermatitis	Personal respiratory allergy	Family and personal atopy	Controls	Ref.
Among atopics, cases of hand eczema						
549 severe		41				84
406 moderate		25				
222 respiratory			5			
199 nonatopic					4	
233 European study (hands/feet)		68				92
130		69				87
586	65			75		89
Atopics among patients with hand eczema						
827		19				15
50				82	28	90
368	39	28				85
20 — 24 years		57				
> 35 years		11				
263	34			67		29

Table 5
NICKEL SENSITIVITY AND ATOPY (%)

Material	Personal atopy only	Personal atopic dermatitis	Family and/or personal atopy	Controls	Ref.
Atopy in nickel-sensitive patients					
37 (36F) (includes urticaria)	14		46		81
172	15	1	37		31
276			31	33	83
53	9				91
166		9			59
108	5.5		60		94
66			44		37
Nickel allergy in atopics with eczema					
173 hand eczema	4.0				85
116 no hand eczema	4.3				
233 with atopic eczema	0 M 4.6 F				92

familial history of atopy of all types,[83] such figures as are available for the incidence of atopy in nickel-sensitive patients are not greatly increased. Atopics probably have a lower incidence of nickel sensitivity (Table 5),[85] although opinions on this differ.[31,32,81,91-93] Of Calnan's series of 400 nickel-sensitive patients, only 4 had had atopic dermatitis. Although only 5.5% of 108 nickel-sensitive patients in another series[94] had a history of personal atopy, 60% had a personal and/or family history — a higher figure than for patients reacting to seven other allergens. There is also some evidence for an increased incidence of atopic disease among relatives of those with allergic contact dermatitis.[96]

''Pompholyx'' (''dyshidrosis'') is not specifically featured in most of these figures, but will be considered below, as will contact urticaria, in which atopy plays a more significant part.[67,96]

VII. SOURCES OF NICKEL-INDUCED HAND ECZEMA

The common sources of sensitization to nickel in both sexes in western Europe are found in clothing accoutrements, jewelry and wristwatches and their buckles or straps. In men, wristwatches account for most cases[5,13] and far exceed industrial causes in general hospital material. Occupational causes are more prominent in males in areas of particular industrial hazard and in eastern Europe[8] (Chapter 12). In nonoccupational cases, the sensitization usually precedes the onset of hand eczema by several years.

Of more direct relevance to this chapter are those sources that may initiate primary hand eczema or elicit it in those already sensitized. For convenience. these may be grouped as follows:

1. Sources capable of causing primary sensitization
   A. Occupational
   B. Environmental
2. Sources more commonly eliciting secondary eczema
   A. Acceptable
   B. Occult or questionable
3. Endogenous sources

Of these 1A and 3 are dealt with in detail in other chapters and will only be discussed briefly here.

A. Sources of Primary Sensitization

1. Occupational

Contact with specific occupational sources is related to particular trades, such as platers and metal workers. This is usually readily diagnosed, subject to compensation and carrying a poor prognosis.[97,98] In contrast, in dealing with women with nickel sensitivity and hand eczema, we are concerned with patients who, whether they work outside the home or not, carry the more general risks of personal and household contacts with nickel. The home is a woman's minifactory,[99] where she combines all manner of wet work with cleaning, polishing, household maintenance, and cooking. If she has young children, the risks are increased — at least for an irritant hand eczema. If she is, at the same time, working outside the home, perhaps as a cashier, in an office, or as a hairdresser, it becomes exceedingly difficult to separate the occupational hazards from those within the home.

2. Environmental

The ubiquitous nature of nickel in everyday life often makes it difficult to identify a specific contact. Clues may be given by the pattern of hand eczema, as with rings, keys, thimbles, etc. Hobbies may be important, e.g., jewelry making and sewing. Because of the degree to which the hands of housewives and workers in "wet" services are subjected to repeated friction, microtrauma,[49] detergents, and cumulative irritants, it is often difficult to separate the effects of nickel from those of irritants. One may be superimposed on the other. This superimposition of nickel allergy on irritant hand dermatitis has been particularly studied in hospital "wet" workers.[73,74,76,77,89] The same situation operates, although less consistently, in housewives.

Occasionally, outside industry, primary nickel sensitization affecting the hands can be diagnosed with confidence on the history, clinical pattern, and patch test. White gold rings are often quoted as an example. Even here, however, the initial dermatitis may have been due initially to an irritant effect of soap or detergent concentration beneath it, or the "primary" nickel allergy may have been the elicitation of a pre-existing and perhaps forgotten metal allergy elsewhere. In our personal questionnaire series,[13] only 2.7% of 354 females considered rings as the primary site of their allergy. Curiously, they were more often a primary source in the older age groups.

Objects that have been cited as causes of nickel hand eczema are given in Table 6. The degree to which they may in reality be primary causes is doubtless dependent on the degree and duration of contact, sweating, pressure, trauma, and friction, as well as the degree to which the metal is released from a particular alloy.[57,60,61]

B. Sources of Elicitation

1. Acceptable

In patients already sensitized to nickel, the risk of acquiring lesions on the hands is obviously greater, especially when the skin is already damaged. All the agents cited as possible primary sensitizers are more understandably able to elicit symptoms on the hands in such patients without necessarily invoking the facilitating factors mentioned above. However, it remains difficult to distinguish irritant (or atopic) dermatitis in a nickel-sensitive patient from elicited nickel dermatitis. Sometimes a change of pattern or a sudden worsening of the condition provide helpful clues, for example, involvement of the volar surfaces of some fingertips or the extended volar pattern described in cashiers and shop assistants handling money.[100] In these and similar cases, the correct answer may depend on the effect of avoidance of the suspected contact and other nickel objects. A marked improvement provides strong supporting evidence of the cause. Even this can be deceptive when, as usually occurs, effective nonspecific treatment for the eczema is given at the same time or, once established, the eczema persists despite removal of the prime causation agent.[101]

Table 6
SOURCES OF NICKEL ALLERGY, WITH SPECIAL REFERENCE TO THE HANDS, SPECIAL OCCUPATIONAL CAUSES EXCLUDED

White gold or cheap metal rings	Kitchen equipment
Scissors	Stainless-steel sinks, pans
Clips, safety pins	Plated cutlery
Letter openers	Kitchen handles
Knitting needles (not now in U.K.)	Metal machinery pipes, handles
Belt buckles	
Metal thimbles	Prams
Coins	Sewing machines
Handbag, case handles	Taps
Purse	Door handles
Lighters	Furniture frames
Pocket knives	Television dial
Lipstick, powder cases	Telephone dials
Ball-point pens (rare)	Window catches
Zip fasteners	Tin openers
Key rings, keys	

Note: Chromium-plated articles may release underlying nickel through microscopic "pores"[103] especially under conditions of sweating or prolonged contact.

Domestic and environmental causes of nickel allergy relating to the hands will obviously vary from country to country with different habits and cultures in the same way that sources of clothing accessories and jewelry do. Cheaper metals and alloys are likely to be more commonly used in less affluent countries, but are being increasingly displaced by plastics. Knitting needles, for instance, cited as a cause of elicitation in 2 of 177 patients in a Polish series,[102] have not been made of metal in the U.K. for 30 years.[5] Other subjects in this series causing "relapses", e.g., elicitation — presumably on the hands — include scissors, keys, ballpoint pens, pins, and needles (Table 6).[5,103] Metal door handles and furniture hinges provoked dermatitis in nine workers and, in other cases, keys themselves and a wide variety of kitchen equipment have all been incriminated. However, clear-cut cases, occurring in subjects with normal skin, are less common than suspected cases in which several other factors or a pre-existing hand eczema with a variable course are also present.

2. *Questionable*

Some sources of nickel have been incriminated as causes of contact eczema of the hands on a theoretical basis — either because they may contain traces of nickel or because they are in common use. Clinical experience does not always support this view. Three of these will be considered.

a. Coins

Most current coinage includes a number of cupronickel pieces (although no longer in the U.S., where they are cuprozinc). The percentage of nickel varies: in Great Britain the proportions are 25% nickel and 75% copper; Swedish and Danish crowns also contain 25% nickel. Release of nickel from the coins was greater in synthetic sweat than in distilled water by a factor of 20 and by a factor of 35 when the temperature was raised from 20 to 35°C.[104] Thus, during 24 h, 691 μg (SD 83.0 μg) was released from a 1-kroner piece in synthetic sweat at 35°C — a considerable amount. However, such prolonged contact at body temperature is unlikely in everyday life, although levels may approach this in cashiers, shop assistants, and obsessional pocket-jinglers. This is borne out by the failure of other workers to provoke a reaction in nickel-sensitive patients with hand eczema of "pompholyx" type after handling kroner coins in water.[105]

Nickel is not confined to cupronickel coins. It can contaminate other copper coins, bank notes, and the hands of those handling copper coins. Shop assistants in a department store were found to have 50 μg Ni on their hands after 4 h of work.[106]

Although coins should thus be a potent source, at least for the elicitation of hand eczema in sensitive subjects, there is no evidence that they are a common cause. Occasionally, cases are certainly recorded. Husain[100] reported five cases in cashiers, a canteen worker, and a tailor. Two further cashiers were affected in 177 patients in whom the source of nickel sensitivity was studied.[102] Other cases were recorded by Cronin.[5] van Ketel[107] described seven female cashiers, presensitized by jewelry, all of whom had regular contacts with coins and banknotes. Three had not developed hand eczema after 2 to 8 years in this work, two developed a mild eczema of the volar aspects of the fingers, and only one had hand eczema "possibly aggravated by contact with nickel coins". He concludes that coins probably do not elicit dermatitis on normal skin, but only contribute to the maintenance of a pre-existing hand dermatitis. Fisher[103] agrees, but van Ketel points out that in these cashiers, there is no concomitant contact with water or wetting agents. He suggests, with Morgan,[108] that continual exposure to coins may produce hardening.

b. Stainless-Steel Sinks and Pans

Although stainless steel does not release nickel easily and gives negative *N,N*-dimethylglyoxime tests, significant amounts may be leached by sweat or household detergents.[109] This release is accentuated by an acid pH and thus in cooking acid fruit and vegetables.[110] However, the clinical relevance, except in presensitized professional cooks, is questionable.

c. Detergents

Traces of nickel in detergent products have in the past been implicated as a possible cause of sensitization,[111,112] but this view has been held on theoretical grounds rather than as a confirmed fact. The addition of ethylene diamine tetraacetic acid (EDTA) to detergents to chelate the free nickel ions had little effect on the incidence of nickel dermatitis in the Netherlands,[113] and the concentration of nickel in detergents is regarded as being, normally, below the eliciting strength.[111,113] With rare exceptions, the level of nickel in fabric washing powders seldom exceeds 1 to 6 ppm,[112,114] giving a normal use concentration of up to 0.06 — perhaps 0.1 ppm. The mere presence of low levels of nickel — usually less than 10 ppm — has been enough to lead to suggestions that this might contribute to the frequency of hand eczema in nickel-sensitive subjects.[112] However, other reasons are certainly more valid.

Allenby and Goodwin[115] found that the minimal amount of nickel required to provoke a positive patch test reaction (usually 0 to 0.5% or 112 ppm) was considerably greater than that present in fabric washing powder solution. Patch testing with the washing powder in a domestic in-use concentration (1%) instead of water did not significantly alter the reaction. However, 1 patch out of 25 reacted to a solution containing 1 ppm nickel (0.0005%). Pretreatment of patch test sites with sodium lauryl sulfate (0.3%) enhanced the allergic patch test reactions. Of the 15 reactors to 112 ppm nickel, 5 had never had hand eczema. Bearing in mind that the negative reaction occurred in 48-h occluded skin, the likelihood of detergents being causative in normal use is remote. However, detergents may facilitate the penetration of nickel through already damaged skin.[116]

VIII. CLINICAL PATTERNS

In the complex situation of hand eczema in patients who are nickel sensitive, it is uncommon to find a pure pattern of primary nickel dermatitis affecting the hands alone, except in those at risk from occupational contacts (Chapter 12). Individual cases of environmental contact (Table 7) may be detected by the alert physician or recognized by the observant

Table 7
NICKEL ALLERGY (%) IN VARIOUS PATTERNS OF HAND ECZEMA

			Ni +		
Material	**Pattern**	**No.**	**M**	**F**	**Ref**
Contact clinic					
All cases	Palmar	115	—	27	29
(263)	("Apron")	18	—	28	
	Dorsal	40	—	30	
	Fingers only	51	—	31	
	All over	57		25	
Non-occupational	Palms/volar	64	—	30	
(138)	Fingers				
	Dorsal	74	—	31	
	Fingers only				
	All over				
Skin clinic	Vesicular	115	0	19	120
	Palms/fingers				
	Fingertip, ring	32	—	15	
	All over	119	—	18	
		113	4	—	
Contact clinic	"Vesicular"	65	1.5	14	
	(All)	88			28
	"Pompholyx"	30	0	19	
		26			
	All primary				
	Hands	159	1.3	—	
		233	—	15	

patient, but are seldom reported except as "curios" or examples of newly discovered causes. Even here it is necessary to distinguish "hard" proven cases from those "soft" cases for which the evidence is circumstantial.[117] The pattern of contact dermatitis induced by contact with keys, clips, pins, purses, etc. usually involves the dominant (first through third) fingers of the master hand. Those caused by scissors, plated cutlery, taps, handles, or knitting needles[102] may be somewhat different, but relevant to contact or usage. A particular pattern involving the palmar aspects of the fingers and palms, seen in cashiers or shop assistants,[100] was thought to be specific and probably overlooked. van Ketel[107] observed a similar pattern in only two of seven nickel-sensitive cashiers and we have seen it in a nickel-sensitive student working as a part-time cashier. However, Cronin, in her study,[29] did not find nickel sensitivity more common in patients with this "apron" type of eczema, which she thought was predominantly endogenous.

The second, and by far the largest, group comprises those patients (or subjects) in whom any pattern of hand eczema is associated with the finding of a positive patch test to nickel. Again, in specific occupational groups, this finding may be highly relevant and in wet workers perhaps partially relevant. In most cases, we are dealing with a multifactorial situation in which irritants, atopy, and other factors already discussed may play an important if not a predominant role. Here, the relevance of the nickel sensitivity is more difficult to assess. In a survey of hospital wet workers,[41] involvment of the interdigital areas, the fingers, and the ring area was far more frequent in the 197 subjects without contact sensitivity than in the 61 with patch test evidence of this. Involvement of the palm was also three times as common. Only one example of a "dyshidrotic" pattern was found.

It has been pointed out that little reference is made in the literature to clinical patterns of nickel sensitivity involving the hand.[118,119] When they do, the patterns are designated in different ways: few seem able to agree on a specific morphology, and some regard this as a useful — if somewhat tentative — exercise.[1] Dyshidrosis (pompholyx) alone is distinguished. Cronin,[29] in a serious and careful attempt to relate allergy and atopy to different areas of involvement of the hand, found no difference in the frequency of positive nickel patch tests in her four anatomical patterns and doubted the relevance of nickel in 18 patients with the volar ''apron'' pattern. She also did not find any significant difference between the total material and those patients without occupational contact. Her rather high overall figure of 28% for nickel sensitivity in 263 women with hand eczema may reflect a selection of material in this major referral clinic. Lower figures were found in material seen in a dermatological department of a general hospital[28] and referred for routine patch testing. In another study of sites affected by particular allergens,[119] a high degree of correlation was found between nickel allergy and the dorsal, palmar, and (particularly) the interdigital aspects of the fingers as well as the periphery of the palm.[37] These findings are not necessarily in conflict with those of Cronin.

Information is even more sparse in regard to specific morphological sites. In a personal series,[120] palmar hyperkeratotic eczema gave a very low yield of positive patch tests to nickel. ''Ring'' and ''fingertip'' patterns in females gave a figure of about 15% and generalized or patchy patterns about 18%. In a more recent study from the same hospital,[28] all primary cases of hand eczema gave a figure for nickel sensitivity of 15% in females. The figure for all cases of contact dermatitis in the total material was 36%. These figures suggest that a diffuse dorsal or generalized pattern of hand eczema is associated with a moderate increase in nickel sensitivity, that these patterns are not predominantly due to nickel alone, and that nickel allergy is only occasionally relevant in ''ring'' and ''fingertip'' eczema. Pompholyx is discussed below.

The importance of factors other than nickel sensitivity as a cause of interdigital, dorsal, ''fingertip'', and ''ring'' eczema is emphasized in surveys of atopic patients.[84,89] The palms were affected alone in only 2% of Rystedt's cases[84] of atopic dermatitis, but were involved, with other areas, in 25%.

Though seldom identified as a clinical entity in relevant reports, discoid (nummular) eczema confined to the hands is not, we believe, significantly related to nickel sensitivity. We agree with Jordan (if he was indeed referring to this pattern) that (the patients) ''are at a loss to explain how nickel produces the patchy, usually sharply demarcated dermatitis on the hand that can in no way be forced to conform to the pattern of contact with a thousand and one hand-held objects.''[117]

A. Pompholyx, Dyshidrotic Eczema, Dyshidrosis

These terms are variously given to describe a characteristic pruritic eruption affecting the sides of the fingers and/or the palms, sometimes in association with similar lesions on the soles. We agree with Calnan[3] that these terms are misleading and better abandoned. We prefer to use the term ''vesicular palmoplantar eczema'' (acute, subacute, recurrent — or episodic — and chronic). We accept that this is somewhat cumbersome and that the term ''pompholyx'' is hallowed by use, at least in Europe, to describe a vesicular hand/foot eczema of multifactorial etiology. This does at least imply a distinctive, clinical appearance and distribution. These are valid for the great majority of cases, although authors differ in their acceptance or rejection of atypical features, vesiculosquamous, desquamative and ''tylotic'' forms,[121] or the degree of extension.[122,123]

Classically, an acute eruption of deep-seated vesicles, with little or no erythema, occurs on the sides of the fingers or the palms, or both. Extension may occur from one site to another, particularly the soles and sometimes the dorsal aspects of the fingers, with subsequent nail dystrophy, each transverse ridge representing an attack of pompholyx.[123] In 41 cases, the fingers were affected in 65 to 80%, the peripheral and medial edges of the palm in 50 to 64%, and the thenar eminence in 35 to 49% of cases.[33] Hyperhidrosis is said to be common.[33] An explosive onset may be followed by desquamation within 2 to 3 weeks, but recurrences are common, perhaps as many as 6 to 12 a year.[123] In some patients, one attack passes into another, producing a mixed picture of vesicles, scaling, hyperkeratosis, and fissuring. Attacks tend to be worse or more frequent in the summer.[122] In atopics, we have sometimes observed attacks at the height of the hay-fever season.

Few figures are available for the frequency of pompholyx in the general population. In a survey of hand eczema in southern Sweden,[15] it accounted for only 51 of 827 cases. However, it is more common in hospital material and represents about 20% of all cases of hand eczema seen in our own clinics. A "palmar" pattern accounted for 44% of Cronin's 263 women with hand eczema,[29] but these would include patterns[51] other than pure pompholyx. It is unlikely to figure prominently in patch test material except in centers where it is the subject of special interest and study.

Vesicular palmoplantar eczema is regarded as a nonspecific reaction with numerous causes, some known and some the basis of speculation. Many provoking agents are recognized. Some of these, although convincing, are anecdotal or based on single case reports. Others, such as fungal infections of the feet (producing the "pompholyx mycid"), are now regarded as less common than used to be supposed.[123,124] We have some reservations on this point. More follow-up studies are required. Allergic reactions of both Type I and Type IV are certainly involved, at least in some cases; Type III may be suspected in others. Vesicular lesions of palms and fingers have been produced by the application of haptens and proteins[123] including nickel[123,125] and proteins (within 20 min)[96] and by oral administration of neomycin in sensitized patients with leg ulcers.[126]

1. Nickel Allergy and Pompholyx

Allowing for differences in definition and selection, and for the relative paucity of detailed material, this form of hand eczema does seem to be more common in nickel-sensitive subjects in some centers (or, perhaps, countries) than in others. It was found in 15 of 32 twin pairs with nickel sensitivity and hand eczema.[38,127] In another series,[33] 51 of 66 nickel-sensitive patients with hand eczema showed this pattern. Christensen[37] quotes other examples.

Figures for the incidence of nickel allergy in patients with pompholyx are also sparse. Of Cronin's 115 female patients with a palmar pattern, 27% were nickel sensitive. However, this figure was no greater than she found in other patterns of hand eczema. Personal figures for nickel sensitivity in pompholyx among women in two series[118,120] were in the region of 19%. In an Italian study of 364 consecutive patients with pompholyx,[124] 30% gave positive reactions to 1 or more allergens, mostly topical medicaments, but metal only accounted for 7 positive tests (1.9%, or less than 5%, if all the patients had been female). Schuppli[128] obtained a figure of 10 to 15% positive tests with eight allergens. These studies are inadequate for valid comparison, but suggest that contact sensitivity in this pattern of hand eczema is less common than in some other forms and is often of secondary or unknown relevance. External provocation to nickel failed to cause any worsening of the dermatitis in 12 patients.[33]

Secondary eruptions (see below) may occur in patients with pompholyx as in other forms of hand eczema. They were particularly common on the soles and eyelids.[33]

2. The Role of Systemic Nickel in Patients with Pompholyx

For many decades there had been a belief, sometimes substantiated by strong circumstantial evidence, that ingested allergens can evoke or cause flares of pompholyx in sensitized subjects. Indeed, these may represent one of a number of reactions that may occur from systemic exposure to ingested allergens.[129]

The strong association between nickel sensitivity and hand eczema, especially pompholyx, led Christensen and Möller[33] to consider that "recurrent pompholyx . . . may be elicited by a mechanism other than external exposure to the metal'' and stimulated the first of several experimental studies on oral provocation. These are fully discussed in Chapter 14. Although the concept is an attractive one, we feel that the proposition has yet to be proven that flares of pompholyx in nickel-sensitive subjects are related to the normal dietary intake of nickel. Not all the trials have been controlled and pompholyx is notable for its capricious behavior. The apparent ease with which flares can be produced in some centers contrasts with the more variable or less conclusive results obtained in others.[130,131] The response appears to be dose related,[132,133] but flares or other reactions are also seen with placebo capsules.[130,132] There remain difficulties, therefore, in assessing the role of the smaller amounts of nickel that are ingested in a normal diet (Chapter 14). It is probable that the reactions evoked in susceptible subjects may not even be immunological, but perhaps dependent on personal variations of metabolism or nonspecific factors which activate this reaction pattern.[129] Further careful studies based on a strict and identical protocol are obviously needed.

B. Secondary Spread

When suspenders were the common initial site of nickel sensitivity in women, secondary involvement of areas such as the eyelids, neck, and elbow flexures was common,[31,134] the hands being involved in 23% of Marcussen's patients.[134] Calnan[31] was unsure of the relationship in his cases in which the hands were involved. Fisher[103] thinks it more likely that most ''secondary'' sites have had direct, although often inadvertent, contact with nickel, in the same way that earlier nickel patch test sites may flare on oral provocation.[129] However, in one report,[135] the course of hyperkeratotic lesions on the elbows paralleled that of pompholyx in seven nickel-sensitive patients. Flares at other sites of the body not in any obvious contact with nickel are, however, more frequently observed in oral provocation trials.[132] In patterns of hand eczema other than pompholyx, direct contact with the metal is obviously more likely to evoke exacerbations.

C. Contact Urticaria

Although contact urticaria on the fingers or hands is common with certain foods and proteins,[96,136] there have been very few reports of nickel causing this reaction. It is possible that it does occur more frequently but is undetected. Osmundsen[137] reported an instructive case of a nickel-sensitive cleaner with a daily short-lived eruption of palms and fingertips evoked by contact with the metal handle of a bucket. Itching, redness, and swelling occurred within 30 min and lasted up to 3 h. The skin of the palms was dry and fissured, but not obviously eczematous. A 20-min nickel test was negative, but a scratch-chamber test on normal skin evoked a marked urticarial response with intense itching. It is not known whether the reaction was immunological.

In the case of proteins and foods, a ''dyshidrotic'' vesicular reaction was evoked 30 min after application of an open test to normal but previously affected skin.[47] Vesicular lesions, rather than an urticarial response, have been produced on the fingers by application of the responsible protein or food.[47,138] This may support the speculation that some component of vesicular eruption of the fingers and palms might be of Type I rather than Type IV in nature.[138]

IX. DIAGNOSIS

Discussion in any detail of the diagnosis of hand eczema is outside the scope of this chapter. For this, refer to standard texts and articles.[1,15,48] The dermatologist is well aware of possible areas of confusion such as psoriasis, fungal infections, and lichen planus. Some errors in diagnosis of cases referred to an occupational dermatitis clinic, most of which involved the hands, have been listed elsewhere.[139] Characteristic and established cases usually present little difficulty, but opinions differ on the validity of including mild and transient cases or those in which dryness, cracking, and superficial fissuring are the only features.

Once a diagnosis of hand eczema has been made, it becomes necessary to consider a secondary diagnosis — that of the etiology. The distinction between exogenous allergens and irritants on the one hand and endogenous (''constitutional'') factors on the other may at times be simple, but can often be extremely difficult, requiring several consultations to resolve. The answer may not become apparent until investigative procedures have been completed and may, even then, require a painstaking elimination of one potential factor at a time — the ''diagnostic continuum''. Again, refer to special texts.[1,2,48,140,141] A suspicion of atopy can often be confirmed by prick testing, although negative tests may not completely rule this out.

The final diagnostic procedure — that of nickel sensitivity — may be suspected by the history and will be confirmed by patch testing[5,24,29,140] (Chapters 8 and 9). This procedure may be supplemented by other specialized test procedures,[141] such as intradermal tests, which may reveal nickel sensitivity in some patients negative to conventional patch tests.[142] For some, oral provocative test doses[142] may be of value (Chapter 14), but the specificity of this procedure remains somewhat in doubt.

There remains one final diagnostic procedure, perhaps the most difficult of all and often disregarded — that of establishing the relevance of any positive patch test that may be found.[27]

X. TREATMENT

In the course of arriving at a conjoint diagnosis of nickel sensitivity and hand eczema, the dermatologist will have assessed the degree of relative importance he attaches to the various factors that may be implicated — atopy, irritants, wet work, nickel sensitivity, and other allergens found on patch testing. His approach to treatment will be greatly influenced by this assessment and his success will be influenced by its accuracy and his skill in conveying it to the patient. This will always be limited by the multifactorial nature of hand eczema, by extraneous factors favoring persistence[101] including the patient's anxiety, the social and economic factors associated with a chronic disabling condition, and the virtual impossibility of protecting hands already damaged from the rigors of daily life. In all but the mildest cases, any patient with recurent or persistent hand eczema will need repeated observation, counselling, advice, and encouragement — the ''therapeutic continuum''. Most patients can only absorb a limited amount of advice at once and it is often more helpful to attempt to remove or modify only one or two adverse factors at a time. This both allows an estimate of their importance and encourages the patient to adjust slowly to new concepts and new modes of life. It is of vital importance that patients realize that functional repair of the horny layer as an integral barrier lags behind apparent healing.

It is inappropriate here to discuss the practical aspects of the management of patients with hand eczema. Each dermatologist will have his own way of tackling the problem. Standard texts provide ample guides.[1,48,52,101] Fact sheets are often useful. Examples are given in the major texts cited. Gloves are valuable if used properly, but have their limitations.[2,143] They should be of vinyl chloride because nickel can penetrate rubber.[63]

The problem of nickel sensitivity requires a separate approach. Although it may be of less relevance to hand eczema than conditions of wet work or household irritants, this cannot be taken for granted. All nickel-sensitive patients should be given extra instruction about the detection and avoidance of this metal. Occupational sources are easier to trace than environmental ones. Prevention of hand eczema is better than ''cure'', which is so difficult to achieve. However, it is not easy to persuade a woman to abstain when the onset of hand eczema in those sensitized may lie 5 or more years in the future. The value of the ''nickel spot test'' has been emphasized by Fisher,[103,144] and we have found it invaluable to provide home test kits to all patients found to be nickel sensitive.[145] The use of lists of nickel-containing objects and general instructions for nickel-sensitive patients[103,140] is also widespread, though considered by some to be the lazy man's substitute for explanation.

Interest in the possible role of ingested nickel in the provocation of pompholyx has led to specific forms of treatment for this condition: a low or, more accurately, restricted nickel diet and the use of chelating agents. Kaaber et al.[146] obtained an improvement in 9 of 17 patients whose eczema had flared on oral challenge; Veien et al.[147] also found a long-term improvement in a follow-up study of 150 patients (Chapter 14), perhaps because it was not continued long enough. They have subsequently extended their study.[148] They point out the difficulty of interpreting the effects of diet alone, since intercurrent factors may obscure the evaluation; they stress that the diet must be maintained for several weeks. A small pesonal series of 26 patients did not show such impressive results, perhaps because of differences in food sources and diets. However, more controlled studies are required.

Because chelating agents are effective in nickel carbonyl poisoning, they have been given, on an experimental basis, to patients who have sustained flares of pompholyx from provocative oral doses of nickel. Tetraethylthiuram disulfide or disulfiram (Antabuse) are split into sodium diethyldithiocarbamate after absorption. This combines with nickel ions to form an inactive nickel complex, nickel-*bis*-diethyldithiocarbamate.[149]

Good results have been claimed with disulfiram at a dose of 50 to 200 mg/d,[149-153] but toxic reactions occurred in 5 of 11 patients treated with higher doses.[151] Very active flare-ups may also occur and it is probably wise to start at a daily dose of 50 mg,[152] increasing slowly, if necessary, to 200 mg/day. This treatment is also limited by the finding of pathological liver function tests or even toxic hepatitis in a proportion of patients that many would find unacceptable.[151-154] Dithiocarbamate was given to one highly sensitive subject[155] and the levels of nickel in the hair and urine were monitored over 2 months. The urinary excretion of nickel increased tenfold, but the slightly negative nickel balance did not exhaust the store of nickel in the organs. The effect of this treatment on the course of the patient's eczema was inconclusive although less local therapy was needed.

As Fisher points out,[103] concomitant sensitivity to thiurams and dithiocarbamates — as from rubber gloves and domestic rubber — must be excluded before this treatment is considered.

The toxic effects mentioned above have led to trials with other and safer chelating agents. Triethylenetetramine (Trientine) was given, in a dose of 300 mg/d, in a double-blind crossover trial, to 23 nickel-sensitive patients with hand eczema.[156] No significant improvement occurred and, surprisingly, there was no detectable increase in urinary nickel excretion. However, this trial was terminated early because of a report[157] suggesting that Trientine caused an increased teratogenicity in rats.

Finally, disodium EDTA as a 10% cream appeared to reduce significantly patch test reactivity to nickel sulfate.[158] Its action in chelating nickel was thought to offer possibilities of reducing the dermatitis-provoking potential of contact with nickel in women in high-risk occupations, as it may also do when incorporated in foods with a high nickel content.[144]

XI. COURSE AND PROGNOSIS

Hand eczema in general has acquired a reputation for being difficult to diagnose accurately, arduous to treat effectively, and impossible to cure completely. Certainly, there is some justification for such a gloomy view of its course and prognosis. Perhaps this is why it is given so little attention compared with other less common diseases and, perhaps also, why there are so few good follow-up studies of unselected cases. We must remember that most of those that are quoted in the literature are drawn from cases referred to hospital or further

Table 8
FACTORS INFLUENCING THE COURSE AND PROGNOSIS OF HAND ECZEMA

General factors
- Failure to detect or avoid allergens at work or at home
- Inadequate counseling
- Inappropriate or inadequate treatment; overtreatment[140]
- Undetected cross-reactive allergens[140]
- Secondary allergens — gloves[41,143]medicaments,[124] fragrance[76]
- Secondary infection
- ''Moonlighting''; D-I-Y work; hobbies
- Hard water; overuse of soap, detergents; excessive shampooing[99,101]
- Young children; no automation
- Wet work,[21,76] friction, wear-and-tear[101]
- Housework[22,29,76]
- Rubbing, ''fiddling'', anxiety, stress
- Atopy, especially atopic dermatitis;[21,45,67,68,84-86,160] previous hand eczema, female sex;[21,45,67,68,84-86,160] (in atopics) dust;[84] dry, itchy skin[45]

With nickel sensitivity
- Pompholyx pattern[160] (see below)
- Immediate-type food reactions[45,160]
- Additional allergens[18,98]
- Coincidental cobalt sensitivity (?)[160,161]
- Geographical, cultural, genetic (?)[97]
- Dietary allergens (?)

Factors that worsen pompholyx (patient's own observations in order of frequency)
- Stress, fatigue; heat; season; low humidity; cold, hot drinks, alcohol

selected for patch testing. These will inevitably consist of the more severe or more persistent cases. In females with hand eczema seen in a multicenter study, 25% had the condition for over 5 years and the same proportion for over 1 year.[25] If the prevalence of hand eczema in the population is of the order of 1 to 2%[15] and assuming that hand eczema accounts for 21 to 30% of all cases of eczema seen in general dermatology clinics, we estimate that less than 1 case in 20 is referred to hospital. Indeed, Agrup found that only a quarter of those affected had seen their general practitioner in the previous year and a similar proportion had not ever done so. Thus, we are dealing with a condition which in its milder forms is common, but disregarded or regarded lightly by the patient. Some of these may, admittedly, become progressively worse, but we do not know what proportion.

In Agrup's field study,[15] two thirds of the subjects' hands healed or improved in 6 to 22 months; in one third the eczema remained or was worse. In another population survey[18] the figures for healing or improvement were 73% for males and 55% for females. In 46% of the females, the condition remained the same or was worse. In a follow-up of hospital wet workers 53 to 66% continued to have periodic or permanent symptoms.[70] However, the prognosis of nonselected mild cases of hand eczema is probably quite good.[18,21]

There are many reasons, of couse, why hand eczema should tend to become chronic or intractable: the nature of eczema itself,[13] the multifactorial nature of the condition, and secondary psychological influences of anxiety over its occupational and social consequences. In many cases, a situation that might be remedied by early and accurate diagnosis, adequate counseling, and appropriate treatment is allowed to develop into one of cumulative damage to the skin that is extremely difficult to reverse.[52] Perpetuating factors are common[101] and often unrecognized by patient and doctor alike (Table 8). Among these are undisclosed ''moonlighting'' or do-it-yourself activities and hobbies.[101] Young children in the house and the absence of mechanical help predispose to an increased severity of the condition, as does wet work itself,[21,67,68,76] although most cases run a mild course.[70] Whole-time housework seems to carry a less favorable prognosis than when it is combined with an occupation outside the home.[22,29,76]

Atopy, particularly a history of atopic dermatitis, is an especially unfavorable factor,[21,76] increasing threefold the risk of hospital workers developing hand eczema.[21]

Rystedt's important and detailed studies on the role of atopy in hand eczema[84-86] should be read in their entirety. In brief, she believes that endogenous factors are more important than exogenous ones in the aggravation of hand eczema. These include a history of atopic dermatitis (especially of the hands), its persistence elsewhere on the body, and its degree of severity. Even a dry, itchy atopic skin is a bad prognostic sign, as are a family or personal history of atopy and respiratory allergy (though to a less extent). She also noted that 85% of those atopics who had protein contact urticaria of the hands had or previously had eczema, in contrast to 37% of those without such reactions.[45,84] This may obviously be of importance in the food industry.

Periods of sick leave are taken relatively infrequently by patients with hand eczema, indicating that most of those affected continue to live and work with their affliction, unless it becomes severe or disabling. Such periods were found to be as low as 0 to 3.4% of subjects in one study,[21] although the figure rose to 7.3 to 8.2% in those with atopic dermatitis and engaged in wet work. However, in another group of hospital workers,[89] 14% out of 44% who had hand eczema had taken sick leave, more commonly those with an atopic diathesis (23%), compared with those with atopic dermatitis (12%) or those without any atopy or family history of this (10%). These differences are difficult to reconcile and may be related to factors other than the hand eczema itself. Sick leave may be taken as much for social reasons or because of physical hard work as for the severity of the skin condition; it was also much more commonly taken by those who were poorly educated.[159] Agrup[15] found that sick leave was taken more commonly by those with allergic rather than irritant dermatitis. However, she commented that most of those with the latter form of dermatitis were housewives who were unable to benefit from taking this course of action.

In a study of subjects applying for permanent disability benefit in Denmark, the authors[97] also felt that social and perhaps geographical factors were of some influence and that an assessment of disability should take these into account.

In contrast to the usually good prognosis of nickel dermatitis at nonoccupational sites of contact, the outlook for patients with hand eczema is generally considered to be far worse, with a marked tendency to persistence and chronicity.[22,160] In his review of repeat patch tests in nickel-sensitive patients, Morgan[54] found that 6 patients out of 20 still showing positive tests had continuing eczema, 4 of them on the hands. They were all housewives. However 18 out of 23 patients whose tests had become negative also continued to have eczema and in 12 of these it only appeared after the initial sensitization. He postulated that the most important factor was "underlying constitutional liability to develop eczema" and that the nickel sensitization was only one incident in an "eczematous career". He believed that such patients would continue to have eczema, whether they retained or lost their sensititivy to their allergen. Nevertheless, it must be accepted that it is very difficult for the housewife to avoid all contacts with nickel and that this must in many cases contribute to the persistence of hand eczema.

It is generally believed that allergic contact dermatitis carries a worse prognosis than does irritant dermatitis.[15,22,97,98] This depends on the allergen involved and the ease with which contact with it may be avoided. In the case of the housewife, irritant and allergic factors are often inextricably intertwined, leading to a poor prognosis[18] for this reason alone.

Although the poor prognosis of hand eczema in nickel-sensitive patients has received much attention, estimates of its chronicity have not always been compared with age- and sex-matched controls or with those sensitive to other common allergens such as perfume or balsam of Peru.

While the prognosis of nonpompholyx hand eczema in nickel-sensitive subjects is probably little different from that of hand eczema in all subjects,[160] despite Fregert's poor prognosis in his occupational patients,[22] the prognosis of pompholyx may be different, although, here again, there are not sufficient studies of control patients who are not nickel sensitive. We believe with Shelley[122] that this not uncommon condition is multifactorial and that too much attention may have been paid to the group of patients with this condition who are nickel sensitive. It may well be correct to regard it as having a worse prognosis than other patterns of hand eczema. In one series,[160] only 18% of 50 patients followed up for 6 years had healed, compared with 10 of 13 (77%) of those with other forms of hand eczema. However, as Menné and Hjorth pointed out,[123] the prognosis is better in population studies,[127] the severity of the attacks being progressively less and the intervals between attacks progressively longer. In a small personal sample followed up by letter,[13] we were surprised to find the prognosis better than we had expected, but the series did include some cases of "pompholyx mycid" which responded rapidly to griseofulvin. The role of diet (Chapter 14) may be important in nickel-sensitive patients,[105,146-148] but this is still a matter of debate.

The factors determining chronicity of hand eczema in nickel-sensitive patients remain unknown and are mostly conjectural (Table 8). We should perhaps mention the bad prognostic significance in those patients who have immediate-type food reactions[45,160] and possibly the co-existence of cobalt sensitivity[161] or multiple allergens on patch testing.[18,97]

Nickel-related hand eczema is an important cause of claims for disability in Denmark.[97,98] Of 564 claimants with skin complaints from 1970 to 1976, over one third had allergic contact dermatitis, almost always involving the hands.[97] Only 31 applicants were classified as having an irritant dermatitis. Of these 222 claimants, $^{2}/_{3}$ of them women, 96 were considered to be due to nickel. Of these, 91 had hand eczema. The mean interval between the onset of hand dermatitis and a decision on permanent disability was 9.1 years (1 to 36 years). These represent, of course, the "hard core" of chronic cases. In only 42% was nickel the only allergen. The rest had one to three extra allergens, usually cobalt, chromium, or rubber chemicals. These were more frequent than in hospital material. The authors found that a surprising number of applicants (64%), on changing jobs, had transferred to another unsuitable occupation — a poor reflection on management and counseling and a further reminder of the possible causes of persistence of their hand eczema.[101]

The effect of changing jobs on the prognosis is discussed in Chapter 13, but Rystedt's observations[51,85] are worth mentioning. The healing rate 3 years after change of occupation was particularly low in atopics, who showed less improvement than nonatopics. However, twice as many patients with hand eczema who were also metal sensitive changed work as those who were not, suggesting that this combination worsens the prognosis.

In summary, therefore, hand eczema of mild or moderate degree in the general population probably carries a reasonable prognosis. This becomes worse with chronicity or in wet work. Those with preceding atopic dermatitis have a worse prognosis. Previous or concomitant nickel sensitivity also appears to worsen the prognosis somewhat, especially in those with pompholyx in whom the role of ingested nickel may at times be important. The prognosis is worse if more than one allergen is present.

REFERENCES

1. **Burton, J. L., Rook, A., and Wilkinson, D. S.,** Eczema, lichen simplex, erythroderma and prurigo, in *Textbook of Dermatology,* Vol. 1, 4th ed., Rook, A., Wilkinson, D. S., Ebling, F. J. G., Champion, R. H., and Burton, J. L., Eds., Blackwell Scientific, Oxford, 1986, 403,
2. **Church, R.,** Hand eczema in industry and the home, in *Essentials of Industrial Dermatology,* Griffiths, W. A. D. and Wilkinson, D. S., Eds., Blackwell Scientific, Oxford, 1985, chap. 8.

3. **Calnan, C. D.**, Eczema for me, *Trans. St. John's Hosp. Dermatol. Soc.*, 54, 54, 1968.
4. **Menné, T., Borgan, Ø, and Green, A.**, Nickel allergy and hand dermatitis in a stratified sample of the Danish female population: an epidemiological study including a statistic appendix, *Acta Derm. Vernereol.*, 62, 35, 1982.
5. **Cronin, E.**, Nickel, in *Contact Dermatitis*, Cronin, E., Ed., Churchill Livingstone, Edinburgh, 1980, chap. 7.
6. **Fregert, S., Hjorth, N., Magnusson, B., Bandmann, H. -J., Calnan, C. D., Cronin, E., Malten, K., Meneghini, C. L., Pirilä, V., and Wilkinson, D. S.**, Epidemiology of contact dermatitis, *Trans. St. John's Hosp. Dermatol. Soc.*, 55, 17, 1969.
7. **Fregert, S.**, Personal communication, 1987.
8. **Schubert, H., Berova, N., Czermelewski, A., Hegigi, E., Jirásek, L., Kohanka, V., Korossy, S., Michailov, P., Nebenführer, L., Praten, E., Rothe, A., Rudzki, E., Stanski, L., Süss, E., Tarnick, M., Temesvari, E., Ziegler, V., and Zschunke, E.**, Epidemiology of nickel allergy, *Contact Dermatitis*, 16, 122, 1987.
9. **Shaw, S.**, Personal communication, 1987.
10. **Menné, T.**, The prevalence of nickel allergy among women, *Dermatosen*, 26, 123, 1978.
11. **Peltonen, L.**, Nickel sensitivity in the general population, *Contact Dermatitis*, 5, 27, 1979.
12. **Prystowsky, S. D., Allen, A. M., Smith, R. W., Nonomura, J. M., Odom, R. B., and Akers, A.**, Allergic contact hypersensitivity to nickel, neomycin, ethylene diamine and benzocain, *Arch. Dermatol.*, 115, 959, 1979.
13. **Wilkinson, D. S.**, Unpublished data, 1987.
14. **Meding, B. and Swanbeck, G.**, Prevalence of hand eczema in an industrial city, *Br. J. Dermatol.*, 116, 627, 1987.
15. **Agrup, G.**, Hand eczema and other hand dermatoses in south Sweden, *Acta Derm. Venereol. Suppl.*, 49, 61, 1969.
16. **Menné, T.**, Nickel allergy — reliability of patch test evaluation in female twins, *Dermatosen*, 29, 156, 1981.
17. **Johnson, M.-L. T. and Roberts, J.**, Skin Conditions and Related Need for Medical Care Among Persons 1 — 74, Education and Welfare, National Center for Health Statistics, Hyattsville, MD, 1978.
18. **Lantinga, H., Nater, J. P., and Coenraads, P. J.**, Prevalence, incidence and course of eczema on the hands and forearms in a sample of the general population, *Contact Dermatitis*, 10, 135, 1984.
19. **Coenraads, P. J., Nater, J. P., and van der Lende, R.**, Prevalence of eczema and other dermatoses of the hands and arms in the Netherlands. Association with age and occupation, *Clin. Exp. Dermatol.*, 8, 495, 1983.
20. **Kavli, G. and Førde, V. H.**, Hand dermatoses in Tromsø, *Contact Dermatitis*, 10, 174, 1984.
21. **Nillson, E., Mikaelsson, B., and Andersson, S.**, Atopy, occupation and domestic work as risk factors for hand eczema in hospital workers, *Contact Dermatitis*, 13, 216, 1985.
22. **Fregert, S.**, Occupational dermatitis in a 10-year material, *Contact Dermatitis*, 1, 96, 1975.
23. **Wilkinson, D. S., Budden, M. G., and Hambly, E. M.**, A 10-year review of an industrial dermatitis clinic, *Contact Dermatitis*, 6, 11, 1980.
24. **Cronin, E.**, Clinical predictions of patch test results, *Trans. St. John's Hosp. Dermatol. Soc.*, 58, 153, 1972.
25. **Wilkinson, D. S., Bandmann, H.-J., Calnan, C. D., Cronin, E., Fregert, S., Hjorth, N., Magnusson, B., Maibach, H. I., Malten, K. E., Meneghini, C. L., and Pirilä, V.**, The role of contact allergy in hand eczema, *Trans. St. John's Hosp. Dermatol. Soc.*, 56, 19, 1970.
26. **Calnan, C. D., Bandmann, H.-J., Cronin, E., Fregert, S., Hjorth, N., Magnusson, B., Maibach, H. I., Malten, K. E., Meneghini, C. L., Pirilä, V., and Wilkinson, D. S.**, Hand dermatitis in housewives, *Br. J. Dermatol.*, 82, 543, 1970.
27. **Wilkinson, D. S.**, Contact dermatitis of the hands, *Trans. St. John's Hosp. Dermatol. Soc.*, 58, 163, 1972.
28. **Shaw, S. and Wilkinson, J. D.**, unpublished data.
29. **Cronin, E.**, Clinical patterns of hand eczema in women, *Contact Dermatitis*, 13, 153, 1985.
30. **Magnusson, B., Fregert, S., Hjorth, N., Hørding, G., Pirilä, V., and Skog, E.**, Routine patch testing. V. correlation of reactions to the sites of dermatoses and the history of the patients, *Acta Derm. Venereol.*, 49, 556, 1969.
31. **Calnan, C. D.**, Nickel dermatitis, *Br. J. Dermatol.*, 68, 229, 1956.
32. **Wilson, H. T. H.**, Nickel dermatitis, *Practitioner*, 77, 303, 1956.
33. **Christensen, O. B. and Möller, H.**, Nickel allergy and hand eczema, *Contact Dermatitis*, 1, 129, 1975.
34. **Fisher, A. A. and Shapiro, A.**, Allergic eczematous contact dermatitis due to metallic nickel, *J. Am. Med. Assoc.*, 161, 717, 1956.
35. **Marcusson, P. V.**, Ecological considerations on nickel dermatitis, *Br. J. Ind. Med.*, 17, 65, 1960.
36. **Wahlberg, J. E. and Skog, E.**, Nickel allergy and atopy. Threshold of nickel sensitivity and immunoglobulin E determinations, *Br. J. Dermatol.*, 85, 97, 1971.

37. **Christensen, O. B.,** Nickel Allergy and Hand Eczema in Females, Thesis, University of Lund, Lund, Sweden, 1981.
38. **Menné, T.,** *Nickel Allergy: The Genetic and Epidemiological Aspects. Relation between Nickel Allergy and Hand Eczema,* Copenhagen, 1983.
39. **Gawkrodger, D. J., Vestey, J. P., Wong, W.-K., and Buxton, P. K.,** Contact clinic survey of nickel-sensitive subjects, *Contact Dermatitis,* 14, 165, 1986.
40. **Menné, T. and Holm, N.-V.,** Hand eczema in nickel sensitive female twins. Genetic predisposition and environmental factors, *Contact Dermatitis,* 9, 289, 1983.
41. **Lammintausta, K., Kalimo, K., and Havu, V. K.,** Occurrence of contact allergy and hand eczema in hospital wet work, *Contact Dermatitis,* 8, 84, 1982.
42. **von Reichenberger, M., Ebke, M., and Patiri, C.,** Zur Nichelsensibilisierung bei Frauen und ihr Relevanz zur beruflichen Tätigkeit, *Berufsdermatosen,* 24, 91, 1976.
43. **Wahlberg, J. E.,** Nickel allergy and atopy in hairdressers, *Contact Dermatitis,* 1, 161, 1975.
44. **Marks, R. and Cronin, E.,** Hand eczema in hairdressers, *Aust. J. Dermatol.,* 18, 123, 1977.
45. **Rystedt, I.,** Factors influencing the occurrence of hand eczema in adults with a history of atopic dermatitis in childhood, *Contact Dermatitis,* 12, 185, 1985.
46. **Meneghini, C. L. and Angelini, G.,** Contact and microbial allergy in pompholyx, *Contact Dermatitis,* 5, 46, 1979.
47. **Hjorth, N. and Roed-Petersen, J.,** Occupational protein contact dermatitis in food handlers, *Contact Dermatitis,* 2, 28, 1976.
48. **Epstein, E.,** Hand dermatitis: practical management and current concepts, *J. Am. Acad. Dermatol.,* 10, 395, 1984.
49. **Wilkinson, D. S.,** Dermatitis from repeated trauma to the skin, *Am. J. Ind. Med.,* 8, 307, 1985.
50. **van Ketel, W. G.,** Clinical factors promoting the development of allergic contact dermatitis, *Berufsdermatosen,* 25, 78, 1977.
51. **Rystedt, I.,** Work-related hand eczema in atopics, *Contact Dermatitis,* 12, 164, 1985.
52. **Malten, K. E.,** Thoughts on irritant contact dermatitis, *Contact Dermatitis,* 7, 238, 1981.
53. **Rycroft, R. J. G.,** Soluble oil dermatitis, in *Essentials of Industrial Dermatology,* Griffiths, W. A. D. and Wilkinson, D. S., Eds., Blackwell Scientific, Oxford, 1985, chap. 7.
54. **Morgan, J. K.,** Observations on the persistence of skin sensitivity with reference to nickel eczema, *Br. J. Dermatol.,* 65, 84, 1953.
55. **Magnusson, B. and Kligman, A. M.,** *Allergic Contact Dermatitis in the Guinea Pig,* Charles C. Thomas, Springfield, IL, 1970.
56. **Silvennoinen-Kassinen, S., Isotalo, H., and Jakkula, H.,** Estradiol enhances nickel-induced blast transformation, *Contact Dermatitis,* 11, 260, 1984.
57. **Hemingway, J. D. and Moldkhia, M. M.,** The dissolution of metallic metal in artificial sweat, *Contact Dermatitis,* 16, 99, 1987.
58. **Fischer, T. and Rystedt, I.,** Cobalt allergy in hand metal workers, *Contact Dermatitis,* 9, 115, 1983.
59. **Menné, T., Brandrup, F., Thestrup-Pedersen, K., Veien, N. K., Andersen, J. R., Yding, F., and Valeur, G.,** Patch test reactions to nickel alloys, *Contact Dermatitis,* 16, 255, 1987.
60. **Cavelier, C., Foussereau, J., and Massu, M.,** Nickel allergy: analysis of metal clothing objects and patch testing to metal samples, *Contact Dermatitis,* 12, 65, 1985.
61. **Jensen, O. and Nielsen, E.,** "Rusters", The corrosive effect of palmar sweat. II. Physical and chemical factors in palmar hyperhidrosis, *Acta Derm. Venereol.,* 57, 139, 1979.
62. **Marzulli, F. N. and Maibach, H. I.,** The use of graded concentrations in studying skin sensitizers. Experimental contact sensitization in man, *Food Cosmet. Toxicol.,* 12, 219, 1972.
63. **Wall, L. N.,** Nickel penetration through rubber gloves, *Contact Dermatitis,* 6, 461, 1980.
64. **Mouridsen, H. T. and Faber, O.,** Penetration of patch tests through protective gloves by allergens and irritants, *Trans. St. John's Hosp. Dermatol. Soc.,* 59, 230, 1973.
65. **Fullerton, A., Andersen, J. R., Hoelgaard, A., and Menné, T.,** Permeation of nickel salts through human skin in vitro, *Contact Dermatitis,* 15, 173, 1986.
66. **Fregert, S.,** Occupational dermatitis in a 10-year material, *Contact Dermatitis,* 1, 96, 1975.
67. **Nillson, E.,** Contact sensitivity and urticaria in "wet" work, *Contact Dermatitis,* 13, 321, 1985.
68. **Nillson, E. and Bäck, O.,** The importance of anamnestic information of atopy, metal dermatitis and previous hand eczema for the development of hand dermatitis in women in wet hospital work, *Acta Derm. Venereol.,* 66, 45, 1986.
69. **Hansen, K. S.,** Occupational dermatosis in hospital cleaning women, *Contact Dermatitis,* 9, 343, 1983.
70. **Lammintausta, K.,** Risk Factors for Hand Dermatitis in Wet Work, Academic dissertation, Turku University, Turku, Finland, 1982.
71. **Dahlquist, I. and Fregert, S.,** Occupational dermatoses in hospital personnel, *Berufsdermatosen,* 18, 261, 1970.

72. **Barregard, L. and Stenberg, B.,** Arbetsrelaterade handeksem hos Sjukhuspersonal, *Läkartidningen,* 79, 1355, 1982.

73. **Singgih, S. I. R., Lantinga, H., Nater, J. P., Woest, T. E., and Kruyt-Gaspesz, J. A.,** Occupational hand dermatoses in hospital cleaning personnel, *Contact Dermatitis,* 14, 14, 1986.

74. **Clemmensen, O. J., Menné, T., Kaaber, K., and Solgaard, P.,** Exposure of nickel and the relevance of nickel sensitivity among hospital cleaners, *Contact Dermatitis,* 7, 14, 1981.

75. **Malten, K. E. and Spruit, D.,** The relative importance of various environmental exposures to nickel in causing contact sensitivity, *Acta Derm. Venereol.,* 49, 14, 1969.

76. **Lammintausta, K., Kalimo, K., and Aantaa, S.,** Course of hand dermatitis in hospital workers, *Contact Dermatitis,* 8, 327, 1982.

77. **Lammintausta, K.,** Hand dermatitis in different hospital workers who perform wet work, *Dermatosen,* 31, 14, 1982.

78. **Rystedt, I.,** Dermatological problems in the work environment following childhood skin diseases, *Acta Derm. Venereol. Suppl.,* 61, 43, 1981.

79. **Rajka, G.,** *Atopic Dermatitis,* W. B. Saunders, London, 1975.

80. **Haniflin, J. M., Cooper, K. D., and Roth, H. L.,** Atopy and atopic dermatitis, *J. Am. Acad. Dermatol.,* 15, 703, 1986.

81. **Caron, G. A.,** Nickel sensitivity and atopy, *Br. J. Dermatol.,* 76, 384, 1964.

82. **Carr, R. B., Berks, M., and Becker, S. W.,** Incidence of atopy in the general population, *Arch. Dermatol.,* 89, 27, 1964.

83. **Peltonen, L.,** Atopia ja Nikkeliallergia, Thesis, Turku, Finland, 1979.

84. **Rystedt, I.,** Hand eczema in patients with history of atopic manifestations in childhood, *Acta Derm. Venereol.,* 65, 305, 1985.

85. **Rystedt, I.,** Atopic background in patients with occupational hand eczema, *Contact Dermatitis,* 12, 247, 1985.

86. **Rystedt, I.,** *Hand Eczema and Long-Term Prognosis in Atopic Dermatitis,* Karolinska Institute, Stockholm, 1985.

87. **Breit, R., Leutgeb, Ch., and Bandmann, H.-J.,** Zum neurodermatitischer Handeckzem, *Arch. Dermatol. Forsch.,* 244, 353, 1972.

88. **Forsbeck, M., Skog, E., and Åsbrink, E.,** Atopic hand dermatitis: a comparison with atopic dermatitis without hand involvement, especially with respect to influence of work and development of contact sensitization, *Acta Derm. Venereol.,* 63, 9, 1983.

89. **Lammintausta, K. and Kalimo, K.,** Atopy and hand dermatitis in hospital wet work, *Contact Dermatitis,* 7, 301, 1981.

90. **Glickman, F. S. and Silvers, S. H.,** Hand eczema and atopy in housewives, *Arch. Dermatol.,* 95, 487, 1967.

91. **Wahlberg, J. E. and Skog, E.,** Nickel allergy and atopy, *Br. J. Dermatol.,* 85, 97, 1971.

92. **Cronin, E., Bandmann, H.-J., Calnan, C. D., Fregert, S., Hjorth, N., Magnusson, B., Maibach, H. I., Malten, K. E., Meneghini, C. I., Pirilä V., and Wilkinson, D. S.,** Contact dermatitis in the atopic, *Acta Derm. Venereol.,* 50, 183, 1970.

93. **Rystedt, I.,** Contact sensitivity in adults with atopic dermatitis in childhood, *Contact Dermatitis,* 12, 1, 1985.

94. **Rudzki, E., Rebandel, P., Grzywa, Z., Napiórkowska, T., Krajewska, D., and Kozlowska, A.,** Atopic disease and allergic contact dermatitis, *Contact Dermatitis,* 11, 44, 1984.

95. **Forsbeck, M., Skog, E., and Herborn, K. H.,** The frequency of allergic diseases among relatives of patients with allergic eczematous contact dermatitis, *Acta Derm. Venereol.,* 46, 149, 1966.

96. **Hjorth, N. and Roed-Petersen, J.,** Occupational protein contact dermatitis in food handlers, *Contact Dermatitis,* 2, 28, 1976.

97. **Menné, T. and Bachmann, E.,** Permanent disability from skin disease. A study of 564 patients registered over a six-year period, *Derm. Beruf. Umwelt,* 27, 37, 1979.

98. **Menné, T. and Bachmann, E.,** Permanent disability from hand dermatitis in females sensitive to nickel, chromium and cobalt, *Dermatosen,* 27, 129, 1979.

99. **Wilkinson, D. S.,** Contact dermatitis in the home. I. The house, in *Current Concepts in Contact Dermatitis,* Verbov, J., Ed., MTP Press, Lancaster, England, 1987, chap. 1.

100. **Husain, S. L.,** Nickel coin dermatitis, *Br. Med. J.,* 2, 998, 1977.

101. **Wilkinson, D. S.,** Causes of unexpected persistence of an occupational dermatitis, in *Essentials of Industrial Dermatology,* Griffiths, W. A. D. and Wilkinson, D. S., Eds. Blackwell Scientific, Oxford, 1985, chap. 111.

102. **Rudski, E., Rebandel, P., Napiórkowska, T., Dynowska, D., and Legiéc, C.,** Nickel objects causing sensitization and relapses, *Contact Dermatitis,* 13, 335, 1985.

103. **Fisher, A. A.,** *Contact Dermatitis,* 3rd ed., Lea & Febiger, Philadelphia, 1986, chap. 41.

104. **Menné, T. and Solgaard, P.**, Temperature-dependent nickel release from nickel alloys, *Contact Dermatitis*, 5, 82, 1979.
105. **Christensen, O. and Möller, H.**, External and internal exposure to the antigen in the hand eczema of nickel allergy, *Contact Dermatitis*, 1, 136, 1975.
106. **Bang Petersen, N., Fregert, S., Brodelius, P., and Gruvberger, B.**, Release of nickel from silver coins, *Acta Derm. Venereol.* 50, 231, 1974.
107. **van Ketel, W. G.**, Occupational contact with coins in nickel-alloy patients, *Contact Dermatitis*, 12, 108, 1982.
108. **Morgan, J. K.**, Observations on persistence of skin sensitivity with reference to nickel eczema, *Br. J. Dermatol.*, 65, 84, 1953.
109. **Katz, S. A. and Samitz, M. H.**, Leaching of nickel from stainless steel consumer commodities, *Acta Derm. Venereol.*, 55, 113, 1975.
110. **Christensen, O. B. and Möller, H.**, Release of nickel from cooking utensils, *Contact Dermatitis*, 4, 343, 1978.
111. **Malten, K. E., Schutter, K., van Senden, K. G., and Spruit, D.**, Nickel sensitivity and detergents, *Acta Derm. Venereol.*, 49, 10, 1969.
112. **Quinones, P. A. and Garcia Minoz, C. M.**, Sensibilisation allergiques de contact au nickel et au chrome: présence de ces elements métalliques dans les détergents commerciaux d'usage domestique, *Ann. Dermatol. Syphilogr.*, 92, 383, 1965.
113. **Malten, K. E. and Spruit, D.**, The relative importance of various environmental exposures to nickel in causing contact hypersensitivity, *Acta Derm. Venereol.*, 49, 14, 1969.
114. **Wahlberg, J. E., Ludstedt, G., and Einarsson, O.**, Chromium, cobalt and nickel in Swedish cement, detergents, oils and cutting oils, *Berufsdermatosen*, 25, 220, 1977.
115. **Allenby, C. F. and Goodwin, B. F. J.**, Influence of detergent washing powders on minimal eliciting patch test concentration of nickel and chromium, *Contact Dermatitis*, 9, 491, 1983.
116. **Kvorning, S. A. and Svendsen, I. B.**, A synthetic detergent as a provocative agent in patch tests, *J. Invest. Dermatol.*, 26, 421, 1956.
117. **Jordan, W. P., Jr.**, Allergic contact dermatitis in hand eczema, *Arch. Dermatol.*, 110, 567, 1974.
118. **Christensen, O. B.**, Prognosis in nickel allergy and hand eczema, *Contact Dermatitis*, 8, 7, 1982.
119. **Edman, B.**, Sites of contact dermatitis in relationship to particular allergens, *Contact Dermatitis*, 13, 129, 1985.
120. **Wilkinson, D. S.**, unpublished data.
121. **Simons, R. D. G.**, *Eczema of the Hands. Investigations into Dyshidrosiform Eruptions*, 2nd ed., S. Karger, Basel, 1966.
122. **Shelley, W. B.**, Dyshidrosis (pompholyx), *Arch. Dermatol.*, 68, 314, 1953.
123. **Menné, T. and Hjorth, N.**, Pompholyx — dyshidrotic eczema, *Semin. Dermatol.*, 2, 75, 1953.
124. **Meneghini, C. L. and Angelini, G.**, Contact and microbial allergy in pompholyx, *Contact Dermatitis*, 5, 46, 1979.
125. **Lombardi, P., Campolini, P., Caccinttolo, R., and Sertoli, A.**, A clinical and experimental approach to nickel dermatitis, 6th Int. Symp. Contact Dermatitis, Tokyo, 1982.
126. **Ekelund, A. G. and Möller, H.**, Oral provocation in eczematous contact allergy to neomycin and hydroxy-quinolines, *Acta Derma. Venereol.*, 49, 422, 1969.
127. **Menné, T. and Holm, N. V.**, Nickel allergy in a female twin population, *Int. J. Dermatol.*, 22, 22, 1983.
128. **Schuppli, R.**, Zur Ätiologie der Dyshidrosis, *Dermatologica*, 108, 393, 1954.
129. **Menné, T. and Hjorth, N.**, Reactions from systemic exposure to contact allergens, *Semin. Dermatol.*, 1, 15, 1982.
130. **Burrows, D., Creswell, S., and Merrett, J. D.**, Nickel, hands and hip prostheses, *Br. J. Dermatol.*, 105, 437, 1981.
131. **Jordan, W. P. and King, S. E.**, Nickel feeding in nickel-sensitive patients with hand eczema, *J. Am. Acad. Dermatol.*, 1, 506, 1979.
132. **Gawkrodger, D. J., Cook, S. W., Fell, G. S., and Hunter, J. A. A.**, Nickel dermatitis: the reaction to oral nickel challenge, *Br. J. Dermatol.*, 115, 33, 1986.
133. **Cronin, E. di Michiel, A. D., and Brown, S. S.**, Oral challenge in nickel-sensitive women with hand eczema, in *Nickel Toxicology*, Brown, S. S. and Sunderman, F. W. Eds., Academic Press, London, 1980, 149.
134. **Marcussen, P. V.**, Spread of nickel dermatitis, *Dermatologica*, 115, 596, 1957.
135. **Kaaber, K., Sjølin, K. E., and Menné, T.**, Elbow eruptions in nickel and chromate dermatitis, *Contact Dermatitis*, 9, 213, 1983.
136. **von Krogher, G. and Maibach, H. I.**, The contact urticaria syndrome — 1982, *Semin. Dermatol.*, 1, 59, 1983.
137. **Osmundsen, P. E.**, Contact urticaria from nickel and plastic additives (butylhydroxytoluene, oleylamide), *Contact Dermatitis*, 6, 452, 1980.

138. **Krook, G.,** Occupational dermatitis from *Lactuca sativa* lettuce and *Cichorium* (endive). Simultaneous occurrence of immediate and delayed allergy as a cause of contact dermatitis, *Contact Dermatitis,* 3, 27, 1977.
139. **Wilkinson, D. S.,** Some causes of error in the diagnosis of occupational dermatitis, in *Essentials of Industrial Dermatology,* Griffiths, W. A. D. and Wilkinson, D. S., Eds., Blackwell Scientific, Oxford, 1985, chap. 5.
140. **Wilkinson, J. D. and Rycroft, R. J. G.,** in *Textbook of Dermatology,* Vol. 1, 4th ed., Rook, A., Wilkinson, D. S., Ebling, F. J. G., Champion, R. H., and Burton, J. L., Eds., Blackwell Scientific, Oxford, 1985, chap. 14.
141. **Veien, N. K., Hattel, T., Justesen, O., and Nørholm, A.,** Diagnostic procedures for eczema patients, *Contact Dermatitis,* 17, 35, 1987.
142. **Veien, N. K. and Kaaber, K.,** Nickel, cobalt and chromium sensitivity in patients with pompholyx (dyshidrotic eczema), *Contact Dermatitis,* 5, 371, 1979.
143. **Wilkinson, D. S.,** Protective gloves, in *Essentials of Industrial Dermatology,* Griffiths, W. A. D. and Wilkinson, D. S., Eds., Blackwell Scientific, Oxford, chap. 9.
144. **Fisher, A. A.,** Possible role of diet in pompholyx and nickel dermatitis, *Cutis,* 22, 412, 1978.
145. **Wilkinson, D. S. and Hambly, E. M.,** Nickel spot testing at home, *Contact Dermatitis,* 4, 114, 1978.
146. **Kaaber, K., Veien, N. K., and Tjell, J. C.,** Low nickel diet in the treatment of patients with chronic nickel dermatitis, *Br. J. Dermatol.,* 98, 197, 1978.
147. **Veien, N. K., Hattel, T., Justesen, O., and Nørholm, A.,** Dietary treatment of nickel dermatitis, *Acta Derm. Venereol.* 65, 138, 1985.
148. **Veien, N. K., Hattel, T., Justensen, O., and Nørholm, A.,** Dietary restrictions in the treatment of adult patients with eczema, *Contact Dermatitis,* 17, 223, 1987.
149. **Christensen, J. D.,** Disulfiram treatment, of three patients with nickel dermatitis, *Contact Dermatitis,* 8, 105, 1982.
150. **Menné, T. and Kaaber, K.,** Treatment of pompholyx due to nickel allergy with chelating agents, *Contact Dermatitis,* 4, 289, 1978.
151. **Kaaber, K., Menné, T., Tjell, J. C., and Veien, N.,** Antabuse treatment of nickel dermatitis. Chelation — a new principle in the treatment of nickel dermatitis, *Contact Dermatitis,* 5, 221, 1979.
152. **Menné, T., Kaaber, K., and Tjell, J. C.,** Treatment of nickel dermatitis. The influence of tetraethyl thiuram disulphide (Antabuse) on nickel metabolism, *Ann. Clin. Lab. Sci.,* 10, 160, 1980.
153. **Christensen, O. B. and Kristensen, M.,** Treatment with disulfiram in chronic nickel hand dermatitis, *Contact Dermatitis,* 8, 59, 1982.
154. **Kristensen, M.,** Toxic hepatitis induced by disulfiram in a non-alcoholic, *Acta Med. Scand.,* 209, 335, 1981.
155. **Spruit, D., Bongaarts, P. J. M, and de Jongh, G. J.,** Dithiocarbamate therapy for nickel dermatitis, *Contact Dermatitis,* 4, 350, 1978.
156. **Burrows, D., Rogers, S., Beck, M., McMaster, D., Merrett, D., and Eedy, D. J.,** Treatment of nickel dermatitis with Trientine, *Contact Dermatitis,* 15, 55, 1986.
157. **Keen, C. L., Cohen, N. L., Lönnerdal, B., and Hurley, L. S.,** Teratogenesis and low copper status resulting from triethylenetetramine in rats, *Proc. Soc. Exp. Biol. Med.,* 173, 598, 1983.
158. **van Ketel. W. G. and Bruynzeel, D. F.,** Chelating effect of EDTA on nickel, *Contact Dermatitis,* 11, 311, 1984.
159. **Kavli, G. and Førde, O. H.,** Hand dermatoses and sick leave: relationship between educational level, physical activity at work and housework as main occupation, *Contact Dermatitis,* 11, 70, 1984.
160. **Christensen, O. B.,** Prognosis in nickel allergy and hand eczema, *Contact Dermatitis,* 8, 7, 1982.
161. **Menné, T.,** Relationship between cobalt and nickel sensitization in females, *Contact Dermatitis,* 6, 337, 1980.

Chapter 14

NICKEL DERMATITIS: ITS RELATIONSHIP TO FOOD AND EXPERIMENTAL ORAL CHALLENGE

Niels K. Veien

TABLE OF CONTENTS

I. Introduction 166

II. Clinical Experience Suggesting that Systemically Induced Nickel Dermatitis is an Entity 166
- A. Secondary Eruptions of Nickel Dermatitis 166
- B. Clinical Observations Associated with the Introduction of Nickel into the Body 167
- C. Disulfiram 167

III. Clinical Reactions Other Than Those Explained by Cellular Immunity 168

IV. Experimental Oral Challenge with Nickel 168
- A. Patients with Positive Patch Tests to Nickel 168
- B. The Interplay between Nickel and Cobalt 171
- C. Patch Test-Negative Patients 172

V. Diet Treatment of Nickel Dermatitis 172

VI. Nickel in Foodstuffs 173

VII. Clinical Relevance of Nickel in Food 173

References 176

I. INTRODUCTION

The primary event in nickel dermatitis is characteristically located in an area of long-term close contact with a nickel-plated item. As fashions have changed, these primary sites have wandered from areas on the thighs in contact with garter clasps to the abdomen, where jeans buttons touch the skin, to the earlobe, a common primary site since the piercing of earlobes has become popular.

II. CLINICAL EXPERIENCE SUGGESTING THAT SYSTEMICALLY INDUCED NICKEL DERMATITIS IS AN ENTITY

A. Secondary Eruptions of Nickel Dermatitis

In his 1939 thesis, Bonnevie described secondary eruptions among his patients with garter clasp dermatitis.[1] These eruptions were frequently located in flexural areas which had no obvious contact with nickel. Calnan[2] examined female nickel-allergic patients and found secondary eruptions in 232 of 312 or 74% of the patients. The most common site were elbow flexures (184), eyelids (61), the sides of the neck (59), and inner thighs (5); there were also generalized eruptions (11). The incidence was found to be related to the activity of dermatitis at the primary site, and the same study showed that nickel, but not rubber, dermatitis gave rise to secondary eruptions. Interestingly, the secondary eruptions recurred from time to time regardless of the activity at the primary site of dermatitis.

Hematogenous spread of the dermatitis was considered to be one possible explanation for the secondary eruptions.

Calnan had 124 patients with other types of eczema not considered to be secondary eruptions. These included 81 patients with hand eczema for which it was difficult to find a cause, since only a few had direct long-term contact with nickel-plated items. Most of these cases were diagnosed as housewives' eczema, but pompholyx was also seen.

Marcussen[3] studied 622 patients with positive patch tests to 5% nickel sulfate (552 were females). The dermatitis of 517 (80%) of the patients had spread to areas other than the site of primary contact. This was seen particularly among patients with strong patch test reactivity. The dermatitis of the women typically spread to the thighs, neck, breast; for males the spread was seen on the hands, face, calves, and feet. Of the 517, 38% had hand eczema. Marcussen thought that contamination with nickel particles, rather than a hematogenous spread, explained the secondary eruptions.

Christensen[4] found secondary eruptions in 34 (52%) of 66 patients with nickel allergy and hand eczema. They were usually seen in the elbow flexures and on the neck, eyelids, and soles of the feet. The chronicity of hand eczema and flares without obvious external contact with nickel led Christensen and Möller to consider the hands a secondary rather than primary site of dermatitis.

Menné[5] found that nickel allergy in females predisposes to the development of hand eczema and that hand eczema predisposes to the development of nickel allergy.

The clinical observations described above indicate that mechanisms other than external physical contact with nickel plated items are operative in some of the rashes seen in nickel-allergic patients.

The symmetry of the secondary eruptions suggests a systemic spread which could be due either to an immunological or a nonspecific irritant reaction to nickel in the tissues or to the nickel excreted in sweat. Such reactivity could be based upon (1) hyperreactivity to normal amounts of nickel in the tissues or (2) dose-dependent reactivity of nickel-allergic patients to marked fluctuations in the levels of nickel in the tissues, particularly the skin.

The preponderance of secondary eruptions among patients with intense primary eruptions and patients who have strongly positive patch tests suggests a hypersensitive state and/or hyperreaction to the normal amounts of nickel present in the tissues, blood, and sweat.

B. Clinical Observations Associated with the Introduction of Nickel into the Body

Sporadic case reports suggest that secondary reactions may occur after changes in the amounts of nickel present in the tissues. Nickel released from a hemodialysis needle was thought to be responsible in one patient for an increase in plasma levels of nickel from <2.1 to 45.2 μg/l. The same patient experienced an associated rash (particularly on the face, neck, flexural folds, and fingers) similar to what she had previously experienced following contact with inexpensive metal jewelry. Patch testing showed nickel allergy.[6] Stoddart[7] described two patients with rashes thought to be caused by infusion cannulas. One of these patients developed hand eczema. Smeenk and Teunissen[8] described seven patients who typically reacted with vesicular eruptions of the hands following infusions, and some of these patients also experienced flares at sites of previous nickel patch tests. It was shown that a few micrograms of nickel were, in fact, released from the infusion cannulas. It would be interesting to demonstrate, under controlled experimental conditions, whether a dose of nickel of that magnitude given intravenously has such a profound clinical effect. This would add weight to the hypothesis that small alterations of the amount of nickel in the tissues may cause otherwise unexplainable fluctuations in the intensity of nickel dermatitis.

Kvorning[9] saw seven female patients with flares of dermatitis after abdominal surgery where nickel-plated retractors were used. The patients typically had flares of dermatitis on the abdomen, thighs, and earlobes and could relate these sites to previous occurrences of nickel dermatitis. All five patch-tested patients reacted to nickel. The flares were assumed to be caused by intra-abdominal release of nickel from surgical instruments.

A pompholyx-like eruption appeared 24 h after a nickel-allergic woman had swallowed a coin; the coin was removed by gastroscopy, and the eruption faded within 2 weeks.[10] Rudzki and Grzywa[11] described a patient who experienced a flare of dermatitis at sites of previous nickel contact after consuming 250 g of margarine. Nickel is used as a catalyst in the production of margarine, and the Polish margarine consumed had a content of up to 0.2mg/kg.

The significance of orthopedic and dental prostheses in relation to nickel dermatitis will be presented elsewhere in this volume.

C. Disulfiram

During the initial phases of disulfiram (diethyldithiocarbamate) therapy, 9 of 11 nickel-allergic patients with hand eczema experienced a flare of dermatitis when a dose of 300 mg/d was given. In several of these patients eruptions were seen in flexures, the ano-genital area, and around the umbilicus. These flares were generally more severe than those seen after oral challenge with 2.5 mg nickel, and they faded in spite of continued disulfiram therapy. Transient leukocytoclastic vasculitis on the legs developed in two patients. Serum nickel levels were determined for three patients, and this level was seen to increase during the first week of disulfiram therapy from 0.4 μg/l or less to 4 to 5 μg/l, while the urinary excretion of nickel increased from pretreatment levels of 1 to 2 μg/24 h to 30 to 60 μg/24 h after 1 week of therapy.[12]

Christensen and Kristensen[13] treated 11 nickel-allergic patients with 200 mg disulfiram per day. Flares of vesicular hand eczema, as well a secondary eruptions at the classical sites and associated increases in serum nickel levels, were seen in 9 of the patients after 1 week of treatment. Christensen[14] noted flares of dermatitis in three nickel-allergic patients during the initial phase of treatment with 50 mg disulfiram per day.

In all three of the above studies, disulfiram was considered to be an effective treatment for nickel hand eczema. In a double-blind, placebo-controlled study of 24 patients, disulfiram was found to be effective in 5 of 11 patients compared with 2 of 13 who received placebo. This difference was not, however, statistically significant for all the parameters used to monitor efficacy.[15]

Disulfiram used in the treatment of patients with chronic alcoholism may give rise to flares of dermatitis if the patient is sensitive to nickel. Of four such patients, three had flares of nickel dermatitis and vesicular eruptions of the fingers, while one developed vasculitis.[16] Menné noted a similar flare in a cobalt-allergic patient treated for alcoholism with disulfiram.[17]

The reactions to disulfiram therapy described above, as well as the associated increase in serum levels and the urinary excretion of nickel, suggest that fluctuations in the nickel levels in the tissues, as well as a hypersensitive state, may be of significance for the patterns of dermatitis in nickel-sensitive patients. This correlation is further underlined by the fact that despite the flares seen in the first days of therapy, disulfiram is an effective therapeutic agent for some nickel-allergic patients.

III. CLINICAL REACTIONS OTHER THAN THOSE EXPLAINED BY CELLULAR IMMUNITY

Interestingly, it is possible to encounter clinical aspects of nickel sensitivity other than classical contact dermatitis. Leukocytoclastic vasculitis was seen in some of the above-mentioned studies, and Hjorth has described vasculitis in nickel-allergic patients.[18] Asthma and rhinitis are other clinical manifestations of nickel allergy.[19-21] Such manifestations indicate that immunological mechanisms other than classical cellular immune reactions may be associated with nickel hypersensitivity. It is possible to raise antibodies to nickel albumin in rabbits, and antibodies to nickel albumin are present in some nickel-sensitive patients.[22]

Immune-complex reactions have been suggested in the etiology of similar reactions seen in guinea pigs sensitized to chromium and then challenged orally.[23]

The possibility of nonimmunological reactions should also be considered, and recurrent vesicular hand eczema should be thought of as a nonspecific clinical reaction pattern until more evidence is given concerning the causes of this dermatitis. The known and suspected causes of this clinical entity have been reviewed by Menné and Hjorth.[24]

IV. EXPERIMENTAL ORAL CHALLENGE WITH NICKEL

A. Patients with Positive Patch Tests to Nickel

The association between recurrent vesicular hand eczema and nickel allergy suspected to be caused by endogenous nickel was explored by Christensen.[4] In a double-blind, placebo-controlled trial, they challenged 12 nickel-sensitive women who had recurrent vesicular hand eczema with an oral dose of 5.6 mg nickel as nickel sulfate. Following ingestion of the active preparation, but not after the placebo, 9 of the 12 experienced a flare of their usual dermatitis. None of the same women experienced flares of dermatitis after intense handling of nickel-contaminated metal objects.

A number of similar studies have been carried out since then. Typically, single-dose challenge schemes or daily doses given for 1 or 2 d have been used. A synopsis of the information provided by these studies is given in Table 1.

These studies show that a flare of recurrent vesicular hand eczema is commonly seen following an oral dose of 2.5 to 5 mg of nickel given as an inorganic salt of the metal. Roduner et al.[33] found that flares among nickel-allergic patients were as frequent in patients with vesicular hand eczema as among those with other types of nickel dermatitis.

It can also be deduced from these studies that there is a dose-response relationship, as relatively few specific reactions were seen when doses of 0.5 mg nickel were used, whereas most patients experienced flares after doses of 5 mg nickel. In addition to the flare of vesicular hand eczema which was the reason for the challenge, secondary eruptions were noted in a number of patients after the challenge. Such secondary eruptions were seen in

Table 1
REVIEW OF THE RESULTS OF ORAL CHALLENGE WITH NICKEL IN PATIENTS WITH POSITIVE PATCH TESTS TO NICKEL

Authors	Method	Types of patients	Amount of nickel (given as $NiSO_4$)	No. of patients	With flare of usual eczema
Christensen[4]	Double-blind, placebo-controlled	Recurrent vesicular hand eczema	5.6 mg	12	9
Kaaber et al.[12]	Single-blind, placebo-controlled	Recurrent vesicular hand eczema	0.6 mg 1.2 mg 2.5 mg	14	1 1 9
Sertoli et al.[25]	Double-blind, placebo-controlled	Not given	2.4 mg	20	13
Kaaber et al.[26]	Single-blind, placebo-controlled + blind evaluation of photographs,	Recurrent vesicular hand eczema	2.5 mg	28	17
Jordan and King[27]	Double-blind, placebo-controlled crossover	Recurrent vesicular hand eczema	0.5 mg daily for 2d	10	1
Burrows et al.[28]	Double-blind, placebo-controlled crossover	Palmar dermatitis	2 mg daily for 2d 4 mg daily for 2d	22	5 to Ni, 5 to placebo, 4 to both 6 to Ni, 2 to placebo, 2 to both
Bedello et al.[29]	Placebo-controlled, repeated if in doubt	Recurrent vesicular hand eczema and other types	2.24 mg	49	31 15 had recurrent vesicular hand eczema 10 periorbital eczema 6 other reactions
Cronin et al.[30]	Open	Chronic hand eczema	0.6 mg 1.2 mg 2.5 mg	5 5 5	1 3 4
Gawkrodger et al.[31]	Double-blind, placebo-controlled	Recurrent vesicular hand eczema	0.4 mg daily for 2 d 2.5 mg daily for 2 d 5.6 mg	10 10 6	3 to Ni, 2 to placebo, 2 to both 2 to Ni, 1 to placebo, 3 to both 4 to Ni, 0 to placebo, 2 to both

Table 1 (continued)
REVIEW OF THE RESULTS OF ORAL CHALLENGE WITH NICKEL IN PATIENTS WITH POSITIVE PATCH TESTS TO NICKEL

Authors	Method	Types of patients	Amount of nickel (given as $NiSO_4$)	No. of patients	With flare of usual eczema
Veien et al.[32]	Double-blind, placebo-controlled	Hand eczema and other types of dermatitis	2.5 mg also 1 mg cobalt	144	55 of 131 with nickel allergy 23 of 131 to placebo or metals + placebo (for details see text)
Roduner et al.[33]	Double-blind, placebo-controlled	Recurrent vesicular hand eczema and other types	2.5 mg	19	5 of 12 with recurrent vesicular hand eczema 3 of 7 with other types of nickel dermatitis

seven of nine of Christensen's patients as a flare at previously healed eczema or nickel patch test sites.[4] Gawkrodger et al.[31] noted patch test flares in three of six patients challenged with 5.6 mg nickel. Patch test flares are usually not seen as clinical problem during fluctuations of nickel dermatitis. This indicates that experimental challenge with 5.6 mg nickel gives rise to greater fluctuations of nickel in the tissue than that encountered clinically. In these studies several patients also had widespread erythematous flares which suggested immediate-type reactions. These appeared within hours of the oral challenge and lasted a few days. These reactions often faded without developing into classical contact dermatitis.[4,26,30]

The studies by Burrows et al.[28] and Gawkrodger et al.[31] present results which are at variance with those from Scandinavia, Italy, and Switzerland in that these authors found that a dose of 2 to 2.5 mg nickel did not give rise to flares of dermatitis more often than a placebo. It would be interesting to gather experience from larger groups of patients in these centers since the above-mentioned studies included only limited numbers of patients.

The clinical variations and the variation in the intensity of the flares seen after oral challenge can be explained by:

1. Variations in
   - A. The amounts of nickel available for the immunologically active cells, caused by variations in the ingestion of nickel (peak intakes probably of greatest significance)
   - B. Binding in the gastrointestinal tract to other compounds
   - C. The absorption of nickel
   - D. The simultaneous ingestion of drugs or other substances which interfere with nickel metabolism by means of competitive inhibition or alterations of absorption and excretion
   - E. The concentrations of nickel in the target organ
2. Variations in the immunological reactivity as suggested by Calnan's observations[2]
3. Simultaneous action of other, possibly nonimmunological, inflammatory mechanisms activated by external primary irritants

Rather than using a dose of nickel comparable to the average daily intake for oral challenge experiments, the dose is probably better determined as one which is sufficient to give a reasonable number of positive reactions in patients who are nickel allergic and have dermatitis suspected to be endogenous, without giving rise to too many reactions in other patients. This was, in fact, the method originally used to determine the proper patch test concentration for nickel-allergic patients.

B. The Interplay between Nickel and Cobalt

A study was carried out to determine whether the ingestion of cobalt might play a role for some nickel-allergic patients. This study included a wide range of clinical morphologies of dermatitis, as all nickel-allergic patients with dermatitis not easily explained by external contact with nickel were included. In this investigation all the patients were challenged with 2.5 mg nickel, 1 mg cobalt, and a placebo given in a random fashion.[32]

The number of reactions to nickel and/or cobalt was statistically significant when compared with the appropriate patch test reactivity. Reaction to cobalt occurred more frequently than predicted by patch test reactivity.

Concomitant nickel and cobalt sensitivity are correlated to hand eczema.[5,34,35] Animal experiments also support the mutual enhancing effect of nickel and cobalt sensitization.[36] It might therefore be of interest to study in greater detail the significance of the daily ingestion of cobalt in food for hand dermatitis in patients sensitive to nickel and/or cobalt.

An *in vitro* study indicated that specifically sensitized circulating lymphocytes are stimulated after oral challenge with 4 mg nickel, but not after patch testing.[37]

In an *in vitro* study of cobalt-sensitive patients it was demonstrated by means of a lymphocyte transformation test that cobalt-allergic patients had specific transformation of lymphocytes after stimulation with cobalt, but not after stimulation with nickel.[38]

C. Patch Test-Negative Patients

Patch testing with 5% nickel sulfate in petrolatum is a reliable way to detect cellular immune reactions to nickel. However, even in sensitized patients, false-negative patch test results have been seen after the use of this method.[39,40] Alternative test methods, such as intradermal testing, may reveal nickel sensitivity despite a negative patch test.[41]

In some situations, an oral challenge procedure can help disclose nickel sensitivity in patients with negative patch tests to 5% nickel sulfate in petrolatum. Reactions to an oral test procedure can be indistinguishable from the reactions seen in patch test-positive patients.[42,43] It is interesting to note that reactions among these patch test-negative patients were seen among patients with recurrent vesicular and hyperkeratotic hand eczema, but not among a large group of patients with other types of hand dermatitis.

Upon close questioning, some of the patch test-negative patients who reacted to oral challenge with nickel did give a history of intolerance to cheap metal jewelry. In four patients, flares of dermatitis at sites of contact with garter clasps or other nickel-plated items were seen in addition to a flare of hand eczema. For such patients, the reaction to the oral challenge procedure is most likely an indication of cellular immunity to nickel, and their negative patch tests are therefore false negative. This is in agreement with histopathological findings made after oral challenge experiments carried out by Christensen et al.[44,45]

In a study involving 16 patients, poor correlation was found between, on the one hand, the results of intradermal skin tests and patch testing after stripping of the skin and, on the other hand, the results of experimental oral challenge.[46]

V. DIET TREATMENT OF NICKEL DERMATITIS

The flares seen after oral challenge with nickel, particularly the reactivation of the clinical problem in patients with recurrent vesicular hand eczema, made it logical to consider dietary depletion studies. In 1978, Kaaber et al.[26] prescribed a nickel-restrictive diet for 17 patients whose vesicular hand eczema had flared after oral challenge with 2.5 mg nickel. The dermatitis of nine patients improved after the diet had been followed for 6 weeks; for seven of these, the dermatitis returned to prediet levels when the diet was no longer followed.

Similarly, Veien et al.[47] instructed 204 patients to reduce nickel intake in their daily diets. These patients either had positive patch tests to nickel and a type of dermatitis which was thought to be endogenous, or their dermatitis had flared after oral challenge with nickel, or both. At a follow-up consultation 1 to 2 months after initiation of the diet trial, the dermatitis of 121 of these patients had improved. After 1 to 5 years, a questionnaire follow-up was carried out. Of the 150 patients responding, 88 (59%) reported that there was long-term improvement if they continued to restrict nickel intake. Diet instructions were still being followed to some degree by 90 of the 150 patients at the time of the questionnaire follow-up, and several noted that they could regulate the severity of the dermatitis by following the instructions more or less strictly. A typical response to the questionnaire was that fewer, less intense eruptions which cleared rapidly were noted during the diet period. The dermatitis cleared completely for only 32 patients. These results suggest that diet treatment has a suppressive rather than a curative effect — similar to treatment with disulfiram.

In a study of the types of dermatitis seen among 7887 patients with eczema, there were 377 cases of nickel dermatitis.[48] A nickel-restrictive diet was prescribed for 73 of these patients, and a questionnaire follow-up study 1 to 3 years after its institution showed that 37 or approximately 10% of the patients with nickel dermatitis had long-term improvement while following the restrictive diet.[49]

VI. NICKEL IN FOODSTUFFS

The limited value of diet treatment could be due to an insufficient reduction of nickel intake. Studies of the nickel content of foodstuffs have given fairly uniform information for the food items most likely to contain significant amounts of nickel. Koivistoinen's comprehensive study[50] of a large number of food items in Finland is in agreement with most other recent studies.[51-56]

High concentrations of nickel are found in tea leaves, cocoa, soy meal, sunflower seeds, oatmeal, hazelnuts, beans and other seeds used for sprouts, peanuts, mussels, and licorice, while most meats contribute little to nickel intake. There is no hard evidence for the introduction of nickel into foodstuffs from industrial processing.[56,57] Under certain conditions, certain water taps may release up to 490 μg nickel into the first 250 ml of water withdrawn after the tap has been closed overnight.[58]

Inhalation of nickel occurs, but the amounts inhaled are probably less than 15 μg/d.[59] Absorption through the pulmonary tissue is, however, much higher than from the gastrointestinal tract — 50% compared with less than 10%.[57]

Absorption through the skin is probably only significant for those heavily exposed to nickel. Experiments with ^{57}Ni indicate that $^{2}/_{3}$ of the nickel placed on the skin will penetrate the skin.[60]

The mean intake of nickel during a 24-h period was determined by Myron[52] to be 165 μg/d. Flyvholm et al.[61] calculated the daily intake in Denmark to be 150 μg nickel per day, which is comparable to the 130 μg found to be the mean intake in a 24-h period in another study.[56] In the latter study, a group of eight volunteers also demonstrated that adherence to a nickel-restrictive diet could reduce the 24-h intake to 70 μg, while conscious ingestion of items with a high nickel content resulted in a mean intake of 250 μg — for one person the intake was as much as 480 μg.

The diet instructions (Table 2) which these volunteers were asked to follow were based on the above-mentioned recent studies of the nickel content of foodstuffs and the responses to questionnaires suggesting food items which upon ingestion caused flares of the dermatitis of nickel-allergic patients.

VII. CLINICAL RELEVANCE OF NICKEL IN FOOD

The intakes listed above are considerably lower than the 300 to 600 μg suggested by Schroeder et al. in a study from 1962,[62] and the discrepancy between the actual daily intake and the challenge doses of 0.5 to 5 mg nickel poses some questions when interpreting the results of oral challenge:

1. Are the flares seen after the experimental oral challenges of clinical significance?
2. How does the challenge with an inorganic nickel salt compare with the ingestion of complex nickel compounds in genuine food?
3. Could it be that we should determine peak intake of nickel rather than mean intakes, due to the eruptive nature of many types of nickel dermatitis?
4. What is the concentration of nickel in the target organ, i.e., the skin, and how does this concentration vary with working conditions, excessive sweating, psychological stress, etc.?
5. Why do patients in Northern Ireland and England not experience flares after oral challenge with 2.5 mg nickel?
6. Is the single-dose challenge a relevant diagnostic method?

Nickel excretion in the urine increased dramatically in a group of persons not hypersensitive

Table 2
DIET INSTRUCTION FOR THE REDUCTION OF NICKEL INTAKE

It is not necessary to restrict intake of the following:

- All kinds of meat and poultry, fish, and eggs; milk, unflavored yogurt, butter, margarine, and cheese
- Vegetables: asparagus, red beets, broccoli, brussels sprouts, white cabbage, cauliflower, champignon, Chinese cabbage, corn, cucumber, dill, eggplant, parsley, green and red peppers, and potatoes
- Grain products: finely milled rye and wheat flour, whole grain rye and wheat bread (in moderation), white rice, and breakfast foods made of rice, cornflakes, and popcorn
- Fruits: peaches, pears, raisins, rhubarb, and all berries *except* raspberries
- Drinks: coffee and tea (not too strong and in moderation), soft drinks, and alchoholic beverages (distilled products and drinks made from these)
- Miscellaneous items including spaghetti and yeast

The following foods have a high nickel content, and intake should be restricted:

- Shellfish such as prawns and mussels
- Vegetables: beans (green, brown, white), kale, leeks, lentils, lettuce, spinach, peas, and sprouts made from beans and lucerne
- Grain products: buckwheat, millet, oatmeal, wheat bran, i.e., bran and fiber products including muesli, bran biscuits, and fiber tablets and multigrain breads
- Fruits: figs, pineapple, prunes, and raspberries
- Drinks: chocolate and cocoa drinks and tea from drink dispensers
- Miscellaneous items including almonds, baking powder (in large amounts), linseed, hazelnuts, peanuts, soya-containing foods such as meat with soya additives, sunflower seeds, sweets containing chocolate, marzipan, nuts, strong licorice, and vitamins containing nickel

Various other food items and drinks can aggravate nickel eczema, even though the nickel content of these foods may be low; included among these are beer, wine (particularly red), herring, mackerel, tuna, tomato, onion, carrot, and certain fruits, in particular, apples and citrus fruits/juice; the vegetables mentioned here can usually be tolerated when cooked

The first liter of water taken from the tap in the morning should not be used in food preparation, as nickel may be released from the tap during the night; nickel-plated kitchen utensils, such as eggbeaters and tea balls, should be replaced

Acid foods such as stewed fruits and rhubarb cooked in stainless-steel utensils should be avoided; the acids in the foods can cause nickel to be released from the utensils; tinned foods should be eaten only in moderation

Although eczema is probably mostly due to the intake of foods containing nickel, it should be pointed out that prolonged physical contact with nickel-plated items can make it worse

It is important to follow the diet instructions for at least 1 to 2 months; it can be difficult to change eating habits for a long period of time, and it is therefore recommended that the instructions be adhered to as closely as possible when eating at home; when eating away from home, the diet can be modified as necessary

The eczema will not necessarily clear during the diet period, but there will ordinarily be fewer and shorter flares, and these will gradually diminish in intensity

If after 1 to 2 months the diet seems to have had a good effect, it is recommended that it be continued for 6 months to 1 year, eating only moderate amounts of those foods with a high nickel content

From Veien, N.K. and Andersen, M. R., *Acta Derm. Venereol. (Stockholm)*, 66, 502, 1986. With permission.

to nickel who were treated with indomethacin.[63] Penicillamin and tetracycline are other examples of drugs which can interfere with nickel metabolism (see Chapter 15).

Drugs like disulfiram and dithiocarbamate, which are the treatments of choice for nickel poisoning, alter nickel metabolism profoundly, as has been shown in the studies of disulfiram for the treatment of nickel dermatitis.

The bioavailability of nickel may vary with the foods ingested. Theoretically, phytin, present for example in oats, could alter the absorption pattern of nickel.[64] The experimental addition of 5 mg nickel as nickel sulfate to a North American breakfast, to a Guatemalan meal, and to various drinks inhibited nickel absorption, whereas the addition of phytic acid did not suppress nickel absorption.[65] In another experiment, less than 5% of nickel added to scrambled eggs in a North American breakfast was absorbed, compared to 8 to 50% of nickel ingested in water after overnight fasting.[66]

The problem of *in vivo* chelation of nickel as a factor in the metabolism of nickel will be dealt with elsewhere in this volume.

A combination of high nickel concentration in the target organ during a period in which the immune system is particularly active and during which nonimmunologically inflammatory mechanisms are active may be responsible for the intermittent flares of dermatitis of some nickel-allergic patients. This could also explain the eruptive nature of the dermatitis of some patients.

One aspect of nickel metabolism which could explain some of the unexpected clinical findings is the possible interaction between nickel and other substances in the tissues. At the level of thc Langcrhans cells there might be interaction with other metals, such as iron, which could alter the sensitization rate as well as the reaction to nickel brought to the skin from the peripheral blood, e.g., via sweat. The questions of interaction between nickel and iron and nickel and zinc have not been resolved.[67,68] Histidine binds nickel strongly, and such binding could alter the bioavailability of nickel from the peripheral blood.

In an open study, an attempt was made in our clinic to treat six patients with nickel dermatitis by giving a ferrous salt (240 mg Fe^{++} per day) as three oral doses. The results were not impressive. One other nickel-allergic patient treated herself with zinc sulfate tablets (200 mg/d in two doses per day) and showed a dramatic improvement in her dermatitis. We then treated seven other nickel-allergic patients who had vesicular hand eczema, or other types of dermatitis thought to be endogenous, with zinc sulfate tablets (100 mg three times daily). No definite improvement was seen after 1 month of therapy (unpublished data).

The following questions could therefore also be asked:

1. What role do chelating compounds in natural foods and those present in drugs play in the flares of nickel dermatitis?
2. Will certain foods and drugs significantly alter the concentration of nickel in the tissues by mechanisms other than chelation?

Until these questions, and possible others, have been answered, we are left with an undetermined etiology of the flares and the current oral challenge technique which is, at best, a rather crude test procedure.

In carrying out a double-blind oral challenge, we routinely use a tablet containing 2.5 mg nickel as nickel sulfate and a placebo given after an interval of 1 week. All the patients are given an instruction sheet describing the challenge, and they determine themselves whether or not a flare has taken place. The code is broken when the patient returns 1 week after the last tablet has been taken. Performed in this manner, the oral challenge procedure is easy to carry out and only one extra consultation is necessary — unless the dermatitis is very active, in which case the oral challenge procedure is postponed until the eczema is relatively quiescent.

The flares seen after this procedure are commonly equivalent to the patient's usual flares of dermatitis and fade within 1 week of the challenge. A few patients experience rather severe reactions which indicates that, for these patients, the dose given is high compared with what is realistically absorbed by and is present in the tissues. The flares seen at old patch test sites or sites of garter belt clasps in our experience appear only after experimental challenge, and they are not seen as natural fluctuations of nickel dermatitis. One type of flare is the "baboon" syndrome,[69,70] which is also most commonly seen after experimental challenge.

We have used the described procedure for 10 years and have found it useful in sorting out the nickel-allergic patients who may have systemically induced dermatitis. It is also useful as an illustration to the patient of the possible endogenous factors causing the dermatitis. Most patients more readily accept a tedious treatment like a restrictive diet or disulfiram, which requires the repeated blood tests to monitor serum levels of liver enzymes, when they have actually seen their hands flare after the ingestion of nickel.

This experience is in contrast to that of Burrows[71] and Gawkrodger et al.[31] who consider it unlikely that the amounts of nickel present in food can influence the course of nickel dermatitis.

Further clinical experience as well as well-controlled clinical trials will be necessary before these questions can be resolved.

REFERENCES

1. **Bonnevie, P.,** *Ätiologie und Pathogenese der Ekzemkrankheiten,* Busck, Copenhagen and Barth, Leipzig, 1939, 600.
2. **Calnan, C. D.,** Nickel dermatitis, *Br. J. Dermatol.,* 68, 229, 1956.
3. **Marcussen, P. V.,** Spread of nickel dermatitis, *Dermatologica,* 115, 596, 1957.
4. **Christensen, O. B.,** Nickel Allergy and Hand Eczema in Females, Thesis, Lund University, Malmö, Sweden, 1981, 28.
5. **Menné, T.,** Nickel Allergy, Thesis, University of Copenhagen, Copenhagen, 1983, 69.
6. **Olerud, J. E., Minako, Y. L., Uvelli, D. A., Goble, G. J., and Babb, A. L.,** Presumptive nickel dermatitis from hemodialysis, *Arch. Dermatol.,* 120, 1066, 1984.
7. **Stoddart, J. C.,** Nickel sensitivity as a cause of infusion reactions, *Lancet,* 2, 741, 1960.
8. **Smeenk, G. and Teunissen, P. C.,** Allergische Reacties op Nikkel uit Infusie-Toedieningssystemen, *Ned. Tijdschr. Geneeskd.,* 121, 4, 1977.
9. **Kvorning, S. A.,** Post-operative dermatitis following non-epidermal nickel contact, *Contact Dermatitis,* 1, 327, 1975.
10. **Boonk, W. J. and van Ketel, W. G.,** Nickel dermatitis after swallowing a nickel coin (Dutch florin), *Contact Dermatitis,* 5, 341, 1979.
11. **Rudzki, E., and Grzywa, Z.,** Exacerbation of nickel dermatitis by margarine, *Contact Dermatitis,* 3, 344, 1977.
12. **Kaaber, K., Menné, T., Tjell, J. C., and Veien, N. K.,** Antabuse® treatment of nickel dermatitis. Chelation — a new principle in the treatment of nickel dermatitis, *Contact Dermatitis,* 5, 221, 1979.
13. **Christensen, O. B. and Kristensen M.,** Treatment with disulfiram in chronic nickel hand dermatitis, *Contact Dermatitis,* 8, 59, 1982.
14. **Christensen, J. D.,** Disfulfiram treatment of three patients with nickel dermatitis, *Contact Dermatitis,* 8, 105, 1982.
15. **Kaaber K., Menné, T., Veien, N., and Hougaard, P.,** Treatment of nickel dermatitis with Antabuse®; a double blind study, *Contact Dermatitis,* 9, 297, 1983.
16. **Veien, N.K.,** Cutaneous side effects of Antabuse® in nickel allergic patients treated for alcoholism, *Boll. Ital. Dermatol. Allergol. Prof.,* 42, 139, 1987.
17. **Menné, T.,** Flare-up of cobalt dermatitis from Antabuse® treatment, *Contact Dermatitis,* 12, 53, 1985.
18. **Hjorth, N.,** Nickel vasculitis, *Contact Dermatitis,* 2, 356, 1976.
19. **Block, G. T. and Yeung, M.,** Asthma induced by nickel, *JAMA,* 247, 1600, 1982.
20. **Niordsen, A.-M.,** Nickel sensitivity as a cause of rhinitis, *Contact Dermatitis,* 7, 273, 1981.
21. **Mitchell, J. C.,** Type I without type III and type IV hypersensitivity reaction to nickel, *Contact Dermatitis,* 7, 270, 1981.
22. **Veien, N. K., Christiansen, A. H., Svejgaard, E., and Kaaber, K.,** Antibodies against nickel-albumin in rabbits and man, *Contact Dermatitis,* 5, 378, 1979.
23. **Polak, L., Turk, J. L., and Frey, J. R.,** Studies on contact hypersensitivity to chromium compounds, *Progr. Allergy,* 17, 145, 1973.

24. **Menné, T. and Hjorth, N.,** Pompholyx — dyshidrotic eczema, *Semin. Dermatol.*, 2, 75, 1983.
25. **Sertoli, A., Lombardi, P., Francalanci, S., Gola, M., Girogini, S., and Panconesi, E.,** Effetto della somministrazione orale di apteni in soggetti sensibilizzati affetti da eczema allergico da contatto, *G. Ital. Dermatol. Venereol.*, 120, 213, 1985.
26. **Kaaber, K., Veien, N. K., and Tjell, J. C.,** Low nickel diet in the treatment of patients with chronic nickel dermatitis, *Br. J. Dermatol.*, 98, 197, 1978.
27. **Jordan, W. P., Jr. and King, S. E.,** Nickel feeding in nickel-sensitive patients with hand eczema, *J. Am. Acad. Dermatol.*, 1, 506, 1979.
28. **Burrows, D., Creswell, S., and Merret, J. D.,** Nickel, hands and hip prostheses, *Br. J. Dermatol.*, 105, 437, 1981.
29. **Bedello, P. G., Goitre, M., Cane, D., and Roncarolo, G.,** Nickel: aptene ubiquitario, *G. Ital. Dermatol. Venereol.*, 120, 293, 1985.
30. **Cronin, E., di Michiel, A., and Brown, S. S.,** Oral Nickel Challenge in Nickel-Sensitive Women with Hand Eczema, abstr. of scientific papers presented at Swansea Conf. on Nickel Toxicology, Swansea, Wales, September 3 to 6, 1980, 91.
31. **Gawkrodger, D. J., Cook, S. W., Fell, G. S., and Hunter, J. A. A.,** Nickel dermatitis: the reaction to oral nickel challenge, *Br. J. Dermatol.*, 115, 33, 1986.
32. **Veien, N. K., Hattel, T., Justesen, O., and Nørholm, A.,** Oral challenge with nickel and cobalt in patients with positive patch tests to nickel and/or cobalt, *Acta Derm. Venereol. (Stockholm)*, 67, 321, 1987.
33. **Roduner, J., Haudenschild-Falb, E., Kunz, E., Hunziker, T., and Krebs, A.,** Perorale Nickelprovokation bei nichtdyshidrosiformem und dyshidrosiformem Nickelekzem, *Der Hautarzt*, 38, 262, 1987.
34. **van Joost, T. and van Everdingen, J. J. E.,** Sensitization to cobalt associated with nickel allergy: clinical and statistical studies, *Acta Derm. Venereol. (Stockholm)*, 62, 525, 1982.
35. **Rystedt, I. and Fischer, T.,** Relationship between nickel and cobalt sensitization in hard metal workers, *Contact Dermatitis*, 9, 195, 1983.
36. **Lammintausta, K., Pitkänen, O.-P., Kalimo, K., and Jansen, C. T.,** Interrelationship of nickel and cobalt contact sensitization, *Contact Dermatitis*, 13, 148, 1985.
37. **Veien, N. K., Svejgaard, E., and Menné, T.,** In vitro lymphocyte transformation to nickel: a study of nickel-sensitive patients before and after epicutaneous and oral challenge with nickel, *Acta Derm. Venereol. (Stockholm)*, 59, 447, 1979.
38. **Al-Tawil, N. G., Marcusson, J. A., and Möller, E.,** In vitro testing for cobalt sensitivity: an aid to diagnosis, *Acta Derm. Venereol. (Stockholm)*, 64, 203, 1984.
39. **Kieffer, M.,** Nickel sensitivity: relationship between history and patch test reaction, *Contact Dermatitis*, 5, 398, 1979.
40. **Marcussen, P. V.,** Specificity of patch tests with 5% nickel sulphate, *Acta Derm. Venereol. (Stockholm)*, 39, 187, 1959.
41. **Marcussen, P. V.,** Comparison of intradermal test and patch test using nickel sulfate and formaldehyde, *J. Invest. Dermatol.*, 40, 263, 1963.
42. **Veien, N. K., Hattel, T., Justesen, O., and Nørholm, A.,** Oral challenge with metal salts. I. Vesicular patch-test-negative hand eczema, *Contact Dermatitis*, 9, 402, 1983.
43. **Veien, N. K., Hattel, T., Justesen, O., and Nørholm, A.,** Oral challenge with metal salts. II. Various types of eczema, *Contact Dermatitis*, 9, 407, 1983.
44. **Christensen, O. B., Lindström, C., Löfberg, H., and Möller, H.,** Micromorphology and specificity of orally induced flare-up reactions in nickel-sensitive patients, *Acta Derm. Venereol. (Stockholm)*, 61, 505, 1981.
45. **Christensen, O. B., Beckstead, J. H., Daniels, T. E., and Maibach, H.,** Pathogenesis of orally induced flare-up reactions at old patch sites in nickel allergy, *Acta Derm. Venereol. (Stockholm)*, 65, 298, 1985.
46. **Veien, N. K. and Kaaber, K.,** Nickel, cobalt and chromium sensitivity in patients with pompholyx (dyshidrotic eczema), *Contact Dermatitis*, 5, 371, 1979.
47. **Veien, N. K., Hattel, T., Justesen, O., and Nørholm, A.,** Dietary treatment of nickel dermatitis, *Acta Derm. Venereol. (Stockholm)*, 65, 138, 1985.
48. **Veien, N. K., Hattel, T., Justesen, O., and Nøholm, A.,** Diagnostic procedures for 7887 patients with eczema, *Contact Dermatitis*, 17, 35, 1987.
49. **Veien, N. K., Hattel, T., Justesen, O., and Nørholm, A.,** Dietary restrictions in the treatment of adult patients with eczema, *Contact Dermatitis*, 17, 223, 1987.
50. **Koivistoinen, P.,** Mineral composition of Finnish foods: N, K, Ca, Mg, P, S, Fe, Cu, Mn, Zn, Mo, Co, Ni, Cr, F, Se, Si, Rb, Al, B, Br, Hg, As, Cd, Pb and ash, *Acta Agric. Scand, Suppl.*, 22, 1980.
51. **Ellen, G., van den Bosch-Tibbesma, G., and Douma, F. F.,** Nickel content of various Dutch foodstuffs, *Z. Lebensm. Unters. Forsch.*, 166, 145, 1978.
52. **Myron, D. R., Zimmerman, T. J., Shuler, T. R., Klevay, L. M., Lee, D. E., and Nielsen F. H.,** Intake of nickel and vanadium by humans. A survey of selected diets, *Am. J. Clin. Nutr.*, 31, 527, 1978.

53. **Brun, R.,** Nickel dan les aliments et eczéma de contact, *Dermatologica,* 159, 365, 1979.
54. **Kirkpatrick, D. C. and Coffin, D. E.,** Trace metal content of various cured meats, *J. Sci. Food Agric.,* 26, 43, 1975.
55. **Thomas, B., Roughan, J. A., and Watters, E. D.,** Cobalt, chromium and nickel content of some vegetable foodstuffs, *J. Sci. Food Agric.,* 25, 771, 1974.
56. **Veien, N. K. and Andersen, M. R.,** Nickel in Danish food, *Acta Derm. Venereol. (Stockholm),* 66, 502, 1986.
57. **Nielsen, G. G., Simesen, M. G., Flyvholm, M.-A., and Veien, N. K.,** *Nikkel,* Akademiet for de Tekniske Videnskaber, Lyngby, Denmark, 1984, 33.
58. **Anderson, K. E., Nielsen, G. D., Flyvholm, M.A., Fregert, S. and Gruvberge, B.,** Nickel in tap water, *Contact Dermatitis,* 9, 140, 1983.
59. *Nickel,* National Academy of Sciences, Washington, D. C., 1975, 268.
60. **Nørgaard, O.,** Investigations with radioactive Ni 57 into the resorption of nickel through the skin in normal and in nickel-hypersensitive persons, *Acta Derm. Venereol. (Stockholm),* 35, 111, 1955.
61. **Flyvholm, M.-A., Nielsen, G. D., and Andersen, A.,** Nickel content of food and estimation of dietary intake, *Z. Lebensm. Unters. Forsch.,* 179, 427, 1984.
62. **Schroeder, H. A., Balassa, J. J., and Tipton, I. H.,** Abnormal trace metals in man — nickel, *J. Chronic Dis.,* 15, 51, 1961.
63. **Spruit, D.,** Increased nickel absorption following indomethacin therapy, *Contact Dermatitis,* 5, 62, 1979.
64. **Vohra, P., Gray, G. A., and Kratzer, F. H.,** Phytic acid-metal complexes, *Proc. Soc. Exp. Biol. Med.,* 120, 447, 1965.
65. **Solomons, N. W., Viteri, F., Shuler, T. R., and Nielsen, F. H.,** Bioavailability of nickel in man: effects of foods and chemically-defined dietary constituents on the absorption of inorganic nickel, *J. Nutr.,* 112, 39, 1982.
66. **Sunderman, F. W., Jr., Hopfer, S. M., Swift, T., Ziebka, L., Marchus, A. H., Most, B. M. and Creason, J.,** Nickel absorption and elimination in human volunteers, in *Trace Elements in Human Health and Disease,* 2nd Nordic Symp., Odense University, Odense, Denmark, August, 1987.
67. **Nielsen, F. H.,** Effect of form of iron on the interaction between nickel and iron in rats: growth and blood parameters, *J. Nutr.,* 110, 965, 1980.
68. **Ling, J. R. and Leach, R. M., Jr.,** Studies on nickel metabolism: interaction with other mineral elements, *Poult. Sci.,* 58, 591, 1979.
69. **Andersen, K. E., Benezra, C., Burrows, D., Camarase, J., Dooms-Goossens, A., Ducombs, G., Frosch, P., LaChapelle, J.-M., Lahti, A., Menné, T., Rycroft, R., Scheper, R., White, I., and Wilkinson, J.,** Contact dermatitis. A review, *Contact Dermatitis,* 16, 55, 1987.
70. **Andersen, K. E., Hjorth, N., and Menné, T.,** The baboon syndrome: systemically-induced allergic contact dermatitis, *Contact Dermatitis,* 10, 97, 1984.
71. **Burrows, D.,** The effect of systemic and implanted metals on metal dermatitis, *Dermatol. Clin.,* 2, 603, 1984.

Chapter 15

NICKEL ALLERGY AND DRUG INTERACTION

Kaare Weismann and Torkil Menné

TABLE OF CONTENTS

I. Chelate and Complex Formation 180

II. The Stability Constant(K) 180

III. Drugs and Metal Binding 180

IV. Disulfiram (Antabuse) and Diethyldithiocarbamate (Dithiocarb) 181

V. Triethylenetetramine (Trientene) 182

VI. D-Penicillamine (Dimethylcysteine) 182

VII. Ethambutol 183

VIII. Halogenated 8-Oxyquinolines (Vioform, Diodoquin) 184

IX. Tetracycline 185

X. Discussion 185

References 186

I. CHELATE AND COMPLEX FORMATION

Chelate formation is a process involving coordinate covalent binding between an electrophilic electropositive atom of a metal, especially those of the second half of the first transition series (Fe, Co, Ni, Cu, and Zn), and nucleophilic atoms of an organic molecule, usually N, O, and S ligands. The result is the formation of a molecule with a complex structure containing the metal and one, two, three, or four molecules of the organic compound. The organometallic complex behaves differently from the uncomplexed molecule and is nonionic. The sequestered metal becomes part of a new structure; it may, for instance, be lipophilic, which completely alters its normal metabolic pathways.

An important implication of chelating is interference with substrates and enzymes containing metals as an integral part or employing them as a cofactor.[1] Inactivation or removal of the metal will change the normal enzyme-substrate reactions and cause metabolic disturbances. The configuration of proteins and nucleic acids is influenced by the presence of bivalent metallic cations which are able to form bonds between phosphate oxygen and the amino groups of bases of nucleic acids.[2] Removal of the stabilizing metal may change nucleic acid synthesis which again may exert profound effects on the organism.

II. THE STABILITY CONSTANT (K)

The potency of complex formation between a metal and an organic compound is characterized by its stability constant (K) defined as

$$K = \frac{(MS)}{(M)\cdot(S)}$$

where M is the metal, S the substrate or organic molecule, and MS the organometallic complex.

K is expressed as log K; for example calcium-ethylenediaminetetraacetic acid (EDTA) has a log K of 3.5 and zinc-EDTA a log K of 16.6. For complexes formed between ligands containing –O, –N, and –S and bivalent metal ions of the second half of the first transition series, K follows the order of the periodic system, except for copper and zinc: Ba < Sr < Ca < Mg < Mn < Fe < Co < Ni < Cu < Zn. For metals to the left of iron, N residues are the preferred electron donor source; to the right of iron, –O is preferred. Thus, Ni, Cu, and Zn will preferably bind to oxygen, e.g., in phosphate.[1]

Nature is full of chelates, beneficial and harmful. As an example of the first, naturally occuring vitamin B compounds are usually complexed with zinc, both elements being indispensable for normal life. Harmful chelates are formed by phytic acid of unleavened grains and calcium, zinc, copper, and iron, whereby these essential elements may be withheld from absorption. In living tissues, important processes depend on chelate formation, e.g., chelation of ferrous iron with the porphyrin ring of hemoglobin is the basis of oxygen transport. Transport of essential trace elements such as copper, zinc, and nickel in the body is mediated by complex formation with serum proteins, primarily albumin, prealbumin, and transferrin.

III. DRUGS AND METAL BINDING

Many drugs, including antibiotics, are capable of forming complexes or chelates with essential bivalent metals, and it is likely that therapeutic mechanisms are related to interference with metal ions. So far, little work has been done to penetrate this complex (in a double sense) world. Some drugs have been used intentionally to interfere with trace element metabolism. D-Penicillamine has been given for Wilson's disease (causing increased excre-

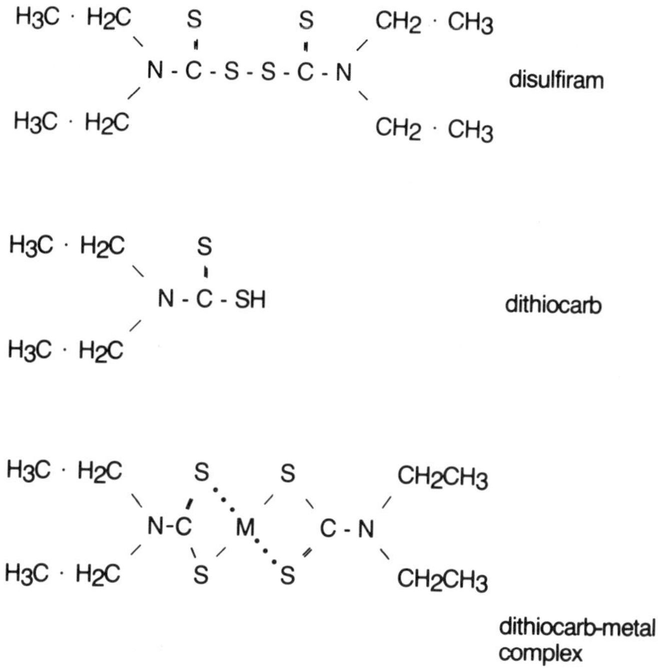

FIGURE 1. Disulfiram and its split product dithiocarb. Complex formation with a divalent metal cation (M) is shown.

tion of high tissue levels of copper), 8-oxyquinoline therapy has been given for acrodermatitis enteropathica, and disulfiram therapy has been given for nickel sensitivity.[3] In these instances, therapy is intended to regulate or normalize the metabolism of essential trace elements. This is to be distinguished from efforts to treat heavy metal poisoning with cadmium, lead, and mercury. In the following, nickel and drug interaction and possible therapeutic applications of chelating therapy for nickel sensitivity are discussed.

IV. DISULFIRAM (ANTABUSE) AND DIETHYLDITHIOCARBAMATE (DITHIOCARB)

Disulfiram (Antabuse) was introduced in 1948 for the supportive management of alcohol addiction. Its effect is due to an inhibition of aldehyde dehydrogenase, whereby ethyl alcohol is metabolized to acetaldehyde which is accumulated in the body. After absorption, disulfiram is rapidly converted to diethyldithiocarbamate (dithiocarb) (Figure 1). Dithiocarb has potential chelating properties and it is believed that this compound is responsible for the alteration in nickel metabolism observed in nickel-sensitive patients[3] (Chapter 2). In human medicine, sodium dithiocarb is an experimental drug used for nickel carbonyl poisoning. The compound and closely related analogs are widely used as fungicides and pesticides in farming, forestry, and gardening. Dithiocarb has strong chelating properties, forming inactive dithiol complexes with heavy metals such as nickel, copper, and zinc. It has been shown in animal experiments to increase intestinal absorption of ^{65}Zn significantly.[4] Nickel studies have not been performed, but it can be expected that nickel absorption is enhanced by its administration. Possibly only intravenous administration can be expected to deplete the body of nickel. One patient with Wilson's disease who was treated with both oral and intravenous sodium dithiocarb showed a negative nickel balance after 4 to 6 d of treatment.[5] The use of the drug for nickel sensitivity has not been evaluated and it cannot be recommended for human use until further studies have been performed.

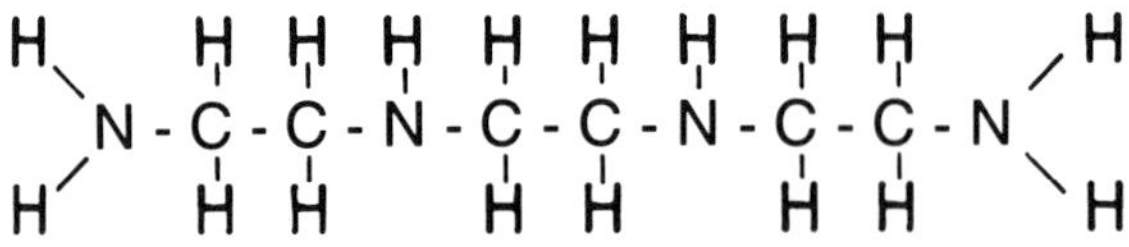

FIGURE 2. Trientene.

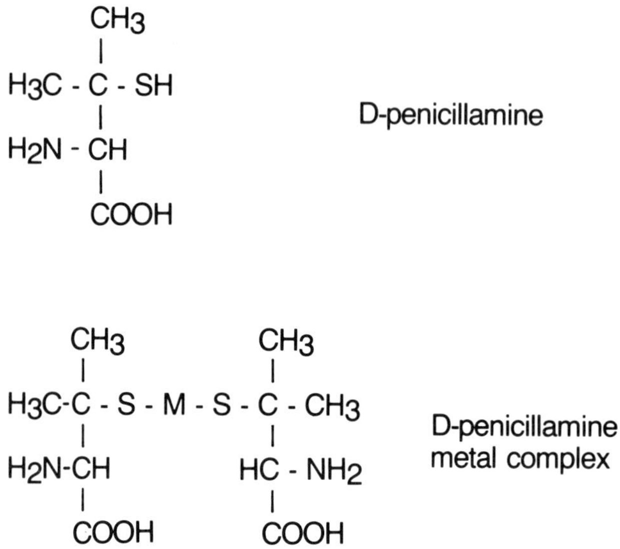

FIGURE 3. D-penicillamine (dimethylcysteine) and complex formation with a divalent metal cation (M).

V. TRIETHYLENETETRAMINE (TRIENTENE)

Triethylenetetramine (trientene) (Figure 2) has been used experimentally in rats to promote nickel excretion.[6] The drug is known as a potential chelator of nickel, among others, and could therefore be expected to improve nickel sensitivity. This was studied in a double-blind crossover trial in 23 nickel-sensitive patients.[7] The patients receiving the active drug took 300 mg trientene daily by mouth for 6-week periods. There was a 4-week drug-free interval before crossover. Urinary excretion was 3.28 ± 3.87 μ/l during trientene and 3.09 ± 2.59 μ/l during the placebo treatment. Thus, there was no indication that the drug possesses any nickeluric effect. This was in accordance with lack of any significant alterations in the skin condition of the treated patients. Trientene had an effect on urinary copper excretion and is therefore a possible candidate for treatment of Wilson's disease. The drug has proven ineffective in zinc-intoxicated rats.[6] Local treatment with trientene for nickel dermatitis has not been reported.

VI. D-PENICILLAMINE (DIMETHYLCYSTEINE)

D-Penicillamine (dimethylcysteine) is a strong metal chelator and was originally used mainly for heavy metal poisoning with lead, mercury, copper, and zinc. The molecule contains one thiol group and one carboxyl group (Figure 3). Two molecules form stable complexes with bivalent heavy metals. Its use for nickel sensitivity has apparently not been studied systematically. Studies have concentrated on the effects of D-penicillamine on copper and zinc metabolism. Due to the similarity of the coordinate chemistry of the heavy metals

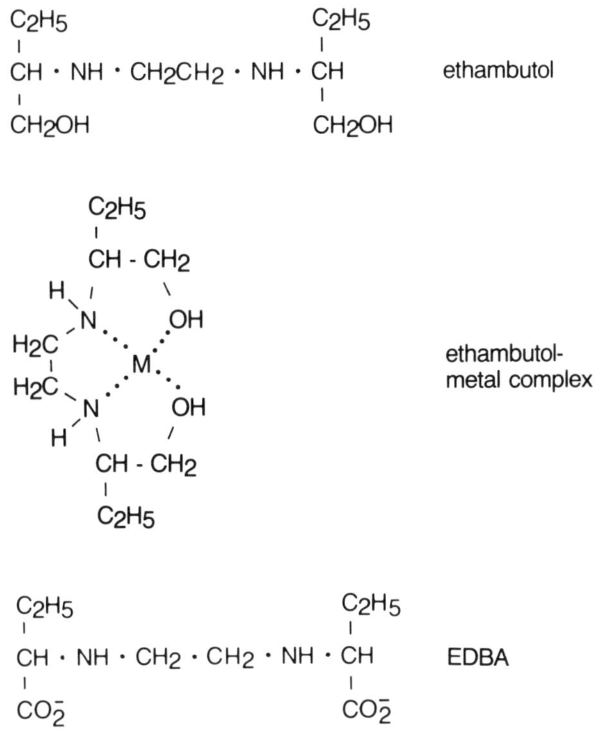

FIGURE 4. Ethambutol and ethambutol-metal (M) complex. The intermediary product 2,2′- EDBA is shown.

of the first transition series, results based on copper and zinc studies may provide a guide, albeit an incomplete one, for the effects on nickel metabolism. D-Penicillamine given orally by gavage to rats increases ^{65}Zn absorption in a dose-dependent way.[8] In animals receiving 10 mg D-penicillamine daily, the mean zinc absorption was elevated by 40%; 20 and 50 mg resulted in average increases of about 70 and 90%, respectively. Higher doses have a decreasing effect on the absorption, probably due to an increase in the urinary excretion of tissue zinc. In patients with generalized scleroderma 500 to 750 mg D-penicillamine resulted in a 66% increase in ^{65}Zn absorption.[9] The serum level of the element remained unchanged. Urinary zinc was not measured. Based on experimental and clinical data, oral D-penicillamine in common doses cannot be expected to promote nickel excretion in man. However, clinical trials to evaluate the drug in nickel-sensitive patients have not been performed.

VII. ETHAMBUTOL

The general ability of antituberculosis drugs to bind bivalent transition metals is well known.[10] Whether or not the metal-binding effect is related to antitubercular activity, the administration of such compounds is likely to cause alterations in heavy metal metabolism. Ethambutol has been found to be biologically superior to other closely related analogs. This has been attributed to the intermediary conversion of the drug to 2,2′-(ethylene-diamino) dibutyric acid (EDBA) which is a much more powerful metal chelator than ethambutol (Figure 4). Ethambutol increases urinary zinc excretion and high doses lead to damage of zinc-rich tissues such as tapetum lucidum in dogs. A zinc-depleting effect is believed to account for visual abnormalities occasionally observed in patients undergoing prolonged therapy with ethambutol.

From a theoretical point of view, ethambutol could be expected to work in nickel sen-

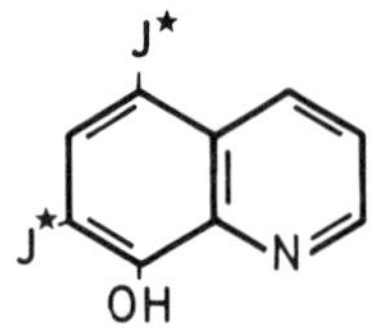

diiodo-hydroxyquinoline
(Diodoquin)

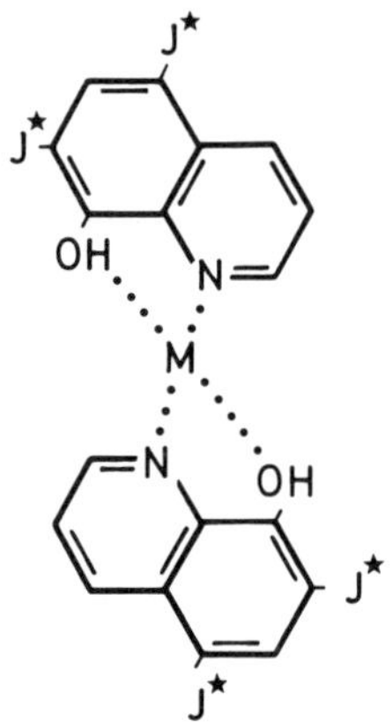

diiodo-hydroxyquinoline-
metal complex

FIGURE 5. 8-Oxyquinoline derivate. Complex formation with a divalent metal cation (M) is shown. Halogen substitutes indicated by ★.

sitivity. Ethambutol probably increases the absorption of nickel, but the excretion of nickel, in analogy with zinc, is possibly further increased by binding of tissue nickel to EDBA which is lost with the urine. In this respect there is an analogy with disulfiram, which is also converted to a potential metal-binding compound after absorption, with a consequent net loss of sequestered metal ions from the tissues. Experimental and clinical studies remain to be performed to evaluate the application of ethambutol for nickel sensitivity.

VIII. HALOGENATED 8-OXYQUINOLINES (VIOFORM, DIODOQUIN)

8-Oxyquinolines, whether or not substituted with halogens (e.g., Vioform containing one molecule of iodine and chlorine, and Diodoquin containing two molecules of iodine), are able to bind bivalent heavy metals due to the presence of neighboring –O and –S groups in the molecule (Figure 5). The complexes are lipophilic, and although only about 10% is absorbed, this promotes absorption of heavy metals, such as zinc, as used therapeutically in acrodermatitis enteropathica. After absorption, 8-oxyquinolines are metabolized in the liver to nonchelating glucoronic acid compounds, which leave the body with urine and feces via the bile. Hereby the complexed metal is released and available for metabolic purposes in the body. Extremely high oral doses or an increased transport of the drug over the intestinal mucosa due to pathologic processes in the gut may exert a metal-depleting effect on the body. Zinc deficiency has been believed to occur in certain zinc-rich tissues such as the optic nerve and the retina, especially the tapetum lucidum of animals. Such events have been suggested to explain subacute myelooptic neuropathy (SMON) as an acute severe zinc deficiency. It occurred in Japan in the 1950s and 1960s where epidemic acute gastroenteritis was treated with large oral doses of halogenated 8-oxyquinolines.[11]

The systemic use of 8-oxyquinolines is banned today. The derivates are only used for local treatment of secondarily infected inflammatory skin disorders, often combined with a potent glucocorticoid. 8-Oxyquinoline acts as an antimicrobiologic agent, probably by se-

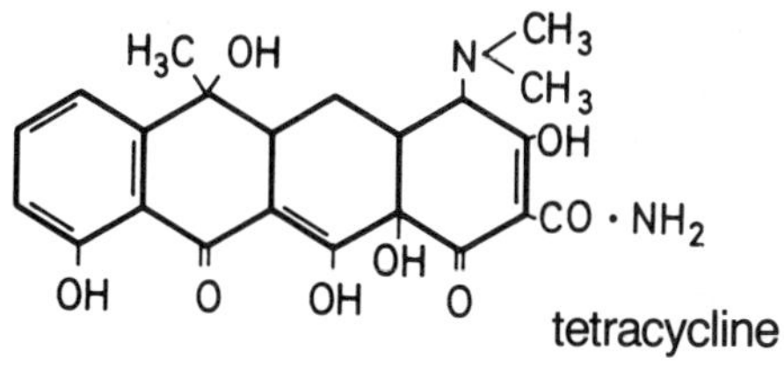

FIGURE 6. Tetracycline. There are several –O and two –N residues where divalent metal cations can bind.

questering essential heavy metals of the pathogenic microorganism, but the exact mechanism of action is not known. Oral 8-oxyquinolines cannot be expected to have beneficial effects on nickel sensitivity, on the contrary. However, local use seems logical due to the sequestering effect of the drug on metals present in and on the epidermis. Prophylactic use of 8-oxyquinolines should be evaluated on bilateral nickel dermatitis of the hands.

IX. TETRACYCLINE

Tetracycline is a large complex molecule which can form stable and poorly absorbable complexes with various metal ions (Figure 6). Well-known examples are tetracycline interactions with calcium, iron, aluminum, zinc, and copper, a process whereby tetracycline absorption is inhibited. The opposite is also the case: tetracycline depresses the absorption of divalent and trivalent metals. Tetracycline is the only drug which as been found to inhibit ^{65}Zn absorption in a dose-dependent way.[12] Rats receiving 25 mg tetracycline daily by gavage had a mean reduction in ^{65}Zn retention of 10%; 50 mg caused a 30% and 100 mg a 35% reduction. In analogy with other members of the first transition series, nickel is complexed as 1:1 and 2:1 complexes with tetracycline. There is a high affinity of tetracycline and heavy metal cations. The overall log K of Fe^{3+} complex is 25.3, Al^{3+} 19.2, Cu^{2+} 12.8, and Ni^{2+} 11.0. The zinc-tetracycline log K is lower, 9.[13] This makes tetracycline a possible candidate for therapy of nickel sensitivity by forming insoluble nickel complexes in the gastrointestinal tract and thereby reducing nickel absorption. Local treatment with tetracyclines should also be considered as an alternative or supportive treatment.

X. DISCUSSION

The ideal metal-chelating drug for nickel sensitivity should possess a high affinity for the metal when inside the body and a low affinity when outside, i.e., in the gut. It should be readily excreted with urine to avoid reabsorption during enterohepatic recirculation. The ability of the drug to act as a lipophilic ionophor across the gastrointestinal mucosal barrier should be low. Furthermore, the prolonged use of the drug should be without any risk of side effects. Apart from disulfiram, chelating drug therapy for nickel sensitivity has only been studied sporadically. Disulfiram has its drawbacks, such as dermatitis and hepatotoxicity, and cannot be used uncritically due to the risk of systemic side effects. There are other possible drugs available, primarily ethambutol and tetracycline.

Local application of nickel-sequestering agents may have a place in the therapy, especially in patients with severe nickel contact dermatitis, and this could be combined with systemic treatment and a diet low in nickel, if possible.

A different hypothetical way of combating nickel sensitivity would be depression of nickel absorption by oral intake of heavy metals such as zinc, with the intention of inhibiting nickel absorption by competition. Oral doses of zinc 5 to 10 times the daily recommendation of 15 mg can be taken for months without any serious risk except for a tendency to lower copper levels. Animal studies should be undertaken to substantiate or reject the hypothetical action of competition inhibition.

A combined therapy with an absorption inhibitor such as tetracycline and a nickel excretion promotor such as ethambutol and possibly D-penicillamine remains a further possibility in the treatment of nickel sensitivity. Careful clinical and biochemical investigations must be performed before we know the positive and negative consequences of such strategies.

REFERENCES

1. **Sigel, H. and McCormick, D. B.,** On the discriminating behavior of metal ions and ligands with regard to their biological significance, *Acc. Chem. Res.*, 3, 201, 1970.
2. **Shirotake, S. and Sakaguchi, T.,** Complexes between nucleic acid bases and bivalent metal ions. II. Complexes formed by guanine or cytosine, and zinc (II). *Chem. Pharm. Bull.*, 26, 2941, 1978.
3. **Menné, T., Kaaber, K., and Tjell, J. C.,** Treatment of nickel dermatitis (the influence of tetraethylthiuramdisulfide (Antabuse) on nickel metabolism, *Ann. Clin. Lab. Sci.*, 10, 160, 1980.
4. **Weismann, K.,** Increased absorption of ^{65}zinc during oral supplementation with diethyldithiocarbamate (Dithiocarb) in rats, *Trace Elements Med.*, 2, 37, 1985.
5. **Sunderman, F. W.,** Nickel and copper mobilization by sodium diethyldithiocarbamate, *J. New Drugs,* 4, 154, 1964.
6. **Basinger, M. A. and Jones, M. M.,** Chelate antidotal efficacy in acute zinc intoxication *Res. Commun. Chem. Pathol. Pharmacol.*, 33, 263, 1981.
7. **Burrows, D., Rogers, S., Beck, M., Kellett, J., McMaster, D., Merrett, D., and Eedy, D. J.,** Treatment of nickel dermatitis with trientine, *Contact Dermatitis,* 15, 55, 1986.
8. **Weismann, K. and Knudsen, L.,** Effects of penicillamine and hydroxyquinoline on absorption of orally ingested ^{65}zinc in the rat, *J. Invest. Dermatol.*, 71, 242, 1978.
9. **Høyer, H., Hvid-Jacobsen, K., and Weismann, K.,** ^{65}Zinc absorption in untreated and D-penicillamine-treated patients with generalized scleroderma: determination by whole-body counting technique, *Acta Derm. Venereol. (Stocklholm),* 62, 438, 1982.
10. **Cole, A., May, P. M., and Williams, D. R.,** Metal binding by pharmaceuticals. I. Copper (II) and zinc (II) interactions following ethambutol administration *Agents Actions,* 11, 296, 1981.
11. **Sturtevant, F. M.,** Zinc deficiency, acrodermatitis enteropathica, optic atrophy, subacute myelo-optic neuropathy, and 5,7-dihalo-8-quinolinols, *Pediatrics,* 65, 610, 1980.
12. **Weismann, K., Knudsen, L., and Høyer, H.,** Effects of tetracycline on the absorption of ^{65}zinc in rats, *Arch. Dermatol. Res.*, 263, 135, 1978.
13. **Albert, A. and Rees, C. W.,** Avidity of the tetracyclines for the cations of metals, *Nature (London),* 177, 433, 1956.

Chapter 16

NICKEL ALLERGY AND ORTHOPEDIC PROSTHESES

John D. Wilkinson

TABLE OF CONTENTS

I. Sensitivity and Cutaneous Reactions to Implanted Nickel or Nickel-Releasing Prostheses 188

II. Hip Prostheses, Metal Sensitivity, and Joint Loosening 190

III. Cutaneous Reactions from Other Sources of Implanted Nickel 191

References 191

I. SENSITIVITY AND CUTANEOUS REACTIONS TO IMPLANTED NICKEL OR NICKEL-RELEASING PROSTHESES

Nickel is a hard, strong metal which is resistant to corrosion and can be polished to give a bright gloss. This and the fact that it forms alloys with many other metals means that it is a common component of many orthopedic prostheses.[1]

Cutaneous sensitivity and systemic reactions to implanted nickel are well documented.[2] The problem appears to relate to nickel more than any other metal.

There have been many reports in the literature of patients proven to be nickel sensitive who have developed widespread eczematous eruptions following insertion of metal implants (Table 1). Various other types of reactions to implanted nickel have also been reported. Most commonly, these are urticarial, bullous, or vasculitic in nature (Table 2). Reactions may be local or generalized.

Many reports of eczema or other reactions associated with implants have failed to confirm the presence of metal sensitivity by patch test.[2-5] In these cases, the association must therefore be regarded as speculative.

Since eczema is such a common condition among elderly, hospitalized, or postoperative patients, the clinical association of eczema and recent implant is alone insufficient to claim an immunological association.

Similarly, with nickel sensitivity being so common in women and being present in approximately 7 to 10% of a preoperative orthopedic population,[6-9] it is obviously important that the clinical evidence associating the rash with the implant be strong. The presence of nickel sensitivity alone is not enough.

Features suggestive of an allergic reaction to a nickel-containing implant include:

1. A chronic eczema or other dermatosis that is resistant to conventional treatment, or an immunofluorescence-negative bullous dermatosis
2. A strong positive patch test to nickel (particularly in men)
3. A dermatosis starting or persisting at the site of the implant
4. Improvement in the eczema or dermatosis following removal of the implant or prosthesis
5. The absence of any other more likely explanation for the dermatosis
6. The demonstration that there was biologically available nickel in the putative implant[10,11]

The possibility that nonimmunological eczemas could be the result of implants or prostheses[3,4] remains speculative.

The delay between implant and resultant dermatosis may vary from as little as 20 d to 8 years in patients with eczematous-type reaction, from 1 d to 2 months in patients with urticarial reactions, and from 3 months to 12 years in those with bullous or other types of reaction.[2] It might be expected that those with preexisting sensitivity to nickel would react more quickly following a prosthetic implant. This, however, does not always appear to be the case because it may take several years for even those who are known to be nickel sensitive to develop a cutaneous reaction.[12]

Most of the eczemas that occur following implantation are both chronic and relatively steroid resistant, and eczema may only develop months or years after the operation. Improvement after removal of the implant may not be immediate due to persisting nickel in the tissues. It is now well established that nickel may be released into the plasma and tissues around implants[11,13,14] and that at least in animals, nickel may be found in other tissues such as lungs, heart, and kidneys.[13]

Nickel sensitivity was associated with seven of eight cases of eczematous-type reaction where the eczema could be shown to be convincingly associated with the implant.[2] Two of

Table 1
ECZEMATOUS REACTIONS FOLLOWING INSERTION OF METAL IMPLANTS IN THOSE SUBSEQUENTLY SHOWN TO BE ALLERGIC TO NICKEL[2]

Sex	Type of implant	Comment	Ref.
Female	Stainless steel	Resolved 2 weeks after removal	5
Female	Not known	Flare of dermatitis followed by partial resolution after removal	5
Female	Stainless steel screws	Cleared within 72 h of removal	25
Female	Stainless steel	Generalized dermatitis, resolved with removal of implant	12
Female	Stainless steel (12.5% nickel)	Rapid resolution after removal	52
Female	Vitallium	Flare followed by resolution within 3 weeks	53
Female	Stainless steel	Rapid resolution followed by flare (incomplete removal)	3
Male	Stainless	Cleared within 1 month of removal; recurrence later	2

Table 2
OTHER TYPES OF REACTION TO IMPLANTED NICKEL

Sex	Type of implant	Type of reaction	Comment	Ref.
Female	Vitallium	Urticarial	Cleared within 48 h of removal	18
Male	Vitallium	Localized and generalized urticaria	Cleared within 3 months of removal	19
Female	Vitallium	Localized bullous dermatosis	None	15
Female	Stainless steel	Localized erythema and blistering	Cleared after removal	
Female	Vitallium	Bullous eruption	Cleared after removal	17
Male	Stainless steel	Vasculitic eczema	Cleared within 2 months of removal	5
Female	Not known	Dermatitis herpetiformis-like vasculitis	Cleared after removal	4
Male	Not known	Prurigo/vasculitis	Cleared after removal	4
Female	Stainless steel	Capillaritis	Prosthesis left *in situ*	20

three cases of localized bullous reaction[15,16] and one case[17] with a more generalized bullous reaction were also shown to be nickel sensitive. Two of three patients with urticarial-type reactions following implants or prostheses were found to be both patch test and scratch test positive to nickel.[18,19] A prurigo or dermatitis herpetiformis-like eruption has also been reported in two patients sensitive to nickel[4] (Table 2). Some reactions have been histologically vasculitic in nature, with a variable clinical morphology ranging from the eczematous[5] or prurigo-like[4] to patients with a dermatitis herpetiformis-like rash[4] or capillaritis.

II. HIP PROSTHESES, METAL SENSITIVITY, AND JOINT LOOSENING

Much of the published work on orthopedic prostheses and metal sensitivity/joint loosening concerns elements other than nickel. Some of the basic principles may, however, also apply.[21-24] In general, nickel has more often been implicated with respect to cutaneous reactions to implanted metal,[18-25] whereas chromium and cobalt have most commonly been discussed in respect to the problem of prosthetic loosening.[15,23,26]

Metal-on-metal prostheses undoubtedly lead to a higher incidence of metal sensitivity than the newer metal-on-plastic prostheses.[24,27,28] Where metal-on-metal prostheses are used or when a prosthesis becomes loose, there may be sufficient metallic material released to produce a flare of dermatitis or urticaria in those who are already sensitized, particularly with respect to nickel. There may also be sufficient metal released to induce sensitivity in a proportion of cases. Where there are problems with a joint or with cutaneous sensitivity reactions, the prosthesis may need to be removed. The fact that a patient is patch test positive to nickel does not, in itself, mean that there is any risk to the joint or that any action has to be taken.

At present, three of the four main alloys used for hip prostheses contain nickel. Stainless steel contains 10 to 14% nickel, cobalt-chromium alloy contains up to 2.5% nickel, and cobalt-chromium tungsten alloy contains 9 to 11%. In patients who are known to be extremely sensitive to nickel or in those in whom there has already been a cutaneous reaction to an implanted nickel-containing alloy, a titanium prosthesis is probably best since this contains no nickel. In the vast majority of patients, however, even those who are nickel sensitive, a nickel-containing metal-on-plastic prosthesis is usually well tolerated without any cutaneous or local reaction.[28]

When metal-on-plastic prostheses are used, far less metallic material is released and there is a correspondingly lower risk of eliciting or inducing sensitivity.[28] Most authors now accept[26,27,29] that metal sensitivity is more likely to be a consequence of prosthetic loosening than a cause of joint loosening. Hip loosening probably occurs due to other mechanical or infective factors or as a result of a nonspecific granulomatous response to released particulate matter around the joint.[29] It is otherwise difficult to explain why, in a sensitized patient with bilateral hip prostheses, one joint should become loose and the other remain stable.

It is now known that metal-on-plastic hips may lead to numerous plastic particles being deposited around the joint,[30] and both metal and plastic dynamic implants may give rise to a foreign body/macrophage-type reaction around the joint with release of proteolytic enzymes and subsequent joint loosening.[29] The newer metal-on-ceramic hips give rise to less wear and may therefore incur less risk of this type of reaction occuring.

The mechanisms by which local cutaneous and systemic reactions occur following implantation of nickel remain somewhat obscure. Some reactions[18,19] appear to be type I in nature. In others there is good evidence of type IV hypersensitivity (Table 1). In some patients, however, type I, type III/Arthus, and type IV reactions seem to co-exist.[4,5,18-20]

In respect to joint loosening, the problem of sensitization should perhaps now be regarded more as an epiphenomenon. The principal mechanism by which loosening occurs appears to be via the action of proteolytic enzymes liberated by macrophages that accumulate around the joint in response to particulate matter or wear particles.[29] As an exception, a localized type II or type IV reaction may occur,[4,29] but generally the role of type IV mechanisms remains unproven.[31,32]

The situation regarding hip replacement and induction of metal sensitivity also remains uncertain.[33] There have been only four studies comparing sensitization rates both before and after prosthetic (metal-on-plastic) hip replacement. In two studies[27,31] the postoperative group excluded those with preexisting sensitivity, and a third study[34] had different prospective and retrospective groups. The incidence of nickel sensitivity was, however, the same in both groups. In only one study were preoperative and postoperative groups the same,[28] and in this study the incidence of nickel sensitivity actually fell postoperatively, suggesting that immune tolerance may have developed. This is consistent with earlier animal work.[35] The conclusion of most studies, however, is that prosthetic implants will on occasion lead to the development of sensitivity.[27,31,36] Kubba et al.[3] remark on the high incidence of positive patch test reactions to nickel and the low incidence of cutaneous or systemic reactions, particularly with metal-on-metal prostheses or where there is joint loosening.[6,15,23,26] In most patients, metal sensitivity appears to remain latent or concealed and only a minority appear to have adverse reactions from their metal prostheses.

III. CUTANEOUS REACTIONS FROM OTHER SOURCES OF IMPLANTED NICKEL

Although orthopedic implants account for the majority of the reported reactions to implanted nickel, there are anecdotal accounts of reactions from other sources. Dermatitis and systemic (immediate and Arthus-type) reactions have occurred following the use of nickel-plated cannulae[37-38] and dermatitis may follow the use of nickel-silver skin closures[39,40] or nickel-plated retractors.[41]

Shrapnel may cause allergic eczema[5] and pompholyx-type eczema has developed following the implantation of a nickel-containing pacemaker.[42] Pacemakers may, in fact, contain up to 35% nickel.[1] A more exotic cause of nickel dermatitis is the use of acupuncture needles.[43] The dividing line between nickel dermatitis due to contact and/or from implanted metal is, however, often somewhat arbitrary.

Local reactions have also occurred in some nickel-sensitive women following use of a Dermojet.[44] This has been shown to have been due to release of nickel from the apparatus into the injection solution.

An interesting report[45] is that of a woman with a preexisting nickel sensitivity who had two unsuccessful mitral valve replacements. Only when a valve containing no nickel was used did the valve continue to work satisfactorily.[46] A similar situation occurred in a nickel-sensitive patient who had a metal jaw osteosynthesis.[47] Resolution of chronic osteomyelitis and healing only occurred when the putative implant was removed.

From these and other reports,[48,49] it is apparent that both local and systemic reactions may sometimes occur following implanted nickel. It is, however, also true that the majority of individuals — even the majority of sensitized individuals — will tolerate low levels of nickel in the tissues without adverse effect.[28,36,50,51]

REFERENCES

1. **Cronin, E.,** *Contact Dermatitis,* Churchill Livingstone, Edinburgh, 1980.
2. **Rostoker, G., Robin, J., Binet, O., and Paupe, J.,** Dermatoses d'intolérance aux métaux des matériaux d'osteosynthèse, *Ann. Derm. Venereol.,* 113, 1097, 1986.
3. **Kubba, R., Taylor, J. S., and Marks, K. E.,** Cutaneous complications of orthopaedic implants, *Arch. Dermatol.,* 117, 554, 1981.

4. **Bazex, A., Parant, M., Bazex, J., and Viraben, R.,** Manifestations cutanées provoquées par le matériel metallique d'ostéosynthése, *Ann. Derm. Venereol.*, 105, 9, 1978.
5. **Laugier, P. and Foussereau, J.,** Les dermites allergiques á distance proroquées par le matériel d'osteosynthése, *Gaz. Med. Fr.*, 10, 3409, 1966.
6. **Dry, J., Leynadier, F., Baux, S., and Kinderman, A.,** Reaction immunoallergologique au materiel metallique des prosthèses: influence du descellement aseptique, *Sem. Hop. (Paris)*, 53, 2449, 1977.
7. **Prystowsky, S. D., Allen, A. M., Smith, R. W., Nonomura, J., Odom, R., Maibach, H. I., Budnik, M., and Akers, W. A.,** Contact sensitivity to nickel, neomycin, ethylene-diamine and benzocaine in a general population. Relationship between age, sex, history of exposure and reactivity to standard patch tests, *Clin. Res.*, 26, 301, 1978.
8. **Peltonen, L.,** Nickel sensitivity in the general population, *Contact Dermatitis*, 5, 27, 1979.
9. **Magnusson, B. and Moller, H.,** Contact allergy without skin disease. *Acta Derm. Venereol.*, 59, (Suppl. 85), 1979.
10. **Fisher, A. A.,** Allergic dermatitis presumably due to metallic foreign bodies containing nickel or cobalt, *Cutis*, 19, 285, 1977.
11. **Samitz, H. and Katz, A.,** Nickel dermatitis hazards from prostheses in vivo and in vitro solubilisation studies, *Br. J. Dermatol.*, 92, 287, 1976.
12. **Pegum, J. S.,** Nickel allergy, *Lancet*, 1, 674, 1974.
13. **Fergusson, A., Laing, P., and Hodge, E.,** The ionisation of the metal implants in living tissues, *J. Bone Jt. Surg. Am. Vol.*, 42, 77, 1960.
14. **Mears, D. G.,** Electron probe microanalysis of tissue and cells from implant areas, *J. Bone Jt. Surg.*, 43, 567, 1966.
15. **Munro-Ashman, D. and Miller, A. J.,** Rejection of metal to metal prosthesis and skin sensitivity to cobalt, *Contact Dermatitis*. 2, 65, 1978.
16. **Cramers, M. and Lutch, V.,** Metal sensitivities in patients treated for tibial fractures with plate of stainless steel, *Acta Orthop. Scand.*, 8, 245, 1977.
17. **Kubba, R. and Champion, R. H.,** Nickel sensitivity resembling Millars Pemphigoid, *Br. J. Dermatol.*, 11, 41, 1975.
18. **McKenzie, A. W., Aitken, C. V. E.,, and Ridstill-Smith, R.,** Urticaria after insertion of Smith-Petersen vitallium nail, *Br. Med. J.*, 4, 36, 1967.
19. **Symeonides, P., Pascha-Loglou, C., and Papageorgiou, S.,** An allergic reaction after internal fixation of a fracture using a vitallium plate, *J. Allergy Clin. Immunol.*, 51, 251, 1975.
20. **Tilsley, D. A. and Rotstein, H.,** Sensitivity caused by internal exposure to nickel, chrome and cobalt, *Contact Dermatitis*, 6, 175, 1980.
21. **Swanson, S. A. V., Freeman, M. A. R., and Heath, J. C.,** Laboratory tests on total joint replacement prostheses, *J. Bone Jt. Surg. Br. Vol.*, 55B, 759, 1973.
22. **Coleman, P. F., Herrington, J., and Scales, J. T.,** Concentration of wear products in hair, blood and urine after total hip replacement, *Br. Med. J.*, 1, 527, 1973.
23. **Evans, E. M., Freeman, M. A. R., Miller, A. J., and Vernon-Roberts, B.,** Metal sensitivity as a cause of bone necrosis and loosening of the prosthesis in total joint replacement, *J. Bone Jt. Surg. Br. Vol.*, 56B, 626, 1974.
24. **Benson, M. K. D., Goodwin, P. D., and Brostoff, J.,** Metal sensitivity in patients with joint replacement arthroplasties, *Br. Med. J.*, 4, 374, 1975.
25. **Barranco, V. P. and Solomon, H.,** Eczematous dermatitis from nickel, *J. Am. Med. Assoc.*, 220, 1244, 1972.
26. **Elves, M. R., Wilson, J. N., Scales, J. T., and Kemp, H. B. S.,** Incidence of metal sensitivity in patients with total joint replacements, *Br. Med. J.*, 4, 376, 1975.
27. **Deutman, R., Mulder, J., Brian R., and Nater, J. P.,** Metal sensitivity before and after total hip arthroplasty, *J. Bone Jt. Surg. Am. Vol.*, 59A, 862, 1977.
28. **Rooker, G. D. and Wilkinson, J. D.,** Metal sensitivity in patients undergoing hip replacement, *J. Bone Jt. Surg. (Br.)* 62B, 4, 402, 1980.
29. **Langlais, F., Postel, M., Berry, J. P., Lecharpentier, Y., and Weill, B. J.,** L'intolerance aux débris d'usure des prothèses, bilan immunologiques et anatomopathologique de 30 cas, *Int. Orthop.*, 4, 145, 1980.
30. **Pidhorz, L., Raguin, J., and Varenne, F.** Les matéraux d'ostéosynthese, *Encycl. Med. Chir. Paris. Tech. Chir.*, 12, 44014, 1975.
31. **Waterman, A. H. and Schrik, J. J.,** Allergy in hip arthroplasty, *Contact Dermatitis*, 13, 294, 1985.
32. **Rappaport, M. J.,** Patch testing in orthopaedic prosthetic patients, *Contact Dermatitis*, 6, 229, 1980.
33. **Burrows, D., Creswell, S., and Merrett, D. J.,** Nickel hands and hip prostheses, *Br. J. Dermatol.*, 105, 437, 1981.
34. **Carlsson, A., Magnusson, B., and Moller, H.,** Metal, sensitivity in patients with metal-to-plastic total hip replacements, *Acta Orthop. Scand.*, 51, 57, 1980.

35. **Battisto, J. R. and Miller, J.,** Immunological unresponsiveness produced in the adult guinea pig by parenteral introduction of minute quantities of hapten or protein antigen, *Proc. Soc. Exp. Biol. Med.,* 111, 111, 1962.
36. **Nater, J. P., Brian, R. G., Duetman, R., and Mulder, T. J.,** The development of metal hypersensitivity in patients with metal-to-plastic hip arthroplasties, *Contact Dermatitis,* 2, 259, 1976.
37. **Stoddard, J. C.,** Nickel sensitivity as a cause of transfusion reactions, *Lancet,* 2, 741, 1960.
38. **Holti, G.,** Immediate and Arthus-type hypersensitivity to nickel, *Clin. Allergy,* 4, 437, 1974.
39. **Booth, R. P.,** Possible allergic skin reaction to nickel clips, *Med. J. Australasia,* 2, 546, 1971.
40. **Nurse, D. S.,** Nickel sensitivity induced by skin clips, *Contact Dermatitis,* 7, 297, 1981.
41. **Kvorning, S. A.,** Post-operative dermatitis following non-epidermal nickel contact, *Contact Dermatitis,* 1, 327, 1975.
42. **Landwehr, A. J. and van Ketel, W. G.,** Pompholyx after implantation of a nickel containing pacemaker in a nickel allergic patient, *Contact Dermatitis,* 9, 147, 1973.
43. **Romaguera, C. and Grimalt, F.,** Nickel dermatitis by acupuncture needles, *Contact Dermatitis,* 7, 156, 1981.
44. **LaChapelle, J.-M. and Tennstedt, D.,** An anatomo-clinical study of delayed skin allergic reactions to nickel following intradermal infections of lidocaine with a Dermojet, *Contact Dermatitis,* 8, 193, 1982.
45. **Lyell, A. and Bain, W. H.,** Nickel allergy and valve replacement, *Lancet,* 1, 408, 1974.
46. **Lyell, A.,** Metal allergy and metallic prostheses, *Int. J. Dermatol.,* 18, 805, 1978.
47. **Roed-Petersen, G., Roed-Petersen, J., and Jørgensen, K. D.,** Nickel allergy and osteomyelitis in a patient with metal osteosynthesis of a jaw fracture, *Contact Dermatitis,* 5, 108, 1979.
48. **Fernandez de Corres, L., Garrustazu, M. T., Soloeta, R., and Escayol, P.,** Nickel contact dermatitis in a blood bank, *Contact Dermatitis,* 8, 32, 1982.
49. **Schriver, W. R., Shereff, R. H., Domnitz, J. M., et al.,** Allergic response to stainless steel wire, *Oral Surg.,* 42, 578, 1978.
50. **Fisher, A. A.,** The safety of stainless steel in nickel sensitivity, *J. Am. Med. Assoc.,* 221, 1279, 1972.
51. **Fisher, A. A.,** Somewhat controversial reports of metal reactions, *Cutis,* 24, 28, 1979.
52. **Grimalt, F. and Romaguera, C.,** Acute nickel dermatitis from a metal implant, *Contact Dermatitis,* 6, 441, 1980.
53. **Oleffe, J. and Wilmet, J.,** Generalised dermatitis from an osteosynthesis screw, *Contact Dermatitis,* 6, 365, 1980.

Chapter 17

NICKEL ALLERGY AND DENTAL TREATMENT PROCEDURES

Arne Hensten-Pettersen

TABLE OF CONTENTS

I. Dental Treatment Procedures 196

II. Extraoral Exposure, Extraoral Reactions 196
A. General Considerations 196
B. Orthodontic Treatment 196

III. Intraoral Exposure, Intraoral Reactions 197
A. Orthodontic Treatment 197
B. Prosthodontic Treatment 198
1. Removable Partial Dentures 198
2. Fixed Prosthodontic Restorations 198
C. Oral Surgery 199

IV. Intraoral Exposure, Extraoral Reactions 200

V. Experimental Intraoral Exposure to Nickel-Containing Alloys 201

VI. Induction of Tolerance by Oral Administration of Nickel 202

VII. Confounding Variables 202

VIII. Conclusions 203

References 203

I. DENTAL TREATMENT PROCEDURES

Lost tooth substance may be replaced by inserting restorative materials that set in the mouth, such as amalgams, cements, or resin-based composite materials. More extensive tooth destruction or tooth loss is restored with custom-made prostheses of ceramics, polymeric materials, and alloys. Combinations of these materials are extensively used. Prevention and treatment of malocclusion is performed by pedodontists and orthodontists who use preformed or custom-made devices of metal alloys, polymeric materials, and rubber compounds. Oral surgery in connection with bone fractures or corrective surgery uses metal splints, wires, and implants of the same materials as in orthopedic surgery.

Dental treatment procedures vary from country to country and change with time. Furthermore, the chemical composition of dental materials is not always disclosed by the manufacturers. The same trade name may be used for alloys of very different composition. For example, a dental casting alloy in Portugal contained 70% nickel, while one of the same name in Norway contained no nickel, only cobalt and chromium.

This review on nickel allergy and dental treatment procedures is based on literature surveys and collaboration with Scandinavian and U.S. dentists over the last 10 years. Many of the cases have not been published for various reasons and are used in the present context to illustrate the complexity of this issue in dentistry.

II. EXTRAORAL EXPOSURE, EXTRAORAL REACTIONS

A. General Considerations

A metal bead necklace with metal clips is used to fasten a paper napkin on dental patients during routine examination and treatment procedures. Both the beads and clips may be made of nickel-plated brass and cause skin eruptions in nickel-sensitive individuals. The skin eruption develops during the afternoon or night after the dental appointment and subsides in a few days. Dental appointments are often spaced with 1- or 2-week intervals, making it difficult to associate such reactions with exposure. Similarly, many dental hand instruments such as explorers, mirrors, and saliva ejectors may be either stainless steel or nickel or chrome-nickel-plated brass or steel. These instruments contact the labial and circum-oral area during the dental treatment and have been implicated as causes of skin and lip eruptions in nickel-sensitive individuals. Again, the reactions are delayed and make the association between exposure and skin reactions somewhat uncertain.

B. Orthodontic Treatment

Corrections of malocclusions and misalignment of teeth often involve moving the first and second maxillary permanent molars to a more distal position. The forces used to move these teeth are fairly great and require the use of extraoral anchorage. Headgears of many designs are available for this purpose. They consist of a textile neck strap or headcap connected by metal or plastic buckles to a semirigid stainless-steel extraoral/intraoral facial bow. The headgears are used for 16 to 24 h/d, especially during the initial 6 to 12 months of orthodontic treatment. The extraoral portions come in intimate contact with the cheeks and may give rise to symmetrical localized contact dermatitis. Orthodontists usually describe these reactions as ''saddle sores''. Several reports indicate that some of these reactions may be related to nickel allergy.[1-3] The lesions are especially prominent in the region where metal buckles connect with the neck strap. Analyses of one product commonly involved in these reactions showed that the intraoral and extraoral portions of the facial bow were of conventional 18/8 stainless steel, whereas the metal buckle was made of nickel-plated brass, i.e., 100% metallic nickel in contact with the skin. Patch testing of seven 12-year-old children with symmetrical extraoral contact dermatitis in connection with the nickel-plated brass

Table 1
CHEMICAL COMPOSITION AND CHARACTERISTICS OF ORTHODONTIC WIRES[21]

	Approximate chemical composition (wt %)					
	Fe	Cr	Ni	Co	Mo	Ti
Stainless steel	68	19	9			
Cobalt-chromium alloy	14	19	14	41	5	
Nickel-titanium alloy				52	3	42
Titanium-molybdenum alloy					11	78

buckles indicated that five reacted to nickel.[3] In addition, one reacted to cobalt and colophony and one to ethyleneglycoldimethacrylate, a cross-linking monomer present in her intraoral appliance. Two patients also had extensive angular cheilitis where the stainless-steel bow contacted. Covering the extraoral buckles and stainless steel with a sintered thermoplastic resin[4] was sufficient to avoid their extraoral reactions and angular cheilitis. Some of the orthodontic patients have remarked that the areas of previous contact dermatitis on their cheeks do not tan during the summer, even 2 to 3 years after the reactions have cleared.

A survey of 366 dental patients treated at the Department of Orthodontics, Dental School, Oslo indicates that about 13% of the girls and 5% of the boys had present or previous dermatoses related to skin contact with metallic objects. Two thirds of them had not found it necessary to seek medical attention for this problem.[3]

A questionnaire survey of material-related problems in orthodontic practice indicates that symmetrical extraoral reactions occur in about 1% of patients using the headgear with nickel-plated buckles.[5]

III. INTRAORAL EXPOSURE, INTRAORAL REACTIONS

A. Orthodontic Treatment

Metallic materials are used for archwires, attachments, bands, ligature wire, and other force-regulating and force-transferring devices. In addition, solders are used to join parts of appliances. In the early era of orthodontics, gold and platinum were the predominant materials, but these were later replaced by base metal alloys like stainless steel. The base metal alloys corrode to some extent, but the rate of deterioration is low. The corrosion rates may be affected by clinical handling. Heating during soldering, welding, or extensive shaping and notching with pliers tend to decrease the corrosion resistance of the materials. The tissue response to a Fe-Cr-Ni and a Co-Cr-Ni orthodontic wire, in the as-received state and with silver soldered joints, has been investigated in nickel-sensitized rabbits.[6] The as-received wires gave no tissue response relative to the polytetrafluoroethylene (PTFE) control. In soldered specimens there were moderate to extreme reactions adjacent to the soldered joint and around the wire portion as well. The soldered Co-Cr-Ni wire elicited the most severe reactions, most pronounced in the nickel-sensitized animals. An agar overlay cell culture test of some retrieved implants showed pronounced cytotoxicity of the soldered specimens, possibly due to leachable toxic components of the silver solder.[6]

Most of the orthodontic alloys contain nickel (Table 1). Exceptions are the titanium-molybdenum alloys and an iron-chromium-molybdenum wire.[7] The manufacturers offer a wide range of factory-made devices, i.e., bands with attachments and hooks, brackets for direct bonding to the tooth surface, etc. Such devices may consist of several parts joined by soldering, brazing, or welding. The detailed composition is to a large extent unknown, but it is commonly stated that 18/8 stainless steel is used. Analyses indicate that this is mostly true, but that there are a few devices on the market with 50 to 80% nickel. The orthodontic solders contain silver, copper, and zinc; tin or cadmium may be added to lower the fusion temperature. If fluoride mouth rinses are used, the rate of deterioration may be increased.[8] The least noble components of the solder, zinc and copper, are most eluted. Cadmium is released rapidly even when present in small amounts.[9]

Intraoral reactions during orthodontic treatment are often in the form of aphtous ulcers[10] and as sores related to mechanical injury of the mucosa. Even in patients with extensive orthodontic appliances and with known nickel allergy, it is rare to observe any intraoral lesions directly related to the alloys per se. A moderate redness of the oral mucosa had been seen in a nickel-allergic girl with metallic orthodontic appliances on 12 teeth, and then only in the areas directly surrounding the orifices of the salivary glands of the soft and hard palates. The patient did not have any discomfort or complaints associated with this reaction. In other nickel-sensitive orthodontic patients with nickel-containing intraoral appliances, clinical examination of the oral mucosa has not indicated any pathological reactions.

A common complaint from orthodontic patients is dry, fissuring, cracking, or crusting lips. Some patients use petrolatum-based ointments or Vaseline® to soften the lips. Nickel-allergic orthodontic patients have experienced extensive swelling of the lips and a perioral dermatitis, roughly confined to the area covered by petrolatum. Removal of the intraoral metallic appliances in two patients resulted in resolution of the cheilitis and perioral dermatitis within 1 week.

B. Prosthodontic Treatment

1. Removable Partial Dentures

Removable partial dentures usually comprise a cast metal framework of cobalt-chromium alloys with a superstructure of resin or porcelain teeth imbedded in poly(methylmethacrylate), either heat- or cold-cured. Stomatitis is fairly frequently seen beneath acrylic and metallic removable partial dentures. Poorly fitting dentures move during mastication, leading to friction and mechanical injury of the contacting oral mucosa. Fungal infections are common. Treatment of infection, correction of occlusion, or making a better-fitting prosthesis with the same materials usually resolves the problem. Denture stomatitis due to allergic reaction is rare and probably occurs in less than 1 to 2% of the patients.[11,12] In a few instances, the stomatitis extends beyond the borders of the prosthesis and may involve the area surrounding the orifices of salivary glands of the hard and soft palates. Also, the portion of the tongue surface that contacts the metal framework during speech, mastication, and deglutination may be desquamated and have erosive lesions. The stomatitis seems to be related to metal contact allergy and it cannot conclusively be regarded as a role of nickel. Some textbooks state that cobalt-chromium alloys for partial frameworks are free of nickel. Analysis in our laboratories indicates that they contain about 0.1 to 0.2% nickel. Whether this amount is sufficient to elicit intraoral reactions in nickel-sensitive individuals is not known. In patients with denture stomatitis where patch testing with nickel, chromium, or cobalt was positive, the patients reacted to two or all three of the metals. None reacted to nickel alone.[13-18] Thus, it was not possible to ascertain whether nickel was of major importance in this respect.

Gold plating the metal framework of the prostheses has also been tried in selected instances, but seems to be a temporary measure. If tooth-supported fixed prostheses made of conventional gold alloys cannot be used, recasting the metal framework in a type IV conventional gold alloy (Au > 75%) is an alternative.

2. Fixed Prosthodontic Restorations

Fixed prosthodontic restorations are used to replace missing teeth or rebuild destroyed tooth structure. The majority of crowns are custom-made for each patient by using the lost wax technique. A multitude of casting alloys are available. Traditionally, the alloys were

Table 2
COMPOSITION OF MAJOR TYPES OF DENTAL CASTING ALLOYS[21]

	Approximate chemical composition (wt %)								
Alloys	Ni	Co	Cr	Mo	Au	Pt	Pd	Ag	Other
Nickel-chromium	82		10	2					6
Cobalt-chromium	<0.1	60	35	5					
<75% noble metals					46	6		40	8
Silver-palladium					2		28	60	10
Gold alloy (white)					87	9	2		2
Gold alloy (yellow)					76	2	2	13	7

based on noble metals, mostly gold alloys. Due to the increasing price of gold on the world market, a series of alternative alloys have been introduced in the last 20 years. At present, more than 600 alloys are on the market in Germany alone.[19] In some countries, especially the U.S., alternative alloys have been used in more than 50% of the cast metal restorations.[20] The dental casting alloys are a heterogenous group and are difficult to put into defined categories.[21] They comprise high gold alloys (Au > 75%), high noble content (palladium-silver based), low gold alloys, and base metal alloys, the latter mainly based on nickel-chromium or cobalt-chromium (Table 2). The nickel content of the casting base metal alloys may vary from 56 to 82% and the corresponding chromium content from 32 to 12%. The nickel content has been a major concern from the outset. Manufacturers are required to declare the nickel content and label the alloy packages with the following statement: ''Caution — contains nickel and shall not be used for patients with known nickel sensitivity''.[22,23] However, this label is on the package as delivered to the dental technicians before casting. Often this information does not seem to reach the dentist or the dental patient, much less the dermatologist.

The health authorities in Sweden and Finland have introduced limitations in the use of cadmium and nickel in dentistry. Dental alloys containing cadmium or more than 1% nickel are not recommended for use in humans.

Preformed steel crowns are extensively used in some East European countries. In the U.S. and Western Europe, preformed metal crowns are used on primary teeth or as part of space retainers for teeth that will erupt at a later stage. Some of these crowns and devices are based on alloys containing 70 to 80% nickel and others on 18/8 stainless steel.

By visual inspection alone, it is almost impossible to note differences between base metal alloys with 70 to 80% nickel and gold-based alloys with a high palladium or platinum content. Although the extensive use of base metal alloys for fixed restoration has been a major issue and of concern to the dental profession, very few case reports are available to substantiate this concern.[24] In fact, allergy to gold crowns seems to be more common than allergic reactions to nickel-containing base metal alloys.[25] One author states that ''after a 14-year experience spanning some 50,000 units, I have yet to see an oral lesion attributable to the restorative work'' (relating to a nickel-chromium-beryllium alloy and nickel-sensitive patients).[47]

C. Oral Surgery

Like orthopedic surgery, oral surgery may entail the use of stainless-steel splints, fixtures, and sutures. Two reports have been published where the surgical treatment was compromised by nickel allergy in the patients.[26,27]

IV. INTRAORAL EXPOSURE, EXTRAORAL REACTIONS

Ingested nickel may be absorbed through the oral mucosa or gastrointestinal tract. In nickel-sensitive individuals, a variety of generalized symptoms may be elicited, as reviewed by Veien in Chapter 14.

The relationship between skin and mucosal reactivity to contact allergens was studied by Nielsen and Klaschka.[28] They used individually fitted allergen carriers made of poly(methyl methacrylate) which permitted the simultaneous use of three to five different allergen concentrations. The epimucosa test was used on 25 persons who had shown positive skin reactions to one of five contact allergens: nickel sulfate, potassium dichromate, benzocaine, turpentine, or *p*-hydroxybenzoic acid ester. Macroscopic reactions of the palatine mucosa were evident only when the allergen concentration was 5 to 12 times stronger than in the skin test. Intraoral testing with potassium chromate and *p*-hydroxybenzoic acid ester did not give rise to any extraoral reactions, whereas tests with nickel sulfate, benzocaine, and turpentine gave rise to various types of generalized urticarial or eczematous reactions or fixed drug eruption, delayed-type reactions. Frykholm,[29] studying the relationship between skin and oral reactions in mercury hypersensitivity, observed a similar relationship. Intraoral exposure to test pieces of dental amalgam only gave a positive mucosal reaction in one of six mercury-sensitive subjects. The others developed a perioral dermatitis or generalized eczematous reactions.

Generalized reactions due to dental alloys have been noted in orthodontic treatment with nickel-containing alloys.[7] One nickel-sensitive boy developed a perianal dermatitis together with symmetrical eczematous reactions at the knee and elbow flexures that appeared lichenoid after a few months. The reactions appeared 1 month after initiating orthodontic treatment with intraoral nickel-containing alloys. The reactions cleared within 1 week of removing the appliances.

Some unpublished case reports are presented here to illustrate the complexity and diversity of reactions. A nickel-sensitive boy developed a severe generalized eczematous reaction on the trunk and extremities after insertion of two stainless-steel bands on the first molars. No intraoral lesions were evident. The reaction subsided within 2 weeks of removing the bands. He declined further orthodontic treatment. One 6-year-old girl had a space retainer placed on one of her primary molars after extraction of the second primary molar. Within 6 h she developed a generalized malaise, high fever, and was hospitalized. The space retainer was removed on suspicion, and the symptoms cleared within 1 d. Her ears had been pierced at an earlier age, and a flare-up in the form of weeping dermatitis of the puncture holes was seen at the same time. Analysis of the space retainer showed that it contained 80% nickel. Two similar cases have been observed in connection with the use of archwires in orthodontics, both in female adults with nickel sensitivity. One of the patients was hospitalized due to general malaise and high fever developing in the afternoon and night after insertion of a stainless-steel wire. She improved when the wire was removed. The orthodontist then replaced the stainless-steel wire 2 weeks later with one of nickel-titanium (52% Ni), and she reacted more extensively the second time. Her orthodontic treatment was discontinued. Another episode with nickel-titanium wire in a nickel-sensitive subject developed as an eczematous reaction on the dorsal surface of one wrist 1 d after inserting the wire. The reaction subsided somewhat during 2 weeks. The wire was exchanged for a new nickel-titanium wire at that time and the patient developed a flare-up of the hand reaction. The wire was removed. The eczematous reaction improved, but it did not clear up completely within 6 weeks. In none of these cases could any untoward intraoral reactions be observed. Some reactions to orthodontic appliances seem to be similar to the baboon syndrome.[30]

V. EXPERIMENTAL INTRAORAL EXPOSURE TO NICKEL-CONTAINING ALLOYS

The concerns about using nickel-containing alloys in nickel-sensitive individuals have been investigated in several human and animal studies. Moffa et al.[31] exposed ten nickel-sensitive individuals and controls to a 80% nickel-12% chromium dental casting alloy. No reactions were seen in the nonsensitive controls. Eight of the nickel-sensitive subjects reacted to the alloy specimens tested on the skin. Three of them developed extraoral eczematous reactions after a 48-h intraoral exposure to alloy specimens, mainly at sites of previous dermatitis, Jones et al.[32] constructed heat-processed poly(methyl methacrylate) maxillary prostheses with two metal inserts contacting the palatal mucosa. The metal insert on one side of the palate consisted of an alloy containing 70% Ni, 16% Cr, 2% Al, and 0.5% Be. The other metal insert was made of a cobalt-chromium alloy, which the authors state does not contain nickel. They exposed five patients who had positive patch tests to 5% nickel sulfate and five patients who did not react. The test prostheses were worn by the patients for 1 week and removed at least 8 h in each 24-h period. There were no positive mucosal reactions either in the patients positive to nickel or the control group. One patient had a slight urticaria and localized dermatitis during the skin patch test, but none had any apparent hypersensitivity reactions to nickel distant from the site of contact.

Magnusson et al.[33] attached test pieces of five different cobalt-chromium dental casting alloys used for removable partial dentures to the skin of ten women with nickel allergy, five of whom were also sensitive to cobalt. Nine of the ten women showed clear-cut or strong patch test reactions to an alloy with 7% nickel and one patient also showed the same reaction to one of the other alloys. In the five patients sensitive to both nickel and cobalt, responses to some of the other alloys were also observed.

Spiechowicz et al.[34] assessed the relationship between oral exposure to a nickel-containing dental alloy in persons with hypersensitive skin reactions to nickel. In ten partially dentate subjects, all with long histories of skin reactions to nickel, dental treatment was performed with fixed prosthetic appliances manufactured of an alloy containing 66% Ni and 12.9% Cr. None of the subjects had mucosal or skin reactions after the oral exposure to the nickel-containing alloy, nor any exacerbations of skin lesions. In one case, clinical improvement was noted in direct relationship to the insertion of four single crowns. A long-standing eczema on the hands and forearms began to heal rapidly. In another patient, the patch test after placement of the prosthesis showed a less intense nickel reaction, a change from 2+ to 1+ reaction. None of the gingival biopsies showed any inflammatory reactions attributable to the alloys.

The human study[34] was preceded by a guinea pig study with similar alloys.[35] Two nickel- and chromium-containing alloys of otherwise unspecified composition were sutured onto the buccal mucosa of ten nickel-sensitized guinea pigs and five control guinea pigs. No macroscopic mucosal reactions were seen during the 3-week test period. Histopathological examination of the contact area also failed to show any morphological difference between control and test sites.

van Loon et al.[36] exposed the buccal mucosa of five subjects with a positive skin reaction to nickel by attaching pure nickel foil specimens to the buccal side of a premolar for 7 d. The five test subjects all developed local allergic contact stomatitis of the mucosa of the cheek adjacent to the metal specimen after 72 h. This was substantiated by the histological appearance of the lesions. Because of the surrounding color, the demarcation of the reaction area was less clearly perceptible than the skin tests. One of the subjects developed small papules around the mouth during the test period. The same oral test with pure metallic palladium gave no oral or other reaction in subjects with a positive skin patch test to palladium.

Axéll et al.[37] performed intraoral patch testing in three subjects with maxillary poly(methyl methacrylate) plates carrying test pieces of a dental casting alloy containing 66% Ni and 12.9% Cr. Two of the patients were skin test positive to nickel, cobalt, chromate, while the third patient did not react to any of the metals by skin testing. The test specimen carriers allowed close contact between the alloy specimens and the palatinal keratinized mucosa and the buccal nonkeratinized mucosa. Biopsies were taken at the contact areas of metal specimens and of the specimen carrier. Biopsies from two of the patients were of a lichenoid nature with lymphcocytes migrating into the epithelium and with fibrin deposits at the basal lamina. This was seen at the buccal test site of all three patients and at the palatinal test site of one patient. Only in one instance did clinical examination reveal a small, symptomless, red patch in the buccal mucosa.

VI. INDUCTION OF TOLERANCE BY ORAL ADMINISTRATION OF NICKEL

In nonsensitized individuals, oral antigenic contacts may induce tolerance rather than sensitization.[38] In a guinea pig model, Vreeburg et al.[39] demonstrated that oral ingestion of nickel and chromium could induce tolerance to the same metals. They were also able to demonstrate the tolerance by placing dental casting alloys containing the same metals intraorally in the guinea pigs for a 30-d period. van der Burg et al.[40] have analyzed the development of hand eczema in junior hairdressers and junior nurses. Information as to previous contact with nickel suggested that development of nickel allergy had been promoted by ear piercing. A lower incidence of nickel hypersensitivity was observed if the ear piercing had been preceded by orthodontic treatment with nickel-containing alloys.

VII. CONFOUNDING VARIABLES

The alloys present in the oral cavity after extensive dental treatment may contain from 5 to 20 different metals, in addition to all the other dental materials used in conjunction with the alloys. Resin-based materials especially are used in large segments of the population. Resin-based unfilled and composite restorative materials, pit and fissure sealants, orthodontic adhesives, glazes, veneers and repair kits for porcelain-fused-to-metal restorations, root canal sealers, and temporary crowns are used extensively.

Depending on the composition, the polymerization is induced by chemical agents or by ultraviolet or visible light sources. The materials are based on monomers of different types: methacrylate monomers (e.g., methyl methacrylate, bis-phenol A diglycidyl dimethacrylate, triethyleneglycol dimethacrylate, ethyleneglycol dimethacrylate, hydroxyethyl methacrylate), urethane-based dimethacrylates, etc.[41] Epoxy bis-phenol resins and ethyleneimine derivatives are also in use. In addition, the chemicals involved in the polymerization processes are present: benzoyl peroxide, hydroquinone, camphoroquinone, phthalates, tertiary aromatic and aliphatic amines, benzoin ethers, UV stabilizers, antioxidants (butylated hydroxytoluene), benzophenones or benzotriazoles, and their derivatives.[42] Even though the materials set to a hard state in the oral environment, the amount of residual unreacted end-groups of dimethacrylates may vary from 15 to 50%,[43] which may lead to a slow release of degradation products such as formaldehyde.[44]

Allergic contact dermatitis and urticaria from dental resin-based products are considered rare.[45,46] However, it is difficult to assess which materials are present in a patient's mouth. Meager information is available about how components released from dental restorations act as sensitizers or cause dermatological sequelae in presensitized individuals. None of the dental materials are used exclusively in dentistry. The formulations have usually been adopted from other usage and adapted to suit dental purposes by adding colors, fillers, etc. Thus, the possibility of sensitization by skin contact prior to dental use is a definite possibility in many instances.

VIII. CONCLUSIONS

In most instances, intraoral exposure to nickel-containing dental materials does not seem to cause any adverse effects in nickel-sensitized individuals. Nickel allergy is of concern in dental treatment procedures. However, the reaction pattern is diverse and does not seem to be predictable. Intraoral exposure to nickel-containing dental materials may elicit intraoral reactions, extraoral reactions, or both. An interesting possibility is the concept that oral exposure to nickel-containing alloys may induce tolerance.

Many of the dental materials used contain nickel in varying amounts. However, they also contain other metals and chemicals which may act as haptens in sensitized individuals. Such confounding effects often lead to inconclusive evaluations in case reports, where nickel may be the most likely causative agent without it necessarily being the reason in all instances.

There is no information available from *in vivo* studies on the actual concentration of nickel in saliva derived from nickel-containing dental materials. With the great diversity of alloys available, it is impossible to predict which alloys will present a problem or perhaps be beneficial in the present context.

Further carefully controlled clinical trials are necessary to assess the problems associated with nickel allergy and dental treatment procedures. Cooperation between dermatologists, dental materials scientists, and dental clinicians will be needed to gain more insight. Most dental school curricula do not include any training in dermatology or contact dermatitis and contact stomatitis. Dentists are, therefore, not trained in associating any allergy problems with their clinical treatment procedures. Most of the cases we have seen seem to be originally diagnosed or suspected by the patients, not by dentists or dermatologists.

REFERENCES

1. **Greig, D. G. M.,** Contact dermatitis reaction to a metal buckle on a cervical headgear, *Br. Dent. J.*, 155, 61, 1983.
2. **Dickson, G.,** Contact dermatitis and cervical headgear, *Br. Dent. J.*, 11, 112, 1983.
3. **Hensten-Pettersen, A., Gjerdet, N. R., Kvam, E., and Lyberg, T.,** Nikkel allergi og kjeveortopedisk behandling, *Nor. Tannlegefor. Tid.*, 94, 567, 1984.
4. **Kvam, E. and Walde, S.,** Sintring av nikkelholdige legeringer, *Tannteknikerbladet,* 44, 19, 1984.
5. **Jacobsen, M. and Hensten-Pettersen, A.,** Occupational health problems and adverse patient reactions in orthodontics, *Eur. J. Orthod.*, in press.
6. **Gjerdet, N. R., Kallus, T., and Hensten-Pettersen, A.,** Tissue reaction to implanted orthodontic wires in rabbits, *Acta Odontol. Scand.*, 45, 163, 1987.
7. **Gjerdet, N. R.,** Orthodontics, in *Dental Materials: Biological Properties and Clinical Evaluations,* Mjör, I. A., Ed., CRC Press, Boca Raton, FL, 1985, 165.
8. **Mueller, H. J.,** Some considerations regarding the degradational interactions between mouth rinses and silver-soldered joints, *Am. J. Orthod.*, 81, 140, 1982.
9. **Berge, M. and Gjerdet, N. R.,** Corrosion of silver soldered orthodontic wires, *Acta Odontol. Scand.*, 40, 75, 1982.
10. **Bondevik, O., Kvam, E., and Gjerdet, N. R.,** Traumatic ulcers and pain during orthodontic treatment, *Commun. Dent. Oral Epidemiol.*, 15, 104, 1987.
11. **Kotilainen, R.,** Stomatitis prothetica and allergy, *Proc. Finn. Dent. Soc.*, 68, 7, 1972.
12. **Franz, G.,** The frequency of allergy to dental materials, *J. Dent. Assoc. S. Afr.*, 37, 805, 1982.
13. **Re, S.,** Allergic reactions to steel prostheses, *Minerva Stomatol.*, 9, 53, 1960.
14. **Brendlinger, D. and Tarsitano, J.,** Generalized dermatitis due to sensitivity to a chrome cobalt removable partial denture, *J. Am. Dent. Assoc.*, 81, 392, 1970.

15. **Levantine, A.,** Sensitivity to dental metal plate, *Proc. R. Soc. Med.,* 67, 1007, 1974.
16. **Wood, J. F. L.,** Mucosal reaction to cobalt-chromium alloy, *Br. Dent. J.,* 136, 423, 1974.
17. **Kaaber, S., Thulin, H., and Nielsen, E.,** Skin sensitivity to denture base materials in the burning mouth syndrome, *Contact Dermatitis,* 5, 90, 1979.
18. **Fernandez, J. P., Veron, C., Hildebrand, H. F., and Martin, P.,** Nickel allergy to dental prostheses, *Contact Dermatitis,* 14, 312, 1986.
19. Forschungsinstitut für die Zahnärztlich Versorgung (FZV), in *Übersicht über die Dental-Edelmetallegierungen und Dental-Nichtedelmetallegierungen in der Bundesrepublik Deutschland,* Deutscher Arzte-Verlag GmbH, Köln, 1986.
20. **Smith, D. L.,** Dental casting alloys: technical and economic considerations in the USA, *Int. Dent. J.,* 33, 25, 1983.
21. **Land, B. R., Morris, H. G., and Razzoog, M. E., Eds.,** Biocompatibility, Toxicity and Hypersensitivity to Alloy Systems Used in Dentistry, International Workshop, University of Michigan School of Dentistry, 1986.
22. American National Standards Institute/American Dental Association Specification No. 41 for Biological Evaluation of Dental Materials, 1979.
23. Federation Dentaire Internationale, Recommended Standard Practices for Biological Evaluation of Dental Materials, 1980; *Int. Dent. J.,* 30, 140, 1980.
24. **Hensten-Pettersen, A.,** Allergiske reaksjoner pâ dentale materialer, *NTFs Tidende,* 94, 573, 1984.
25. **Izumi, A. K.,** Allergic contact gingivostomatitis due to gold, *Arch. Dermatol. Res.,* 272, 387, 1982.
26. **Roed-Petersen, B., Roed-Petersen, J., and Jørgensen, K. D.,** Nickel allergy and osteomyelitis in a patient with metal osteosynthesis of a jaw fracture, *Contact Dermatitis,* 5, 108, 1979.
27. **Schriver, W. R., Shereff, R. H., Domnitz, J. M., Swintak, E. F., and Civjan, S.,** Allergic response to stainless steel wire, *Oral Surg.,* 42, 578, 1976.
28. **Nielsen, C. and Klaschka, F.,** Testudien an der Mundschleimhaut bei Ekzemallergikern, *Dtsch. Zahm Mund Kieferheilkd.,* 57, 201, 1971.
29. **Frykholm, K. O.,** On mercury from dental amalgam. Its toxic and allergic effects and some comments on occupational hygiene, *Acta Odontol. Scand.,* 15 (Suppl. 22), 1957.
30. **Andersen, K. E., Hjorth, N., and Menné, T.,** The baboon syndrome: systemically induced allergic contact dermatitis, *Contact Dermatitis,* 10, 97, 1984.
31. **Moffa, J. P., Beck, W. D., and Hoke, A. W.,** Allergic response to nickel containing dental alloys, *J. Dent. Res.,* 56, B78 (Abstr. 107), 1977.
32. **Jones, T. K., Hansen, C. A., Singer, M. T., and Kessler, H. P.,** Dental implications of nickel hypersensitivity, *J. Prosthet. Dent.,* 56, 507, 1986.
33. **Magnusson, B., Bergman, M., Bergman, B., and Söremark, R.,** Nickel allergy and nickel-containing dental alloys, *Scand. J. Dent. Res.,* 90, 163, 1982.
34. **Spiechowicz, E., Glantz, P.-O., Axéll,T., and Chmielewski, W.,** Oral exposure to a nickel-containing dental alloy of persons with hypersensitive skin reactions to nickel, *Contact Dermatitis,* 10, 206, 1984.
35. **Spiechowicz, E., Nyquist, G., Goliszewska, E., and Chmielewski, W.,** Experimental investigations on sensitivity to nickel present in alloys used in dentistry carried out in guinea pigs previously sensitized to this metal, *Swed. Dent. J.,* 7, 39, 1983.
36. **van Loon, L. A. J., van Elsas, P. W., van Joost, T., and Davidson, C. L.,** Contact stomatitis and dermatitis to nickel and palladium, *Contact Dermatitis,* 11, 294, 1984.
37. **Axéll, T., Spiechowicz, E., Glantz, P.-O., Andersson, G. and Larsson, Å.,** A new method for intraoral patch testing, *Contact Dermatitis,* 15, 58, 1986.
38. **Tomasi, T. B.,** Oral tolerance, *Transplantation,* 29, 353, 1980.
39. **Vreeburg, K. J. J., de Groot, K., von Blomberg, M., and Scheper, R. J.,** Induction of immunological tolerance by oral administration of nickel and chromium, *J. Dent. Res.,* 63, 124, 1984.
40. **van der Burg, C. K. H., Bruynzeel, D. P., Vreeburg, K. J. J., von Blomberg, B. M. E., and Scheper, R. J.,** Hand eczema in hairdressers and nurses: a prospective study. I. Evaluation of atopy and nickel hypersensitivity at the start of apprenticeship, *Contact Dermatitis,* 14, 275, 1986.
41. **Ruyter, I. E. and Sjøvik, I. J.,** Composition of dental resin and composite materials, *Acta Odontol. Scand.,* 39, 133, 1981.
42. **Bowen, R. L.,** Compatibility of various materials with oral tissues. I. The components in composite restorations, *J. Dent. Res.,* 58, 1493, 1979.
43. **Ruyter, I. E. and Svendsen, S. A.,** Remaining methacrylate groups in composite restorative materials, *Acta Odontol. Scand.,* 36, 75, 1978.
44. **Ruyter, I. E.,** Release of formaldehyde from denture base polymers, *Acta Odontol. Scand.* 38, 17, 1980.
45. **Nathanson, D. and Lockhart, P.,** Delayed extraoral hypersensitivity to dental composite material, *Oral Surg.,* 47, 329, 1979.

46. **Tinkelman, D. G. and Tinkelman, C. L.,** An unusual etiology of urticaria, *Pediatrics,* 63, 339, 1979.
47. **Weiss, P. A.,** State-of-the-art metal ceramics. Utilizing nickel chromium superalloy frameworks, in *Dental Ceramics, Proc. 1st Int. Symp. Ceramics,* McLean, J. W., Ed., Quintessence, Chicago, 1983, 231.

Chapter 18

SOCIAL AND DEMOGRAPHIC ASPECTS OF NICKEL CONTACT ALLERGY

Björn Edman and Lars Janzon

TABLE OF CONTENTS

I. Introduction 208

II. DALUK 208

III. Patch Tests 208

IV. Demographic Data 209

V. Department of Occupational Dermatology 209

VI. Age and Sex Distribution 209

VII. Time Trends in Incidence 210

VIII. Geographical Variation in Morbidity 210

IX. Occupation 211

X. Site of Eczema 211

XI. Atopy 212

XII. Comments and Conclusions 212

References 213

I. INTRODUCTION

The prevalence of nickel allergy among women in Scandinavia has in some studies been assessed to be as high as 10 to 12%;.[1,2] About 40% of all patients with nickel allergy develop hand eczema.[3,4] Because of this, many of these patients are forced to change jobs. The medical and social consequences of nickel allergy are obvious. It is therefore important to learn more about factors influencing the risk of sensitization.

The ear-piercing procedure is the most common cause of nickel sensitization among women in Sweden.[5] Factors influencing the use of this procedure are not well known. We also have inadequate knowledge about other modes of sensitization. Furthermore, many of those exposed to nickel do not develop contact allergy to nickel. This motivates further studies to evaluate factors which modify the risk of sensitization. It has been inferred that the risk of sensitization is increased as well as decreased by atopy.[6]

Malmö is a southern Swedish city of about 230,000 inhabitants. The age and sex distribution, as well as the number of people living alone, number of people with foreign background, and the number of people unemployed and living on social welfare, varies considerably among the six main areas within the city. From this perspective, the city of Malmö is well suited for studies regarding a relationship between nickel allergy and these different social and demographic factors. With only one hospital and one department of dermatology in the city, it is easy to identify cases belonging to the population being studied. About 65% of the patients patch tested in Malmö are treated in the Section of General Dermatology.[7] About 20% of the patients are treated at the Unit of Occupational Dermatology within the same department. Less than 15% of the patients with nickel allergy are diagnosed and treated by dermatologists in private practice.[7] In an earlier study from Malmö,[8] contact allergy to nickel was shown to increase. Therefore, further study on the trend is essential.

The aims of this present review are to:

1. Assess the age- and sex-specific incidence of nickel allergy in the city of Malmö
2. Study time trends in the incidence of nickel allergy
3. Study geographical differences in the incidence of nickel allergy within the city of Malmö and describe demographic characteristics of areas with high and low incidence
4. Study the relationships between nickel allergy and occupational activity
5. Study relationships between nickel allergy and atopy

II. DALUK

Results from the patch tests are stored in the database DALUK at the Department of Dermatology.[9] Information regarding the patient's sex, age, occupation, living area, atopy, eczema sites, and duration of eczema is also stored with the test results. All this information is collected when tests are applied. Patients living outside the city of Malmö are excluded from this study.

III. PATCH TESTS

During the time period 1982 to 1985, 2276 patients — 1593 females and 683 males — were tested using the standard series recommended by the International Contact Dermatitis Research Group with the technique described in a previous study.[7] Contact allergy to nickel was found in 351 patients, 15.4% of all tested. Of these, 335 (95.4%) were females. Thus, the incidence of nickel allergy among tested females was 19.6% and among males 2.6%.

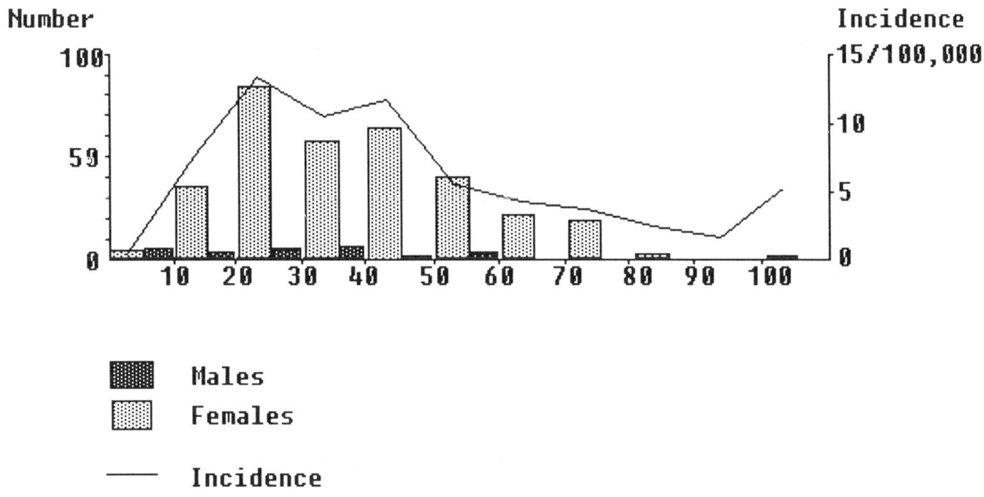

FIGURE 1. Age and sex distribution and incidence.

IV. DEMOGRAPHIC DATA

Data in local registers (from the National Central Bureau of Statistics) were used for the sociodemographic characterization of high- and low-incidence areas of nickel contact allergy.

V. DEPARTMENT OF OCCUPATIONAL DERMATOLOGY

In Malmö, 90% of all patch tests are performed at the Department of Dermatology. Unequivocal cases of occupational dermatitis, about 20% of all patch tests, are referred directly to the Unit of Occupational Dermatology. With this referral system, patients at the Section of General Dermatology cannot be considered a random sample from the population of patients with nickel allergy. In order to assess the degree of selection, we have compared the age and sex distribution and occupational activities among patients treated at the Unit of Occupational Dermatology and patients treated at the Section of General Dermatology. The percentage of positive tests for nickel allergy was about the same among patients in the two sections of the department. The age and sex distributions did not differ when these two groups of patients were compared. Among the patients referred to the Section of General Dermatology, the majority belonged to the occupational groups "retired", "office work", and "cleaning work". Most of the patients referred to the Unit of Occupational Dermatology were employees in either the metal industry or within the health care system.

VI. AGE AND SEX DISTRIBUTION

The age and sex distributions together with the incidence per 100,000 per year are shown in Figure 1. The incidence is defined as the number of new cases diagnosed in the Section of General Dermatology per year. The mean age in both sexes was 40 ± 18 years (SD). Most of the female patients were 20 to 30 years old, while most of the male patients were 30 to 40 years of age. The highest incidence was in the age group 20 to 40 years. The incidence was lower in subsequent age groups.

Because of the low frequency in male patients, only female patients will be included in the following analysis.

Table 1
NUMBER OF OBSERVED CASES IN RELATION TO THE EXPECTED NUMBER OF CASES 1982

	1983	1984	1985	1986
Section of General Dermatology	49/35 (1.4)	78/37 (2.1)	79/36 (2.2)	85/35 (2.4)
Unit of Occupational Dermatology	17/12 (1.4)	22/13 (1.7)	13/13 (1.0)	19/12 (1.6)

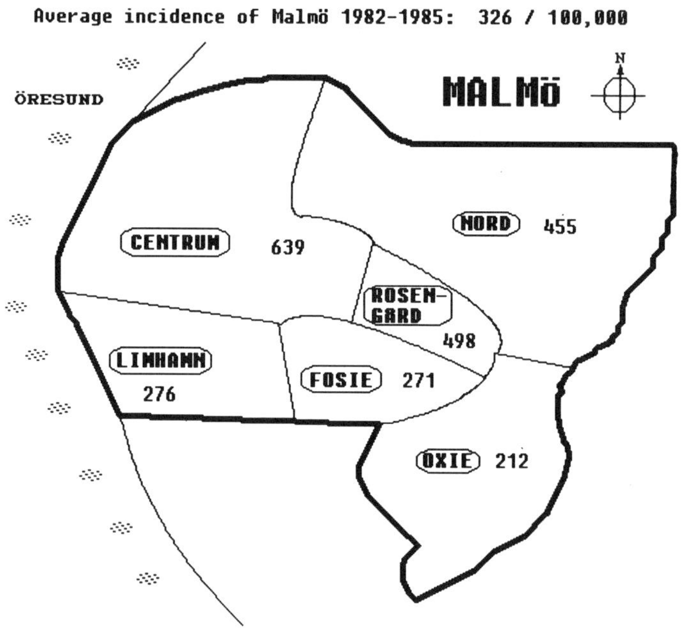

FIGURE 2. Incidence per 100,000 in Malmö, 1982 to 1985.

VII. TIME TRENDS IN INCIDENCE

The time trend in incidence was analyzed for the period 1983 to 1986. The number of new cases per 100,000 per year increased during this time from 35 to 85 (Table 1), while the age and sex distribution within the city did not significantly change.

VIII. GEOGRAPHICAL VARIATIONS IN MORBIDITY

The age-standardized incidence of nickel allergy in the six main areas within the city varied in comparison with the city average of −35% to +96% (Figure 2). These areas were compared in terms of percentages of one-person households, people being unemployed, people belonging to social class III (blue-collar workers),[10] people with foreign background, foreigners (people without Swedish citizenship), and people living on social welfare. Areas with the highest incidence of nickel allergy had, in comparison with areas of lowest incidence, greater than expected percentages of one-person households, unemployment, foreigners, and people living on social welfare (p <0.05). Corresponding factors in areas with the lowest incidence were below the city average (p <0.05) (Figure 3).

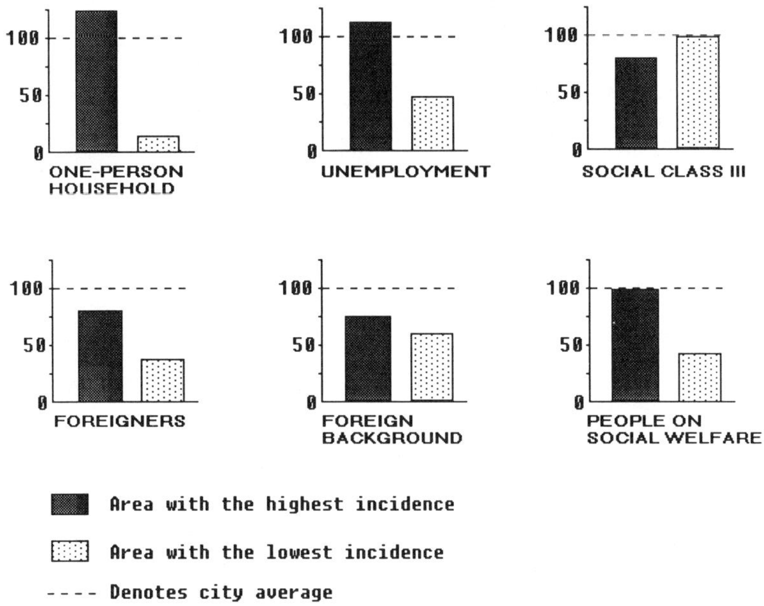

FIGURE 3. Percentage of some demographic characteristics in areas with the highest and lowest incidence of nickel allergy.

Table 2
OCCUPATIONS IN FEMALES WITH NICKEL CONTACT ALLERGY

Rank	Occupation	Percentage	Cumulative
1	Clerk	14	14
2	Retired	13	27
3	Cleaning work	13	40
4	Hospital work	11	51
5	Food work	9	60
6	Transport work	3	63
7	Cashier	3	66
8	Textile work	3	69
9	Beauty work	3	72
10	Underage	1	73
11	Others	27	100

IX. OCCUPATION

The occupational activities among the females tested in the section of general dermatology are shown in Table 2. Of the females, 14% were clerks, 13% were retired, 13% housewives or similar, 11% were employed within the medical care system, and 9% within food works. Compared to the occupational activities among women without nickel allergy (i.e., tested with negative result or positive to other allergens) there was a significant overrepresentation ($p < 0.05$) of clerks and women employed at the hospital among nickel-allergic patients. The percentage of retired people was lower than expected from the city average (Figure 4).

X. SITE OF ECZEMA

Among retired nickel-allergic patients, the most frequent eczema site was the lower leg, possibly without relevance. Among housewives the most common eczema site was the center

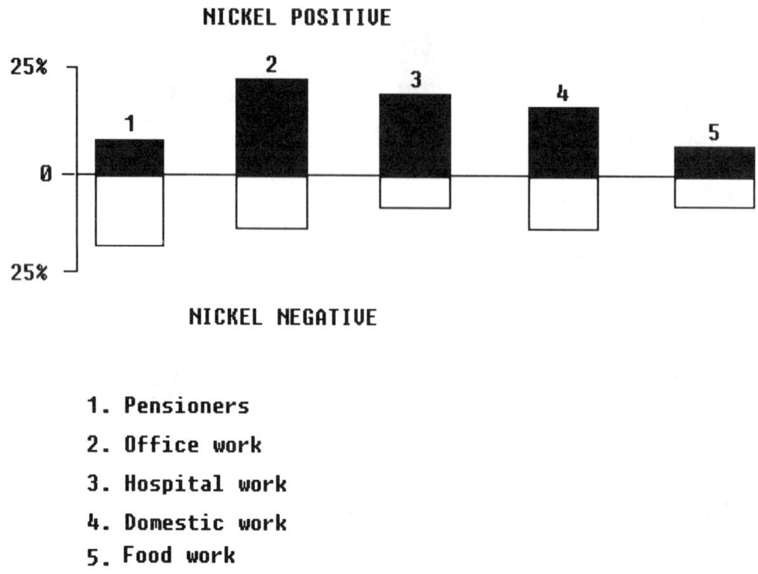

FIGURE 4. Distribution of occupations in nickel-allergic females.

of the palm and the area between the fingers. Patients employed as clerks most often had eczema at the center or the periphery of the palm and between fingers, which shows resemblance with pompholyx[11] and the so-called apron pattern.[12] Patients employed at the hospital likewise had eczema in the center of the palm, but also at the dorsal part of their hands.

XI. ATOPY

About 10% of all patients with contact allergy to nickel had atopy, which was defined as a personal history of asthma, allergic rhinitis, or atopic dermatitis.

XII. COMMENTS AND CONCLUSIONS

This study confirms that contact allergy to nickel is a problem that almost exclusively is restricted to females. The frequency of nickel allergy has continued to increase since a previous study during the period of 1969 to 1979.[8] One study from Sweden[5] has shown a causal relationship between the ear-piercing procedure and nickel allergy. In that study, the prevalence of nickel allergy among women wearing earrings was highest among the youngest girls. This might indicate that the risk of sensitization is related to age.[1] Hand eczema in these sensitized subjects usually develops 5 to 10 years later.[11] The fact that the highest incidence in our study was found in the age group 30 to 40 years might be explained by patients not seeking medical advice until they have developed a hand eczema. Since the ear-piercing procedure now has also been adopted by men, one would have expected a greater proportion of male patients with contact allergy to nickel. It might be that the exposure time has been too short, but it might also reflect a potential influence of hormonal and genetic[1] background on the risk of sensitization.

The observed high- and low-incidence areas within the city could be explained by different care-seeking behaviors or by differences in the incidence of the disease. Few of the city's patients with contact allergy to nickel are tested by physicians and dermatologists in private practice.[7] It is inconceivable that low-incidence areas are areas where people seek medical advice outside the Department of Dermatology. Hence it is reasonable to conclude that the observed differences in the morbidity reflect real differences in the incidence of the disease.

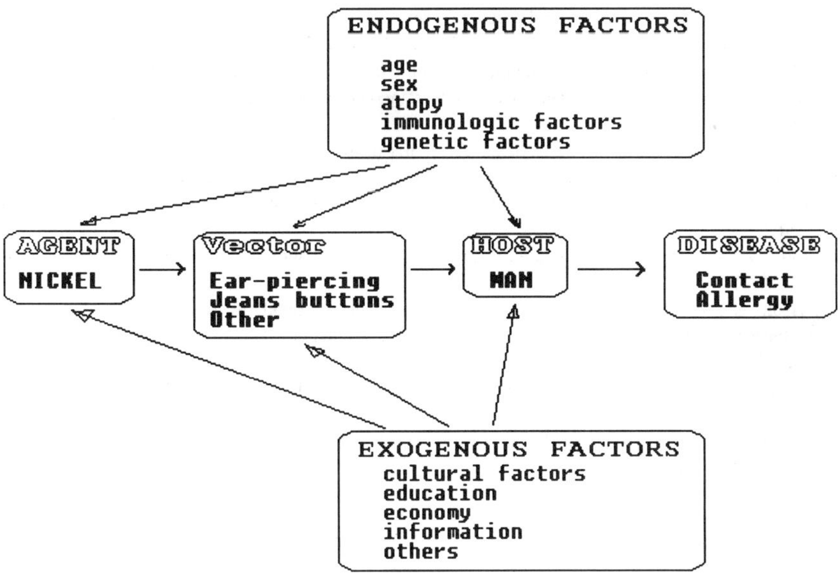

FIGURE 5. The epidemiologic model of nickel contact allergy.

Observed social and demographic characteristics of high-incidence areas support the hypothesis that the risk of becoming exposed to nickel is influenced by cultural, educational, and economic factors. These ecological relationships should be further evaluated in case-control studies. With the observed geographical differences in the disease and with the identification of high-risk groups, it is possible to focus health-education activities and prophylatic measures in order to reduce the risk of sensitization. Such measures may include the introduction of new materials and techniques for ear piercing, which is probably easier than changing popular habits. The need for such activities is strengthened by the rapid increase of nickel allergy, 140% during the time period 1982 to 1985.

Whether the observed overrepresentation of clerks and cleaning workers reflects jobs with an increased risk of contact with nickel or women with an increased risk due to their lifestyles cannot be clarified with available data but should merit further study.

We have no data on the prevalence of atopy in the population of Malmö. In a population study from Finland,[13] the prevalence of atopy was 13.5%. The observed 10% in our patients with nickel allergy is thus lower than expected ($p < 0.05$). This is in accordance with previous studies reporting a lower risk of contact allergy among patients with atopy.[14]

The present study illustrates the value of integrating epidemiology in clinical practice (Figure 5). By describing the disease in terms of time, place, and person we can learn more about risk factors and factors modifying the risk. This is of great importance for the design of preventive measures and also, of course, of great value for proper planning and allocation of medical resources.

REFERENCES

1. **Menné, T., and Holm, N. V.,** Nickel allergy in a female twin population. Genetic predisposition and prevalence, *Int. J. Dermatol.*, 22, 22, 1983.
2. **Carlsson, Å. S., Magnusson, B., and Möller, H.,** Metal sensitivity in patients with metal-to-plastic total hip arthorplasties, *Acta Orthop. Scand.*, 51, 57, 1980.

3. **Peltonen, L.,** Nickel sensitivity in the general population, *Contact Dermatitis,* 5, 27, 1979.
4. **Menné, T., Borgan, Ö.,** Nickel allergy and hand dermtitis in a stratified sample of the Danish female population: an epidemiologic study including a statistic appendex, *Arch. Derm. Venereol.,* 2, 35, 1982.
5. **Larsson-Stymne, B. and Widström, L.,** Ear-piercing — a cause of nickel allergy in schoolgirls?, *Contact Dermatitis,* 13, 290, 1985.
6. **Ho, V. C. and Johnston, M. M.,** Nickel dermatitis in infants, *Contact Dermatitis,* 15, 270, 1985.
7. **Edman, B. and Möller, H.,** Medicament contact allergy, *Berufs-Dermatosen,* 34, 139, 1986.
8. **Edman, B. and Möller, H.,** Trends and forecasts in a 12-year patch test material, *Contact Dermatitis* 8, 95, 1982.
9. **Edman, B.,** The usefulness of detailed information to patients with a contact allergy, *Contact Dermatitis,* 19, 43, 1988.
10. **Carlsson, G.,** *Social Mobility and Class Structure,* Gleerups, Lund, Sweden 1958.
11. **Christensen, O. B. and Möller, H.,** Nickel allergy and hand eczema, *Contact Dermatitis,* 1, 129, 1975.
12. **Cronin, E.,** Clinical patterns of hand eczema in women, *Contact Dermatitis,* 13, 153, 1985.
13. **Luoma, R. and Koivikko, A.,** Occurrence of atopic diseases in three generations, *Scand. J. Soc. Med.,* 10, 49, 1982.
14. **Jones, H.E., Lewis, C. W., and McMarlin, S. L.,** Allergic contact sensitivity in atopic dermatitis, *Arch. Dermatol.,* 107, 217, 1973.

Chapter 19

NICKEL DERMATITIS — A PREVENTABLE HEALTH PROBLEM

T. Menné and H. Maibach

Prevention of diseases is not only rational medicine, but a realistic approach.

Numerous studies during the last 60 years have classified nickel dermatitis as a common and, for some patients, a disabling disease. This volume documents the magnitude of the problem and provides evidence that prevention of nickel sensitization is possible, at least from a theoretical viewpoint.

The understanding of the concentration-dependent sensitization risk makes it possible to estimate a concentration level for haptens below which sensitization is unlikely to occur. It is practically possible for the industry to produce nickel alloys which release minimal nickel under use conditions.

Producers of costume jewelry must be encouraged to discontinue the use of alloys which corrode easily. Corrosive-resistant alloys should be mandatory for ear piercing.

Regulation by law for nickel alloys designed for prolonged skin contact is presently under consideration in Denmark.

Nickel sensitization is a preventable condition — it is our hope that this book provides the basis for planning preventive strategies.

INDEX

A

Absorption, 27
  dermal, 20—21
  intestinal, 21—22
  pulmonary, 22
Acid digestion technique, 4
Acrodermatitis enteropathica, 181
Allergens, mechanism of, 56—57
Alloys, 10, 19, see also specific alloys
Ammonium sulfate, 3
Analytical techniques, 3
Angry back syndrome, 80
Animal sensitization assays, 66—72
Ankylosing spondylitis, 103
Antabuse, interactions of, 181
Antibiotics, in complex and chelate formation, 180
Antigenic complex, 80
Atopy, 127, 139—141
  in hand eczema, 144—146, 157
  latent, 144
  in nickel sensitivity, 144—146, 212
  prevalence of, 144, 146
Auto mechanics, 19

B

Baboon syndrome, 175—176, 200
Bacteriostatic agents, 124
Biliary excretion, 24
Biological monitoring, 4—5, 25
Blood, nickel concentration in, 23
Blood clotting mechanisms, 84
Body fluids
  corrosiveness of, 46
  nickel concentrations in, 5—6, 25
Bronchial epithelial cells, 24
Buccal mucosa, keratinized, 201—202
Building industry, 124
Butchers, 126

C

Carcinogenicity, 36
  classification of, 38
  in foundry work, 37
  future research on, 41
  mechanisms of, 38—40
  of metal trades, 37—38
  of nickel alloys and stainless steels, 37
  of nickel-cadmium battery manufacture, 38
  of nickel plating, 37
  of nickel refining, 36—37
  regulatory requirements for, 40—41
Cell migration inhibition, 81—82
Cellular distribution, 24
Cellular immune response, afferent and efferent limbs of, 56—57
Cement, 118—119, 124
Chalk, 126
Cheilitis, 198
Chelates, 180
Chelating agents, 23, 155, 182—183
Chelating drugs, 185—186
Chemical compounds, 3
Chemiluminescent nickel carbonyl detector, 4
Chemobiokinetics, 20—24
Chloride, in nickel corrosion, 46
Chrome exposure, 36
Chromium carbide, formation of in alloys, 52
Chromium-containing alloys, 49—50
Chromosomal aberrations, 38
Cigarette smoking, nickel exposure through, 18
Cleaning
  in hand eczema and nickel sensitivity, 143—144
  occupations involving, 125
Clean Water Act, 41
Coal gasification workers, 19
Coal slag, 19
Cobalt
  allergy to, 124
  in implants, 51—52
  sensitization to, 120, 126
Cobalt-chromium dental casting alloys, 198, 201
Cobalt-nickel allergy, 125
Cobalt-nickel relationship, 171—172
Coinage, 125
  alloys in, 11
  composition of, 49
  cupronickel, 48—49
  hand eczema with handling of, 148—149
  nickel content of, 48—49, 119
  nickel release from, 50
Complex formation, 180, 182—186
Concanavalin A, 82
Construction industry, 124
Consumer Products Safety Commission regulations, 41
Copper/nickel alloys, solubility of, 48
Corrosion, 51—52
Corrosion fatigue, 51—52
Corrosion inhibitors, 124
Corrosion resistance, 10
Corrosive environment, 46
Cosmetics, nickel content in, 95
Cronstedt, Axel Frederik, 2
Cross-sensitization tests, 76
Cupronickel, 10, 48—49
Cutaneous exposures, prevention of, 26
Cutting fluids, 124, 126

D

DALUK, 208
DDC, 23
Demographic data, 208—213

Dental casting alloys, 199—203
Dental materials, 202—203
Dental treatment procedures, 196
  experimental intraoral exposure in, 202—203
  extraoral reactions to extraoral exposure, 196—197
  extraoral reactions to intraoral exposure, 200
  intraoral reactions to intraoral exposure, 197—200
Dentures, removable partial, 198
Denture stomatitis, 198
Dermal absorption, 20—21, 27
Dermatitis, see also Eczema
  allergic contact
    with dental resin-based products, 202
    common clinical picture of, 92
    with nickel sensitivity, 140—141
    prognosis of, 157
    testing for, 92
  with chromium-containing alloy exposure, 50
  contact, 80
    advances in immunology for, 60—61
    current view on development of, 56—57
    of hands, 139
    history of, 56
    nickel as hapten in, 57—60
  diet treatment of, 172
  flares of and nickel levels in organs, 175—176
  fulminant, 121
  hand, 92, 139, 141—142
  irritant, 157
  with nickel implants, 191
  nickel levels in tissues and, 167
  nickel-related, 12, 46
    background of, 120—122
    clinical expression of, 121—122
    epidemiology of, 110—114
    history of, 110
    occupational, 118—127
    patch test studies of, 110—111
    population studies of, 111—114
    prevention of, 126—127
    sensitizing and eliciting dose of, 120—121
  nickel sources of, 11
  with orthodontic appliances, 200
  perioral, 198
  prevention of, 21, 215
  relationship of to food
    clinical reactions, 168
    clinical relevance in food, 173—176
    diet treatment, 172
    experimental oral challenge, 168—172
    general, 166
    nickel in foodstuffs, 173
    systemically induced nickel dermatitis, clinical experience suggesting, 166—168
  secondary eruptions of, 166
  from skin contact with nickel-plating, 52
Dermojet, 191
Detergents, 118, 119
  hand eczema with, 143—144, 149
  in nickel sensitivity, 120, 143—144
  used in hairdressing, 125
Diagnostic tests, 25
Diet
  daily nickel intake in, 13—17
  low-nickel, 26
  nickel concentrations and, 173
Diethyldithiocarbamate, interactions of, 181
Diet treatment, 172, 174
Differential pulse absorption voltametry (DPAV), 3
Dimethylcysteine, 182—183
Dimethylglyoxime, 93, 96
Dimethyl-sulfoxide, 93
2,4-Dinitrochlorobenzene, 56
Diodoquin, 184—185
Disodium EDTA, 155
Disulfiram, 24
  drawbacks of, 185
  in hand eczema treatment, 26
  interactions of, 81
  nickel dermatitis and, 167—168
  in nickel poisoning treatment, 174
Dithiocarb, interactions of, 181
Dithiocarbamate, 155, 174
DNA, 39—40
DNCB sensitivity, 102, 104—105
DPAV, see Differential pulse absorption voltametry
Draize Repeat Insult Test, 76
Drinking water, nickel content of, 17
Drug interactions, 180—186
Dyshidrosis, 135, 144, 151—152
Dyshidrotic eczema, 122, 151—152

E

EAAS, see Electrothermal atomic absorption spectrophotometry
Ear piercing, 60, 112, 120, 124, 125, 202, 208, 213
Eczema, 80, see also Dermatitis
  apron pattern of, 150, 151
  dyshidrotic, 122, 151—152
  fingertip, 151
  of hand, 12
    age of onset of, 139—140
    chronic, intractable nature of, 156—158
    clinical patterns of, 149—153
    course and prognosis for, 155—158
    diagnosis of, 153—154
    dorsal, 121
    factors involved in nickel-sensitive patient, 141—143
    incidence of nickel sensitization in, 121
    nickel allergy and, 105, 134, 136—141, 143—146
    occupational association of, 122—126
    oral challenge and, 168—171
    palmar, 124
    prevalence of, 135—136
    prevention of, 154
    sex distribution of, 136, 139
    sources of elicitation of, 147—149
    sources of nickel-induced, 146—149
    time relationship of with nickel sensitivity, 139—

140
treatment of, 26, 154—155
typing of, 122
volar, 121—122
with metal sensitivity from implanted nickel, 188—190
minimum nickel dose to produce, 120
mini patch of, 86
nickel sensitization and, 110—111
oral nickel intake and, 18
with orthodontic appliances, 200
ring patterns of, 151
secondary spread of, 153
with shrapnel, 191
site of, 211—212
treatment of, 23
Eczematous diathesis, 142
Electrodeposited coatings, 48, 52—53
Electroforming, 118, 123, 124
Electroforming plant workers, 12
Electroplating operations, 19, 123
Electrothermal atomic absorption spectrophotometry (EAAS), 3—4
Environmental exposures, 10—11, 147
Epicutaneous sensitivity method, 66—67, 71, 104
Epidermal painting sensitization tests, 71—72
Epoxy bis-phenol resins, 202
Ethambutol, 183—186
Ethyleneimine derivatives, 202
Excretion, 24
Exposure monitoring, 27

F

Family studies, 103—104
Fats, nickel content in, 16
FCA, see Freund's complete adjuvant
Fecal nickel excretion, 24
Fibroblasts, 24
Fibrosarcomas, 39
Fiddler's neck, 126
Fisher's dimethylglyoxime spot test, 11
Follicular test reactions, 94
Food and Drug Administration, 41
Foods
nickel absorption from, 21—22
nickel content of, 12—18, 27, 119
nickel dermatitis and, 166—176
Foundry work, carcinogenicity in, 37
Fretting corrosion, 51—52
Freund's complete adjuvant (FCA), 66, 69
Fungal infections, 198

G

Gastrointestinal absorption, 27
Gastrointestinal exposure, 18
Genetic markers, 102—103
Genetic predisposition, 142
German silver, 10, 11
Gersdorffite, 2
Gloves, 126, 154
Gold alloys, in prosthodontics, 199
Gold plating, in dental prostheses, 198
GPMT, see Guinea pig maximization test
Groundwater, nickel content of, 17
Guinea pig maximization test (GPMT), 66—71
Guinea pig sensitization, 69

H

Hairdressers, 12, 113
Hairdressing, 119, 124—125
Haptens, 56
in allergic reaction, 80
nickel as, 57—60
Hard metal manufacturing, 124, see also Metal manufacturing
Hard-metal workers, 12
Hazardous Waste Disposal rule, 40
Hip prostheses, metal sensitivity with, 190—191
Histocompatibility antigens, major, 102
HLA, see Human leukocyte antigen
HLA-DR antigens, 80—81, 84
HLA genes, 102—103
Hormones, 142
Hospital wet work, 125
Housework, 121—122, 125
Human leukocyte antigen (HLA), 102—103
Human nickel exposures
chemobiokinetics of, 20—24
diagnostic tests for, 25
prevention of, 26—27
sources of, 19—20
Human sensitization assays, 76—78
Humidity, 142—143
Hygienic measures, 126
Hyperhidrosis, 152
Hypersensitivity, see also Dermatitis, contact
cell-mediated, 84
with coin handling, 48
in vitro testing in, 80—86

I

IARC, see International Agency for Research on Cancer
Iatrogenic exposures, 19—20
Immune-complex reactions, 168
Immune modulation, 84
Immunology, 80—81
Implants, nickel-containing, 19
corrosion of, 51—52
leaching of, 19
sensitivity and cutaneous reactions to, 188—191
Indomethacin, 174
Inorganic carcinogens, 39
Inorganic nickel, pulmonary exposure to, 18—19
Intercellular fluid, nickel levels in, 23
Interferon, 84—85
Interleukin 2, 57, 80—81
International Agency for Research on Cancer

(IARC), 4, 36, 38
International Union of Pure and Applied Chemistry (IUPAC), 4
Intestinal absorption, 21—22
Intradermal sensitization methods, 66, 67, 95
In vitro sensitivity testing, 80—86, 96
IUPAC, see International Union of Pure and Applied Chemistry
IUPAC/IARC reference procedure, 4

J

Janitors, 12
Jewelry
  corrosive alloys in, 215
  nickel sensitization from, 80, 113, 120
Joint loosening, 190—191

K

Kidney, nickel levels in, 23
Kitchen utensils, nickel leaching from, 17
Kligman studies, 76—77

L

Langerhans cells, 21, 24
  allergen-carrying, 56
  in allergic reaction, 80
Laryngeal cancer, 36
Leaching, 17—19
Leukocyte aggregation test, 96
Leukocyte inhibition factor (LIF), 81—82
Leukocyte migration testing, nickel-protein complexes in, 96
Leukocyte procoagulant activity, 84
Leukocytoclastic vasculitis, 168
LIF, see Leukocyte inhibition factor
Lipophilic organic nickel compounds, 20
Liver, nickel levels in, 24
LTT assays, 85
Lungs, 22, 23, 36
Lymphocyte transformation test, 82—84, 96, 104
Lymphokines, in contact sensitivity, 81

M

Macrophage activating factor, 85
Malocclusion correction, 196
Margarine production, nickel in, 126
Mechanical engineering, 118
Metal
  binding, 180—181, 185
  tolerance to, 26
Metallic nickel, 2
Metal manufacturing, 37—38, 118, 124
Metal workers, 19
MIF, see Migration inhibition factor
Migration inhibition, 81—82
Migration inhibitor factor (MIF), 81, 83, 85
Mixed lymphocyte response (MLR), 82
MLR, see Mixed lymphocyte response
Molybdenum alloys, 51—52
Monel 400 alloy, 11
Motor oils, 19
Musicians, 126

N

National Toxicology Program studies, 41
NDMA sensitivity, 102, 104
Ni^{3+}/Ni^{2+} couple, 2
Nickel, 10
  absorption of, 20—22, 27
    depressed, 185—186
    dermal, 27
  allergy to, 40, 124, 127
    age and sex distribution of, 209
    atopy and, 212
    dental treatment procedures and, 196—203
    dermatitis with, 92
    drug interactions and, 180—186
    epidemiologic model of, 213
    geographical variations in, 210
    incidence of, 134—135
    noneczematous symptoms of, 93
    occupational variations in, 211
    with orthopedic prostheses, 188—191
    pompholyx and, 152
    prevalence of, 112
    prevention of, 215
    relationship of with hand eczema, 136—141
    site of eczema and, 211—212
    social and demographic aspects of, 208—213
    sources of, 148
    time trends in incidence of, 210
  analytical techniques for, 3, 5—6
  atomic number and weight of, 2
  bioavailability of, 174—175
  carcinogenicity of, 36—41
  cellular distribution of, 24
  chemical compounds of, 3
  chemobiokinetics of, 20—24
  crystalline phases of, 2
  concentrations of, 5—6, 167, 175
  discovery of, 2
  electrothermal atomic absorption spectrophotometry of, 3
  excretion of, 24
  exposure sources of
    environmental, 10—11
    iatrogenic, 19—20
    occupational, 11—12
    oral, 12—18
    pulmonary, 18—19
  exposure to, 118, see also Occupational exposure
    diagnostic tests for, 25
    in nickel manufacturing, 118
    occupational, 118—119
    prevention of, 26—27
    through skin, 119—120
  in foodstuffs, 173—176

immunological effects of, 21
induced tolerance to by oral administration, 202
industrial uses of, 47
ingested, 200
isotopes of, 2
leaching of, 17—18
manufacturing of, 118, 123
metabolism of, 175
organ distribution of, 23
oxidation states of, 2
physicochemical properties of, 2
radioisotopes of, 2—3
reference values of, 5
sensitivity to, 10—11
clinical reactions of, 168
friction and, 142
genetic factors related to, 142
in hand eczema patients, 136—138
hand eczema with, 138—140
hormonal factors related to, 142
humidity and temperature in, 142—143
occlusion and, 143
sweating and, 142
specimens of for biological monitoring, 4—5
systemic
distribution of, 23
pompholyx and, 152—153
toxicity of, 69
weight loss after marine environment exposure of, 48
Nickel-57, in nuclear medicine, 2—3
Nickel-63, 3
Nickel acetate, 3, 40
Nickel-alloy coatings, 46—53
Nickel alloys
benefits of, 46
carcinogenicity of, 37
chromium-containing, 49—50
in corrosive environment, 46
loss of passivity of, 50—51
non-chromium-containing, 49
occupational exposure to, 46
passivity of, 49—50
in prostheses, 51—52
release of nickel from, 46—53
solubility of, 46—51
Nickel-associated skin disease, 12, see also Dermatitis; Eczema
Nickel-binding proteins, 23
Nickel-cadmium batteries, 19, 38
Nickel carbonate, 3
Nickel carbonyl
analysis of, 4
carcinogenicity of, 38
decay of, 2
in nickel manufacturing, 118
regulation of, 41
treatment of poisoning from, 181
Nickel chloride, 3
Nickel-chromium platers, 36
Nickel-cobalt relationship, 171—172
Nickel compounds, phagocytization of, 24
Nickel hydroxide, 3
Nickel-iron alloy, 11
Nickel monoxide (NiO), 3
Nickel nitrate, 3
Nickeloplasmin, 23
Nickel oxide, 4
Nickel platers, skin disease of, 12
Nickel plating, 37, 118—120, 123—124
Nickel refinery workers, cancer rates of, 39
Nickel refining, carcinogenicity with, 36—37
Nickel salts, 69, 93
Nickel-sequestering agents, 185
Nickel spot test, 154—155
Nickel subsulfide, carcinogenicity of, 39
Nickel sulfate, 3
effects of on epithelial cell growth and differentiation, 39
patch testing with, 93—94
Nickel sulfides, 3
Nickel urticaria, 122
Nicrobraze LM alloy, 11
Nuclear medicine, 2—3

O

Occlusion, 143
Occupational Dermatology, Department of, 209
Occupational exposure, 118—119
carcinogenicity of, 36
common occupations involving, 122—126
to nickel alloys, 46
pulmonary, 19
from rare and curious sources, 126
regulatory requirements for, 40—41
of skin, 11—12
Occupational nickel dermatitis
association between nickel-sensitization hand eczema and work, 122—126
background of allergic nickel dermatitis, 120—122
general, 118
nickel exposure, 118—120
prevention, 126—127
Occupational Safety and Health Administration, U.S., permissible nickel levels of, 40
Occupational sensitization, 147
incidence of, 121
in women, 120
Open sensitivity testing, 95
Optimization sensitivity test, 71
Oral challenge, 168—173, 175
Oral nickel exposure, 95
Oral surgery, 199
Organ distribution, 23
Orthodontic alloys, 197—198
Orthodontic appliances, reactions to, 200
Orthodontic treatment, 196—197
Orthodontic wires, chemical composition of, 197
Orthopedic prostheses, 188—191
Oxazalone, 85
Oxidation states, 2

Oxide film, 48, 50—52
Oxygen, in nickel corrosion, 46
8-Oxyquinoline, 181, 184—185

P

Paint and pigment workers, 19
Passivity, 49—51
Patch testing, 86, 208
  concentrations in, 94
  concept of, 80
  with different nickel salts, 93
  distortion of, 137
  false- and unexpected negative reactions in, 95
  false-positive reactions in, 94
  importance of, 93
  intraoral, 201—202
  negative reaction to, 172
  for nickel dermititis, 110—111
  occlusion time in, 94
  positive reaction to, 120, 168—171
  results of, 112
  test site for, 94
  vehicles in, 93—94
Penicillamin, 174
D-Penicillamine, 180—182, 186
Perfume, 157
Petrolatum, in patch testing, 93, 94
PHA, see Phytohaemagglutinin
Phrapnel, 191
Phytin, 174—175
Phytohaemagglutinin (PHA), 82—85
Picrylchloride, 56
Polak-Turk test, 71
Polymerization processes, 202
Pompholyx, 122, 136, 151—152
  age of onset of, 139—140
  with coin handling, 148
  definition of, 135
  nickel-allergic, 95, 141, 152
  nickel-negative, 92
  prognosis of, 158
  systemic nickel role in, 152—153
Pompholyx mycid, 158
Population studies, of nickel allergy, 111—114
Pottery workers, 126
PPD, 84
Prednisone, 83
Preventive strategies, 26
Prick test, 93
Procarcinogens, 39
Procoagulant activity, 84
Prostheses, 19
  corrosion of, 51—52
  dental, 198
  metal-on-metal, 190
  metal-on-plastic, 190
  nickel-releasing, 188—191
  orthopedic, 188—191
Prosthodontic restorations, fixed, 198—199
Prosthodontic treatment, 198—199
Protein-precipitation technique, 4
Pulmonary absorption, 22, 27
Pulmonary exposure, 18—19
  deposition and clearance of, 22
  monitoring and limitation of, 27
  prevention of, 27
PUVA, 95

R

Radiotracers, 2
Reference values, 5
Regulatory requirements, 40—41
Reiter's disease, 103
Resin-based dental materials, 202—203
Resource Conservation and Recovery Act, Hazardous Waste Disposal rule of, 40
Rhabdomyosarcomas, 39
RNA synthesis, inhibition of, 39

S

Sandblasters, 19
Sarcomas, 39
Schwarze-Peck tests, 76
Seafood, nickel content of, 16
Sensitivity testing
  animal, 66—72
  clinical concepts of, 92—96
  human, 76—78
  indications of, 92—93
Sensitization, 10, see also Nickel, allergy to; Nickel, sensitivity to
  dermal, 21
  with ear piercing, 112, see also Ear piercing
  from environmental exposure, 147
  family studies of, 103—104
  genetic aspects of, 102—106
  genetic markers for, 102—103
  of guinea pig species, 69
  in hairdressers, 113
  immunology of, 56—61
  incidence of, 113
  individual susceptibility to, 102
  with jewelry, 113
  method of inducing, 66—69
  methods of, 66—72
  from nickel-plated suspenders, 110
  occupational association of, 122—126, 147, see also Occupational sensitization
  prevention of, 26
  sources of, 146—147
  twin studies of, 104—105
  from wristwatches, 110
Serum nickel concentrations, 23, 25
Sewing needles, 126
Shelanski-Shelanski Test, 76
Sinonasal cancer, 36
Skin
  absorption through, 20—21, 27
  damaged, 120

environmental exposures of, 10—11
nickel contact with, 119—120
nickel levels in, 23
occupational exposures of, 11—12
prevention of exposures to, 26
Softisan, in patch testing, 93—94
Stability constant, 180
Stainless steel
carcinogenicity of, 37
hand eczema with handling of, 149
nickel content of, 120
nickel release from, 50
prostheses of, 52
solubilization of nickel from, 51
Steel crown, 199
Stethoscopes, nickel in, 126
Stomatitis, 198
Stress corrosion, 51—52
Subacute myelooptic neuropathy (SMON), 184
Sulfides, crystalline, 3
Surface water, nickel content of, 17
Sweat, nickel content of, 24
Sweating, 142
Systemic distribution, 23
Systemic exposure, 95

T

T-cell factor, 85
Temperature, 142—143
Tetracycline, 174, 185—186
Thioglycol acid, 119
Thioglycolates, 125
Thiuram sulfides, 23
TINA test, 69, 71
T-lymphocytes
in allergic reaction, 57
interleukin 2 release from, 80—81
nickel presentation to, 24
nickel-specific, 57—61
Tooth loss, 196
Toxicity, 69
Trientene, 182
Triethylenetetramine, 155, 182
Tubulin-microtubule system, 24
Twin studies, 104—105

U

UBV, 95
Urinary nickel excretion, 4—5, 21—22, 24—25
Urticaria, contact, 122, 153, 202

V

Vandenberg/Epstein study, 76, 77
Vioform, 184—185
Viral plaque inhibition technique, 85
Vitallium, 19

W

Waste incineration, nickel exposure with, 18
Water, nickel content of, 119
Welders, 19
Welding, 37—38, 124
Wet environments, 125, 143—144
Wilson's disease, 180—182

Z

Zinc, 184—185